I0055513

Microwave and Millimeter Wave Phase Shifters

Volume II

Semiconductor and Delay Line
Phase Shifters

For a complete listing of the *Artech House Microwave Library*, turn to the back of this book

Microwave and Millimeter Wave Phase Shifters

Volume II

Semiconductor and Delay Line Phase Shifters

Shiban Koul
Barathi Bhat

Artech House
Boston • London

Library of Congress Cataloging-in-Publication Data

(Revised for vol.2)
Microwave and millimter wave phase shifters.
    Includes bibliographical references and index.
    Contents: v. 1. Dielectric and ferrite phase shifters-v. 2 Semiconductor and delay line phase
shifters.
    1. Phase shifters. 2. Microwave devices. 3. Millimeter wave devices. I. Koul, Shiban K. II. Bhat,
Bharathi. III. Title: Phase shifters.
    TK7872.P39M53 1991   91-26686
    621.38'5   CIP
    ISBN 0-89006-319-2 (v. 1)
    ISBN 0-89006-585-3 (v. 2)

© Artech House.
685 Canton Street
Norwood, MA 02062

All rights reserved. Printed and bound in the United States of America. No part of this book may
be reproduced or utilized in any form or by any means, electronic or mechanic including photocopy-
ing, recording, or by any information storage and retrieval system, without permission in writing
from the publisher.

International Standard Book Number: 0-89006-585-3
Library of Congress Catalog Card Number: 91-26686

10   9   8   7   6   5   4   3   2   1

To
**Jaykishori, Bansi Lal, Veena, and Aumendra Koul
and
Jaya and the late Gopal Bhat**

Contents of Volume I

Contents of Volume II

Foreword

In 1973, Doctor Bharathi Bhat joined the Centre for Applied Research in Electronics, then called the School of Radar Studies, fresh from Harvard, and she was assigned the task of developing phase shifters for use in radar systems. Since that time, Doctor Bharathi Bhat has concentrated her research and development work on this very interesting branch of microwave technology. She has established a group at the Centre that has been successful in developing a wide variety of phase shifters. Therefore, this book that she and her colleague, Professor Shiban Koul, have written stems from an extensive experience in the analysis, design, and construction of phase shifters. Their intimate knowledge of the subject makes this book an authoritative contribution in the field.

Normally, design engineers do not have adequate opportunity to teach, and academic scholars rarely have the experience of developing devices and systems to the ultimate point of seeing them tested in the field and accepted in service. Both Professor Koul and Professor Bhat combine these two distinct experiences in a very unusual manner and have done so for a number of years. I am sure that with such background and variety of experience, this book will be of great value both as a textbook for graduate courses and a reference book for designers of microwave circuits.

P.V. Indiresan
Professor
Centre for Applied Research
    in Electronics, I.I.T. Delhi
INDIA

Preface

The phase shifter, as a general-purpose microwave component, finds use in a variety of communication and radar systems, microwave instrumentation and measurement systems, and industrial applications. Prior to the advent of electronically variable phase shifters in the 1950s, almost all phase shifters, both fixed and variable, were mechanical. Electronic phase shifters assumed special significance because of their potential utility and volume requirement in phased array antenna systems for inertialess scanning. With the first demonstration of the Reggia-Spencer ferrite phase shifters for phased array scanning in 1957, there began a new era of ferrite phase shifter technology. During the mid-1960s another important class of phase shifters employing *p-i-n* diodes as electronic switches for phase shift control came into existence. The next two decades saw a period of intense research and development activity in both ferrite and *p-i-n* diode phase shifters, leading to a variety of practical designs in both waveguide and planar configurations. In the meantime, particularly since the 1980s, several other types of electronic phase shifting devices have emerged; the most important ones being the GaAs FET active phase shifters and the *magnetostatic wave* (MSW) time-delay phase shifters. With the advent of *monolithic microwave integrated circuit* (MMIC) technology, MMIC phase shifters employing MESFETs and varactors as electronic control elements have been made possible as well. In particular, advances in printed antennas, on the one hand, and microwave GaAs technology, on the other, have opened up new challenges for developing fully monolithic phase shifter modules. Yet another emerging development is in the area of optically controlled bulk semicondutor phase shifters particularly suited for the millimeter-wave frequency range.

Although the development of electronic phase shifters to its present level of sophistication was propelled primarily to cater to large size phased arrays, their realizability in many forms has opened up new vistas of application areas requiring small arrays such as traffic control, vehicular communication, and so forth. Even though the phase shifter revolution of the 1970s and 1980s culminated in a wide variety of practical phase shifters, its evolution toward miniaturization and utilization in higher millimeter-wave frequency range continues. As we move into the 1990s, the time is appropriate to consolidate under a single cover the entire spectrum of phase shifter development for the benefit of microwave community.

This book and its companion volume provide up-to-date coverage of almost all types of phase shifters, both mechanical and electronic, hitherto reported in the open literature. Divided into fifteen chapters, these books offer under different categories, technical details of various phase shifters, starting from the conventional coaxial line stretcher to the modern MMIC and optically controlled millimeter-wave phase shifters. Chapters 1–7, which discuss dielectric and ferrite phase shifters, are included in Volume I, whereas Chapters 8 to 15, which discuss technical details of various semiconductor and delay line phase shifters, are included in Volume II.

For digital, reciprocal operation, *p-i-n* diode phase shifters in MIC configuration is a well-proven technology. Chapter 8 provides basic circuit forms that can be used in conjunction with microwave switching devices to function as digital phase shifters. Chapter 9 is specifically devoted to the analysis and design aspects of *p-i-n* diode phase shifters in hybrid MIC form. The scope for realizing a *p-i-n* diode phase shifter in a finline circuit is included.

GaAs FET phase shifters are relatively new. Chapter 10 provides a review of the current technology for different hybrid versions of FET phase shifters, with special emphasis on active phase shifting. Hybrid MIC analog phase shifters employing voltage-controlled varactor diodes are described in Chapter 11.

Monolithic realization of phase shifters is an emerging trend. Chapter 12 covers the basic monolithic techniques and realization of various phase shifters using FETs and varactors.

Most of the ferrite and semiconductor device phase shifters currently realized are meant for microwave applications. The scope for extending these to millimeter-wave frequencies and their limitations are included in Chapter 13. This chapter also covers miscellaneous millimeter-wave phase shifters, such as those employing bulk semiconductor and integrated dielectric guides.

Surface acoustic wave (SAW) and magnetostatic wave (MSW) delay lines offer phase shifting in the form of true time delays. Chapter 14 describes microwave delay-line phase shifters employing SAW and MSW techniques. The final chapter enumerates broad criteria for the selection of phase shifters for system applications.

This book and its companion volume are primarily intended as reference texts for microwave engineers working in R&D establishments and industry. The chapters are organized so that all theoretical and practical aspects of each phase shifter type are covered in the same chapter. Selected portions of the book therefore can be directly included as part of the postgraduate course material in the area of microwave and millimeter-wave circuits and components.

We are grateful to Professor P.V. Indiresan for initiating and encouraging the activity on phase shifters at the Centre for Applied Research in Electronics, I.I.T., Delhi, and to the Directorate of Training and Sponsored Research, Defence Research

Development Organization (India) for sponsoring developmental programs in this area for more than a decade and a half. We would also like to thank the Defence Research Development Laboratory (Hyderabad) for funding a major project on ferrite phase shifters and to Electronics and Radar Development Establishment (Bangalore) for the interaction and interest. This book is the result of our involvement in these programs and the opportunity that we have had for making an in-depth study of the area.

Shiban K. Koul
Bharathi Bhat

Chapter 8
Phase-Shifter Circuits Using Switching Devices

8.1 INTRODUCTION

An ideal switching device basically possesses two states: a short circuit and an open circuit. In practice, these two states are realized rather approximately by means of a microwave semiconductor control device, such as a *p-i-n* diode or a *metal semiconductor field-effect transistor* (MESFET). For example, a *p-i-n* diode can approximate the desired short-circuit and open-circuit conditions at microwave frequencies under forward and reverse bias states, respectively. A variety of digital phase shifters can be configured by mounting one or more diodes in different types of transmission line circuit elements. Irrespective of the type of switching device used, such phase-shifter circuits can be broadly classified into two groups—namely, the *reflection type* and the *transmission type*. The reflection-type phase shifter is basically a one-port device in which there is reflection of a microwave signal at the termination of a transmission line. The magnitude of the reflection coefficient should ideally be unity, and the phase shift is given by the change in the phase of the reflection coefficient between the two switching states. An ideal transmission-type phase shifter is a two-port network in which the phase of the transmission coefficient through the network is altered by means of a switch, while its magnitude remains unity in both the states. The phase shift is given by the change in the transmission phase through the network.

The above classification applies equally well to analog phase shifters. The circuit configuration in each case remains the same except that the switch is replaced by a variable reactance device. For example, continuously variable phase shift can be obtained by using a varactor diode (as a variable capacitor) or a dual-gate MESFET by varying the dc bias.

Depending on the circuit configuration employed, both reflection- and transmission-type phase shifters can be further classified as *constant time-delay circuits* or *constant phase-shift circuits*. Constant time-delay circuits provide large, instantaneous bandwidths, and their phase shift increases linearly with frequency. With suitable modification in the circuits, they can also be changed to constant phase-shift circuits.

Both reflection- and transmission-type phase shifters have been reported extensively in the literature [1–3]. In this chapter, we present the general operating features and relative merits and demerits of various configurations falling under these two categories. The switches are assumed to be either ideal or nearly ideal and the transmission line circuits are assumed to have their ideal responses.

8.2 REFLECTION-TYPE CIRCUITS

Basic Power-Handling Theorem

A reflection-type circuit can be represented as a passive network terminated in an ideal on-off switch, as shown in Figure 8.1(a). A fundamental theorem governing such an elementary reflective network has been derived by Hines [4]. If Γ_{s0} and Γ_{s1} are the voltage reflection coefficients at the input port of the network when the switch is open circuited and short circuited, respectively, then the theorem states that

$$\Gamma_{s0} - \Gamma_{s1} = \frac{1}{2}\left(\frac{V_{s0}}{V_L}\right)\left(\frac{I_{s1}}{I_L}\right) \tag{8.1}$$

where V_{s0} is the open-circuit RF voltage and I_{s1} is the short-circuit RF current at the switch, assuming a constant resistive-impedance generator attached to the input port

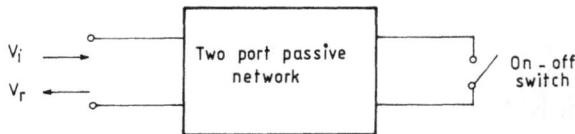

(a)

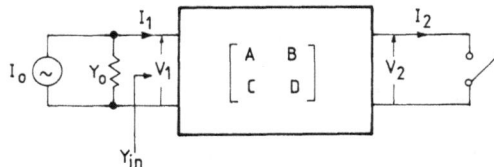

(b)

Figure 8.1 (a) Schematic of a reflection-type network; (b) circuit for determining power rating of the network.

whose short-circuit current is $2I_L$ and whose open-circuit voltage is $2V_L$. If the network is lossless, then the magnitude of the reflection coefficient is unity and we can write

$$\Gamma_{s0} = e^{j\phi_{s0}} \tag{8.2a}$$

$$\Gamma_{s1} = e^{j\phi_{s1}} \tag{8.2b}$$

Using (8.2) in (8.1), we get the following expression for the differential phase shift ($\Delta\phi$):

$$\Delta\phi = |\phi_{s0} - \phi_{s1}| = 2 \sin^{-1} \left| \frac{V_{s0}I_{s1}}{4V_L I_L} \right| \tag{8.3}$$

If the maximum rated rms RF voltage that the switch can withstand when it is open is V_{m0}, and the maximum rated rms current that it can carry when closed is I_{m1}, then the maximum RF power P_{max} and phase shift $\Delta\phi$ are related by the equation (derived from Eq. 8.3)

$$P_{max} = \frac{V_{m0}I_{m1}}{4 \sin \dfrac{\Delta\phi}{2}} \tag{8.4}$$

If n switches are used to produce a phase shift of $\Delta\phi$, then the maximum power limit is given by

$$P_{max} = \frac{nV_{m0}I_{m1}}{4 \sin \dfrac{\Delta\phi}{2}} \tag{8.5}$$

It has also been pointed out by Hines [4] that in order to realize the maximum power limit, it is necessary to design the network suitably in order to approach the maximum rated voltage V_{m0} and the maximum rated current I_{m1} in the switch.

Maximum Average Power Rating

The power rating of the reflective-type phase shifter shown in Figure 8.1(a) can be expressed in terms of the *ABCD* parameters of the passive network. Garver [1] has derived suitable expressions by considering a circuit model, as shown in Figure 8.1(b). The source is a constant-current generator having a peak RF current I_0 and an internal admittance Y_0. For the two-port network, the input voltage and current (V_1 and I_1) are related to the output voltage and current (V_2 and I_2) as

$$V_1 = AV_2 + BI_2 \qquad (8.6a)$$

$$I_1 = CV_2 + DI_2 \qquad (8.6b)$$

The maximum power available from the generator to the network is

$$P_i = \frac{1}{2}\left(\frac{I_0^2}{4Y_0}\right) \qquad (8.7)$$

Since the generator is a constant-current source, we have

$$I_0 = V_1(Y_0 + Y_{in}) \qquad (8.8)$$

where Y_{in} is the input admittance of the network. Substituting for I_0 in (8.7), we obtain

$$P_i = \frac{V_1^2}{8Y_0}(Y_0 + Y_{in})^2 \qquad (8.9)$$

When the switch is closed, $V_2 = 0$. From (8.6), we get

$$V_1 = BI_2 \qquad (8.10a)$$

$$I_1 = DI_2 \qquad (8.10b)$$

so that

$$Y_{in} = \frac{I_1}{V_1} = \frac{D}{B} \qquad (8.10c)$$

Substituting for V_1 and Y_{in} from (8.10) in (8.9), we obtain the following expression for the maximum average power that the reflection phase shifter can handle [1]:

$$\bar{P}_i = \frac{I_2^2}{8Y_0}(BY_0 + D)^2 \qquad (8.11)$$

Peak Power Rating

In order to derive the peak power rating of the reflective phase shifter, we consider the circuit shown in Figure 8.1(b) with the switch open circuited. Setting $I_2 = 0$ in (8.6) gives

$$V_1 = AV_2 \tag{8.12a}$$

$$I_1 = CV_2 \tag{8.12b}$$

so that

$$Y_{in} = \frac{C}{A} \tag{8.12c}$$

Substituting for V_1 and Y_{in} from (8.12) in (8.9), the expression for the peak power rating of the phase shifter is obtained as

$$\hat{P}_i = \frac{V_2^2}{8Y_0} (AY_0 + C)^2 \tag{8.13}$$

Examples of Reflective Terminations

Figure 8.2 shows some practical schemes of achieving switchable impedances. In all these circuits, PP' is the reference reflecting plane and Z_0 is the characteristic impedance of the feeding transmission line. If $\bar{V}_i$ and $\bar{V}_r$ denote the incident and reflected voltage signals normalized with respect to the characteristic impedance Z_0 of the input line, then the voltage V across the line and the current I flowing through it are given by

$$V = \bar{V}_i(1 + \Gamma) \tag{8.14a}$$

$$I = \bar{V}_i(1 - \Gamma) \tag{8.14b}$$

where Γ is the input voltage reflection coefficient and is given by

$$\Gamma = |\Gamma|e^{j\phi} = \frac{\bar{V}_r}{\bar{V}_i} = \frac{\bar{Z} - 1}{\bar{Z} + 1} \tag{8.15}$$

In (8.15), ϕ is the phase difference between the incident and reflected signal and $\bar{Z}$ ($= Z/Z_0$) is the normalized terminating impedance. Let $\bar{Z}_1$ and $\bar{Z}_2$ be the values of

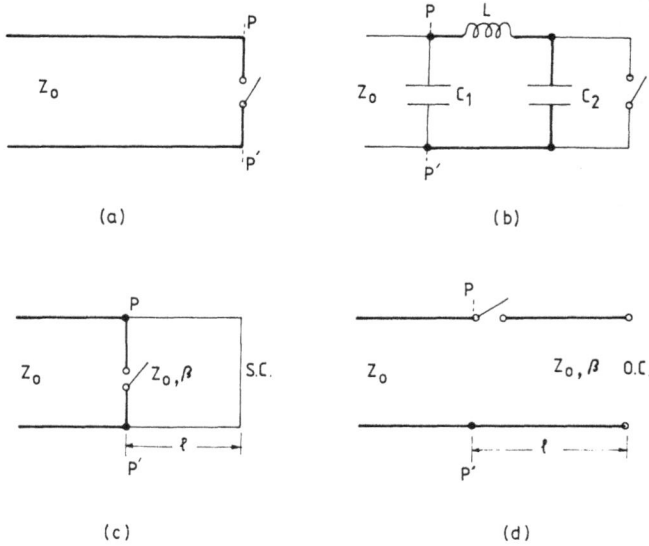

Figure 8.2 Examples of reflection-type circuits: (a) transmission line terminated in a switch; (b) transmission line terminated in lumped-element network and a switch; (c) short-circuited transmission line with a shunt switch; (d) open-circuited transmission line with a series switch.

the terminating impedances corresponding to the two switching states. Adding the subscript 1 or 2 to all the associated quantities and using (8.14), we can write [5]

$$|V_2 I_1 - V_1 I_2| = 2\bar{V}_i^2 |\Gamma_1 - \Gamma_2|$$

$$= 2\bar{V}_i^2 \left[(|\Gamma_1| - |\Gamma_2|)^2 + 4|\Gamma_1||\Gamma_2| \times \sin^2\left(\frac{\phi_1 - \phi_2}{2}\right) \right]^{1/2} \quad (8.16)$$

We now impose the condition that losses in both the switching states must be the same. Setting $|\Gamma_1| = |\Gamma_2| = |\Gamma|$, in (8.16), we obtain

$$|V_2 I_1 - V_1 I_2| = 8P_i |\Gamma| \sin\frac{\Delta\phi}{2} \quad (8.17)$$

where $P_i = \bar{V}_i^2/2$ is the power of the incident signal and $\Delta\phi = (\phi_1 - \phi_2)$ is the phase shift.

Wideband Phase-Shifting Networks

A transmission line terminated in an ideal switch, as shown in Figure 8.2(a), produces a phase shift that is completely independent of frequency. When the switch is switched between short circuit and open circuit, the impedance changes from 0 to ∞. Correspondingly, the reflection coefficient Γ changes from -1 to 1, thus giving a phase shift of 180°.

Figure 8.2(b) shows a switchable lumped-element terminating network. It has been shown by Garver [6] that any desired phase shift with wideband response can be achieved by appropriately choosing the elements of the network. The design formulas are obtained as [6]

$$Z_0' = k\,Z_0 \tag{8.18a}$$

$$L = Z_0'/\omega_0 \tag{8.18b}$$

$$C = 1/Z_0'\omega_0 \tag{8.18c}$$

$$C_1 = (1 - a)C \tag{8.18d}$$

$$C_2 = aC \tag{8.18e}$$

where Z_0 is the characteristic impedance of the feeding transmission line and ω_0 is the angular frequency corresponding to the lower end of the operating band. The parameters k and a depend on the desired phase shift. They are (1) $k = 0.47$, $a = 0.7$ for $\Delta\phi = 90°$; (2) $k = 0.3$, $a = 0.83$ for $\Delta\phi = 45°$; (3) $k = 0.19$, $a = 0.91$ for $\Delta\phi = 22.5°$; and (4) $k = 0.115$, $a = 0.952$ for $\Delta\phi = 11.25°$. With this choice of parameters, Garver [6] has reported an equiripple phase response within a phase error of $\pm 2°$ over more than an octave bandwidth for all phase bits. The smaller the phase shift, the larger the bandwidth is.

Time-Delay Networks

Schemes for achieving switched reactances using a distributed network are shown in Figures 8.2(c) and (d). In both cases, the differential phase shift $\Delta\phi$ is equal to $2\beta l$, where β is the phase constant and l is the length of the switched line section. These are basically time-delay networks giving a phase response that is a linear function of frequency. Time-delay phase shifters provide instantaneous wide bandwidth, which is needed in pulsed phased-array radars [7]. They can, however, be converted to constant phase-shift circuits by incorporating suitable modification in the circuit. This aspect is discussed in Chapter 9.

Reflection-type phase shifters are commonly employed in reflect-array radars. Figure 8.3 shows the schematic of a 2-bit, shunt-mounted, reflection-type phase shifter. The switches are spaced $3\lambda_0/8$ apart, where λ_0 is the guide wavelength in the transmission line at the center frequency. By using different combinations of the three binary switches, four phase steps can be obtained in increments of 90°.

In practice, the incident and reflected signals of the one-port, reflection-type network are separated by using the network to terminate a circulator. Such circuits are called *circulator coupled phase shifters.* Of more common usage are the *hybrid coupled phase shifters,* in which a pair of identical reflective networks terminate two ports of a coupler. General aspects of these two-port reflective-type phase shifters are discussed in subsequent sections (Sections 8.4 and 8.5).

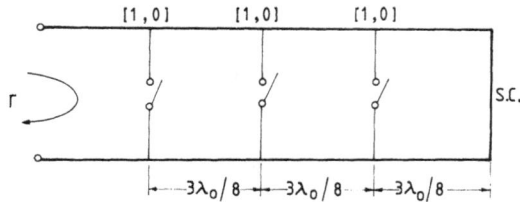

Figure 8.3 Two-bit reflection-type phase shifter; 1: short circuit, 0: open circuit.

8.3 TRANSMISSION-TYPE CIRCUITS

Figure 8.4(a) shows a general schematic of a transmission-type phase-shifting network. The passive network may be either lumped or distributed, and the switch may be series or shunt mounted. Switching between the two states is equivalent to passing the RF signal through two different circuit paths. The change in the transfer phase of the network between the two switching states gives the phase shift.

Maximum Power-Handling Limit [3]

In order to determine the maximum switchable power for a transmission-type phase shifter with a single switch, we consider a simple circuit consisting of a shunt-mounted switch in series with a capacitive susceptance, as shown in Figure 8.4(b). When the switch is open, the current I through the circuit is $I = V/Z_0$. The transmission coefficient $T = 1$. When the switch is closed, the reflection coefficient Γ and the transmission coefficient T are given by

$$\Gamma = \frac{-jB/Y_0}{2 + jB/Y_0} \tag{8.19}$$

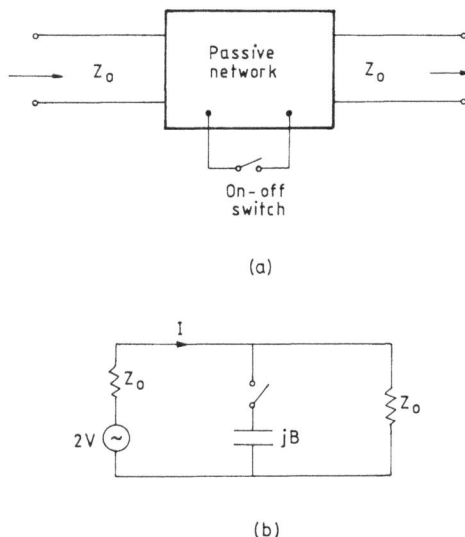

(a)

(b)

Figure 8.4 (a) General schematic of a transmission-type phase shifter; (b) an example of a transmission-type phase shifter with a shunt switch.

$$T = 1 + \Gamma = \frac{2}{(2 + jB/Y_0)} = \left| \frac{2}{\sqrt{4 + (B/Y_0)^2}} \right| e^{-j\tan^{-1}(B/2Y_0)} \qquad (8.20)$$

The phase shift is given by

$$\Delta\phi = -\tan^{-1}\left(\frac{B}{2Y_0}\right) \qquad (8.21)$$

In order to obtain an expression for maximum switchable power (P_m), we consider the voltage V_{m0} across the switch when it is open, and the magnitude of the current I_{m1} passing through it when it is closed. In order to minimize the input voltage standing wave ratio (VSWR) when the switch is closed, we choose $Z_0 \ll 1/jB$. We can then write

$$V_{m0} \simeq V \qquad (8.22a)$$

$$I_{m1} \simeq IB/Y_0 \qquad (8.22b)$$

$$V_{m0}I_{m1} = VIB/Y_0 \qquad (8.22c)$$

Substituting for B/Y_0 from (8.21), we obtain

$$V_{m0}I_{m1} = 2VI \tan \Delta\phi \qquad (8.23a)$$

or

$$P_{max} = \frac{V_{m0}I_{m1}}{2 \tan \Delta\phi} \qquad (8.23b)$$

This equation indicates that in a transmission-type phase shifter with a single switch, the maximum phase shift is limited to 90°, as $P_{max} \rightarrow 0$ when $\Delta\phi \rightarrow 90°$. Secondly, in order to keep a reasonable transmission match over a desired bandwidth, $\Delta\phi$ must be kept small. Thus, for achieving larger phase shift, a number of identical reactive elements (each of small value) can be cascaded, spaced by about a quarter-wavelength apart. With quarter-wave spacing, symmetric reflections cancel mutually at the input, resulting in low-input VSWR. The elements are chosen such that the transmission phases add up. This type of transmission phase shifter is called the *loaded line phase shifter*.

Figure 8.5 shows another general schematic of a transmission-type phase shifter. This one requires a minimum of two switches per phase bit. Phase shift is obtained by switching between the two passive networks. For any reciprocal symmetric passive network, the scattering parameters S_{12} (transmission coefficient) and S_{11} (reflection coefficient) can be obtained in terms of its *ABCD* parameters. They are

$$S_{12} = \frac{2}{A + \dfrac{B}{Z_0} + CZ_0 + D} = S_{21} \qquad (8.24)$$

$$S_{11} = \frac{\dfrac{B}{Z_0} - CZ_0}{A + \dfrac{B}{Z_0} + CZ_0 + D} = S_{22} \qquad (8.25)$$

where Z_0 is the characteristic impedance of the input and output lines. Referring to Figure 8.5, if $S_{21} = |S_{21}|e^{j\phi_1}$ and $S'_{21} = |S'_{21}|e^{j\phi_2}$ are the transmission coefficients of networks 1 and 2, respectively, then the phase shift $\Delta\phi = \phi_2 - \phi_1$. The insertion losses in dB in the two switching states are $20 \log_{10}|S_{21}|$ and $20 \log_{10}|S'_{21}|$.

In practice, passive networks can be sections of transmission lines of different lengths. Such phase shifters are basically time-delay networks and are called *switched line phase shifters*. Another configuration is the *switched filter phase shifter* in which

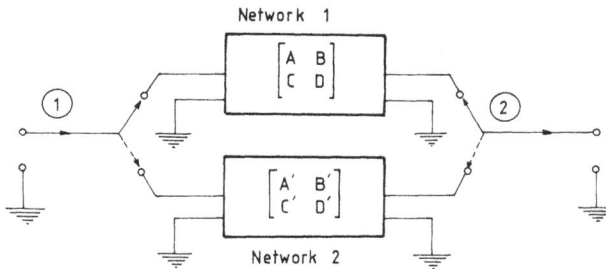

Figure 8.5 Schematic of a transmission-type phase shifter with two switched networks.

the two networks are lumped-element high-pass and low-pass filters. Details of these phase shifters are presented in subsequent sections of this chapter.

8.4 CIRCULATOR COUPLED PHASE-SHIFTER CIRCUITS

Any of the reflective-type circuits shown in Figure 8.2 can be used to terminate a circulator. Figure 8.6 shows the general schematic. The function of the circulator is to separate the incident and reflected signals of the reflective network, thus converting the one-port reflective network into a transmission device. If the circulator is ideal, then the transmission coefficient T between the input and output ports is equal to the reflection coefficient Γ of the reflective network, except for an additional phase delay, which is common to both the switching states. Thus the phase shift of the device remains the same as that offered by the reflective network, which is equal

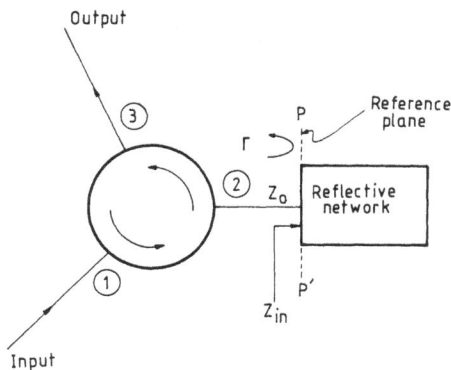

Figure 8.6 Schematic of a circulator coupled phase shifter.

to the difference in the phase factors of Γ between the two switching states. The insertion loss of the circulator coupled phase shifter is given by

$$\alpha(\text{dB}) = 20 \log_{10}\left(\frac{1}{|\Gamma|}\right) \tag{8.26}$$

where

$$\Gamma = \frac{Z_{in} - Z_0}{Z_{in} + Z_0} \tag{8.27}$$

While using the circulator, it is important to ensure that there is no mismatch intervening between the circulator and the terminating impedance. For example, if Γ_c is the reflection coefficient between port 2 of the circulator and the terminating impedance, the phase error introduced is given by [6]

$$\varepsilon_\phi = \pm\left(1 + 3|\Gamma| - \frac{1}{4}\sin \pi|\Gamma|\right) \sin^{-1}\left|\frac{\Gamma_c}{\Gamma}\right| \tag{8.28}$$

For lossless phase shifters, $|\Gamma| = 1$. The phase error then reduces to

$$\varepsilon_\phi = \pm 4 \sin^{-1}|\Gamma_c| \tag{8.29}$$

In terms of VSWR due to the intervening mismatch, the phase error is approximately given by

$$\varepsilon_\phi \simeq \pm 100(\text{VSWR} - 1) \text{ degrees} \tag{8.30}$$

Another source of error is due to the finite isolation of the circulator. It has been reported by Garver [6] that a circulator having isolation factors of 20, 30, and 40 dB gives a maximum phase error of $\pm 22.8°$, $\pm 7.2°$, and $\pm 2.3°$, respectively.

Power-Handling Capability of Time-Delay Phase Shifter [1]

Consider a circulator coupled phase shifter with a time-delay reflective network as shown in Figure 8.7(a). The peak power-handling capability is calculated by considering the equivalent circuit when the switch is open (Figure 8.7(b)). The switch is assumed to be a perfect open sustaining a peak voltage V_{s0} across it. From Figures 8.7(a) and (b), we can write

$$V_{s0} = \frac{V \cdot jX}{Z_0 + jX} \tag{8.31}$$

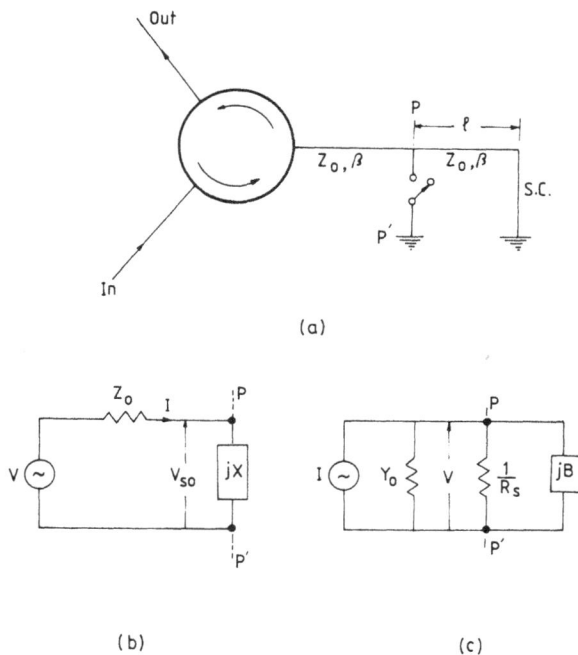

Figure 8.7 (a) Circulator coupled time-delay phase shifter; (b) equivalent circuit with switch open; (c) equivalent circuit with switch closed.

where V is the peak voltage of the generator and

$$X = Z_0 \tan \beta l = Z_0 \tan \frac{\Delta \phi}{2} \tag{8.32a}$$

$$\Delta \phi = 2\beta l \tag{8.32b}$$

Using (8.32) in (8.31) and simplifying, we get

$$|V_{so}| = V \sin \frac{\Delta \phi}{2} \tag{8.33}$$

The maximum available power $\hat{P}_i$ from the generator is given by

$$\hat{P}_i = \frac{1}{2} \left(\frac{V}{2} \right)^2 \frac{1}{Z_0} \tag{8.34}$$

Substituting for V from (8.33), we obtain

$$\hat{P}_i = \frac{V_{s0}^2}{8Z_0 \sin^2\left(\dfrac{\Delta\phi}{2}\right)} \tag{8.35}$$

This is the maximum peak power that the time-delay reflection phase shifter can control.

For determining the average power capability, we consider the equivalent circuit (shown in Figure 8.7(c)) when the switch is shorted. The switch is assumed to offer a small resistance R_s. Referring to Figure 8.7(c), the voltage V is given by

$$V = \frac{I}{Y_0 + \dfrac{1}{R_s} + jB} \tag{8.36}$$

where

$$B = -Y_0 \cot\frac{\Delta\phi}{2} \tag{8.37}$$

Power dissipated in the resistance R_s is given by

$$\bar{P}_d = \frac{|V|^2}{2R_s} \tag{8.38}$$

Substituting for V from (8.36) and using (8.37), we get

$$\bar{P}_d = \frac{I^2}{2Y_0 \left| 2 + \dfrac{1}{Y_0 R_s} + \dfrac{Y_0 R_s}{\sin^2 \dfrac{\Delta\phi}{2}} \right|} \tag{8.39}$$

The maximum average power that the reflection phase shifter can handle is given by

$$\bar{P}_i = \frac{1}{2}\left(\frac{I}{2}\right)^2 \frac{1}{Y_0} \tag{8.40}$$

Substituting for I^2 from (8.39), we obtain

$$\bar{P}_i = \frac{\bar{P}_d}{4}\left(2 + \frac{Z_0}{R_s} + \frac{R_s}{Z_0 \sin^2 \dfrac{\Delta\phi}{2}}\right) \tag{8.41}$$

The insertion loss is calculated using the formula

$$\alpha(\text{dB}) = 10 \log_{10}\left(\frac{\bar{P}_i}{\bar{P}_i - \bar{P}_d}\right) \tag{8.42}$$

Power-Handling Capability of Wideband Phase Shifter

For the wideband terminating network given in Figure 8.2(b), Garver [6] has derived expressions for peak power $\hat{P}_i$ and average power $\bar{P}_i$ using its *ABCD* matrix parameters. Using the same symbols as given in the design equations (8.18), the expressions for $\hat{P}_i$ and $\bar{P}_i$ are

$$\hat{P}_i = \frac{V_{s0}^2}{8Z_0}\left[\left\{1 - a\left(\frac{\omega}{\omega_0}\right)^2\right\}^2 + \left\{\frac{1}{k} - (1 - a)\left(\frac{\omega}{\omega_0}\right)\right\}^2 \left(\frac{\omega}{\omega_0}\right)^2\right] \tag{8.43}$$

$$\bar{P}_i = \frac{\bar{P}_d}{4R_sY_0}\left[k^2\left(\frac{\omega}{\omega_0}\right)^2 + \left\{1 - (1 - a)\left(\frac{\omega}{\omega_0}\right)^2\right\}^2\right] \tag{8.44}$$

where V_{s0} is the voltage across the ideal switch when open circuited and R_s is the small resistance the switch is assumed to offer when it is closed.

8.5 HYBRID COUPLED PHASE-SHIFTER CIRCUITS

The hybrid coupled phase shifter [7] makes use of a 3-dB, 90° hybrid coupler with two of its ports terminated in symmetric phase-controllable reflective networks. Figure 8.8 shows the general schematic of the phase shifter. Any reflective network such as the ones shown in Figure 8.2 can be used in symmetric pairs as terminations. The coupler divides the input signal (fed to port 1) equally between the two output ports (ports 3 and 4) but with a phase difference of 90°. Signals reflected back from the two symmetric terminations add up at port 2 and no signal returns to port 1. The hybrid coupler thus offers a matched transmission behavior for the phase shifter bit.

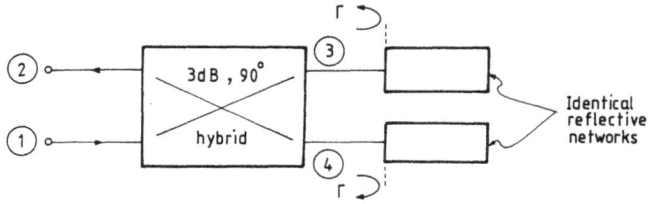

Figure 8.8 Schematic of a hybrid coupled phase shifter.

While the circulator coupled phase shifter uses one reflective termination, the hybrid coupled phase shifter requires two reflective terminations (minimum of two switches) per phase bit. However, from the point of view of insertion loss, size, ease of fabrication, and cost, the hybrid coupled phase shifter is preferred. As in the case of the circulator coupled phase shifter, any phase shift can be obtained by suitably designing the terminating circuit.

There are three commonly used planar circuit forms through which 3-dB, 90° coupler properties can be realized. They are (1) *branchline coupler,* (2) *rat-race hybrid coupler,* and (3) *parallel-coupled backward-wave coupler.* Figure 8.9 shows phase-shifter circuits using these couplers and one type of reflective termination. If the couplers are ideal, then the phase shift $\Delta\phi$ in all three cases is $2\beta l$, where β is the propagation constant and l is the length of the switched transmission line section.

The bandwidth of the phase-shifter bit is governed by the bandwidth of the coupler as well as the reflective network. The coupler bandwidth is assessed from its characteristics in terms of power split, phase relationship between the output ports, input VSWR, and isolation as a function of frequency. Considering a combination of all these factors, the useful bandwidth of the branchline coupler, the rat-race coupler, and the parallel-coupled backward-wave coupler as 3-dB hybrids is approximately 10%, 20%, and 35%, respectively. It may be noted that the branchline and parallel-coupled backward-wave couplers are inherently 90° hybrids, whereas the rat-race coupler is a 180° hybrid. In order to obtain the desired 90° phase difference between the output signals, the reference plane of one of the two output ports is extended by a quarter-wavelength as shown in Figure 8.9(b). The relative advantages and disadvantages of these couplers in realizing phase shifters are discussed in Chapter 9.

The procedure for determining the power-handling capability of a hybrid coupled phase shifter is the same as that of a circulator coupled phase shifter. Because the hybrid coupled phase shifter uses two terminations, its peak power handling capability is twice that of the corresponding circulator coupled phase shifter. For the hybrid coupled phase shifter terminated in time-delay networks as shown in Figure 8.9, the expression for peak power can be written from (8.35) as

$$\hat{P}_i = \frac{V_{s0}^2}{4Z_0 \sin^2\left(\dfrac{\Delta\phi}{2}\right)} \tag{8.45}$$

where V_{s0} is the peak voltage across the open-circuited switch and Z_0 is the characteristic impedance of the transmission line feeding the reflective network. The maximum average power that the hybrid coupled phase shifter can handle is the same as that of the circulator coupled phase shifter (as given by (8.41)).

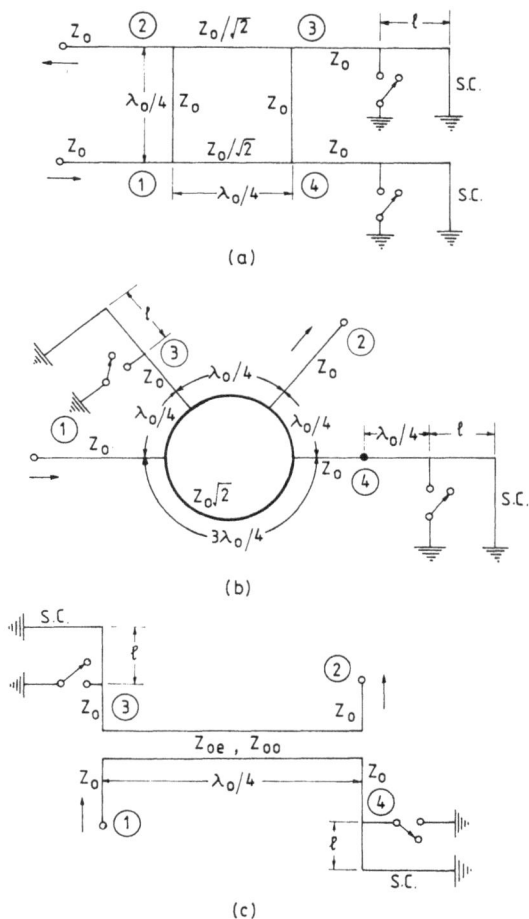

Figure 8.9 Hybrid coupled phase-shifter circuits using (a) branchline coupler, (b) rat-race coupler, and (c) backward-wave coupler.

8.6 SWITCHED LINE PHASE-SHIFTER CIRCUITS

The switched line phase shifter is basically a time-delay circuit in which phase shift is obtained by switching between two transmission line sections of different lengths. Two basic schematic circuits are shown in Figure 8.10, one with series-mounted switches and the second with shunt-mounted switches. In the series configuration (Figure 8.10(a)), when the switches S_1 and S_1' are closed and S_2 and S_2' are open, the transmission is through path length l_1. When the switching states are reversed, transmission is through the upper path of length l_2. The phase shift $\Delta\phi$ between the two switching states is simply $\beta(l_2 - l_1)$ where β is the propagation constant of the transmission line. It is assumed that all transmission lines are identical having the same value of β and characteristic impedance Z_0. In practice, one disadvantage of this phase shifter when nonideal diodes are used is that at a frequency at which the length of the off-path is a half-wavelength, the line becomes resonant. That is, the phases add up in such a manner that they reflect the incident power back to the generator. This problem is somewhat overcome in the shunt-mounted switched line configuration shown in Figure 8.10(b). When the shunt switches S_1 and S_1' short the ends of the transmission line section of length l_1, and switches S_2 and S_2' are open circuited (as shown in Figure 8.10(b)), the signal passes through the upper line. Looking into the lower path from the input and output ports A and B, the signal sees an infinite impedance because of the $\lambda_0/4$ sections terminated in short circuits. When the switching states of the switches in the lower and upper paths are interchanged, transmission is through the lower transmission line. As in the case of series configuration, the phase shift $\Delta\phi$ is given by $\beta(l_2 - l_1)$.

An important feature of this phase shifter is that when practical switching devices are used, the insertion loss is nearly the same in both the switching states and it is nearly independent of the phase shift. As compared with other types of phase shifters, the switched line phase shifter uses the maximum number of switches (four per phase bit).

Peak Power Capability

The peak power-handling capacity of a switched line phase shifter is given by

$$\hat{P}_i = \frac{V^2}{2Z_0} \tag{8.46}$$

where V is the peak RF voltage rating of the switching device (when open) and Z_0 is the characteristic impedance of the transmission line.

(a)

(b)

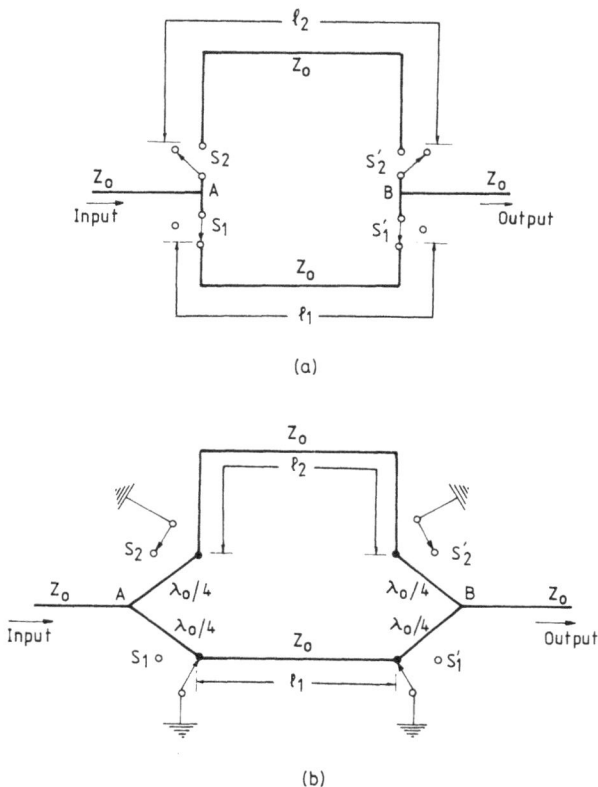

Figure 8.10 Switched line phase-shifter circuits with (a) series-mounted switches and (b) shunt-mounted switches.

Average Power Capability

For determining the average power capability of a series switched line phase shifter, we consider the equivalent circuit shown in Figure 8.11. The switch is assumed to have a small resistance R_f when it is closed. If V is the peak RF voltage of the source, the average power dissipated in R_f is given by

$$\bar{P}_d = \frac{1}{2} \frac{V^2 R_f}{(R_f + 2Z_0)^2} \tag{8.47}$$

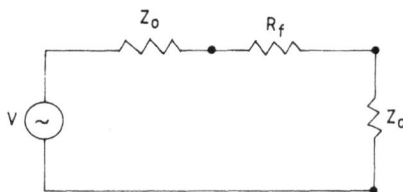

Figure 8.11 Equivalent circuit for calculating average power capability of switched line phase shifter with series switch.

The average incident power is given by

$$\bar{P}_i = \frac{1}{2}\left(\frac{V}{2Z_0}\right)^2 Z_0 = \frac{V^2}{8Z_0} \tag{8.48}$$

Substituting for V^2 from (8.47) in (8.48), the average power capability can be expressed in the form

$$\bar{P}_i = \frac{\bar{P}_d Z_0}{R_f}\left(1 + \frac{R_f}{2Z_0}\right)^2 \tag{8.49}$$

For the shunt-mounted switch, the average power capability is given by replacing R_f by G_f and Z_0 by Y_0. Thus,

$$\bar{P}_i = \frac{\bar{P}_d Y_0}{G_f}\left(1 + \frac{G_f}{2Y_0}\right)^2 = \frac{\bar{P}_d Z_0}{4R_f}\left(1 + \frac{2R_f}{Z_0}\right)^2 \tag{8.50}$$

8.7 LOADED LINE PHASE-SHIFTER CIRCUITS

The loaded line phase shifter [6, 8, 9] makes use of a transmission line loaded with a symmetric pair of switchable reactive elements, as shown in Figure 8.12. The spacing between the reactive elements is chosen equal to about a quarter-wavelength such that the reflections from the reactive elements cancel at the input terminal at the design frequency. Of the two types of loaded line phase shifters shown in Figure 8.12, the shunt-loaded type is the most commonly employed in practice because of simpler biasing circuit requirements.

The loaded line phase shifter can be conveniently analyzed by using the *ABCD* matrices [6]. Considering the shunt-loaded circuit with switches connected to susceptances B_1 as shown in Figure 8.12(a), the overall *ABCD* matrix of the network is obtained as

(a)

(b)

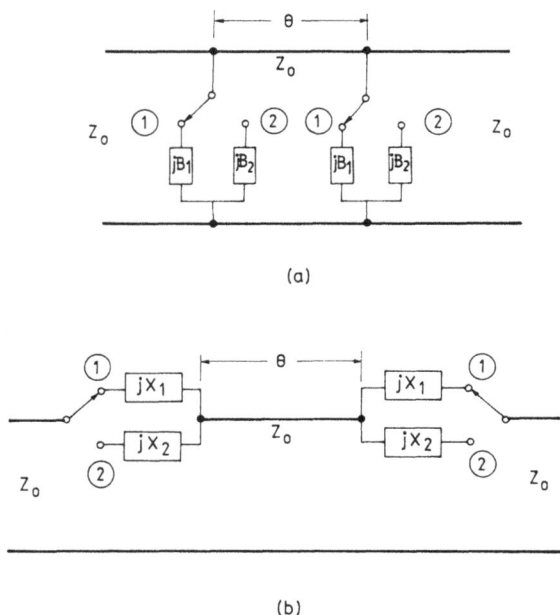

Figure 8.12 Schematic of loaded line phase-shifter circuits with (a) switchable shunt susceptances and (b) switchable series susceptances.

$$\begin{bmatrix} A & B \\ C & D \end{bmatrix} = \begin{bmatrix} 1 & 0 \\ jB_1 & 1 \end{bmatrix} \begin{bmatrix} \cos\theta & jZ_0\sin\theta \\ jY_0\sin\theta & \cos\theta \end{bmatrix} \begin{bmatrix} 1 & 0 \\ jB_1 & 1 \end{bmatrix}$$

$$= \begin{bmatrix} \cos\theta - Z_0B_1\sin\theta & jZ_0\sin\theta \\ j(2B_1\cos\theta + (Y_0 - Z_0B_1^2)\sin\theta) & \cos\theta - Z_0B_1\sin\theta \end{bmatrix} \quad (8.51)$$

Applying the formula for transmission coefficient given by (8.24), we get

$$S_{21} = |S_{21}|e^{j\phi_1}$$

$$= \frac{2}{\left[(\cos\theta - B_{N1}\sin\theta) + j\left\{ B_{N1}\cos\theta + \left(1 - \dfrac{B_{N1}^2}{2}\right)\sin\theta \right\} \right]} \quad (8.52)$$

where B_{N1} is the normalized susceptance given by

$$B_{N1} = B_1 Z_0 = B_1/Y_0 \quad (8.53)$$

The transmission phase ϕ_1 is given by

$$\phi_1 = \tan^{-1}\left| -\frac{B_{N1} + \left(1 - \dfrac{B_{N1}^2}{2}\right)\tan\theta}{(1 - B_{N1}\tan\theta)}\right| \qquad (8.54)$$

The magnitude of the input reflection coefficient is

$$|S_{11}| = [1 - |S_{21}|^2]^{1/2} = \left[1 - \frac{1}{1 + B_{N1}^2(\cos\theta - 0.5B_{N1}\sin\theta)^2}\right]^{1/2} \qquad (8.55)$$

In the second switching state, the switches are connected to susceptances of value B_2. The transmission coefficient S_{21}', its phase factor ϕ_2, and the magnitude of input reflection coefficient $|S_{11}'|$ are given by Equations (8.52), (8.54), and (8.55), respectively, by replacing B_{N1} by B_{N2} ($= B_2 Z_0$). The phase shift $\Delta\phi$ is then equal to $(\phi_2 - \phi_1)$.

It has been reported [6] that the widest bandwidth is achieved when $\theta = \pi/2$ and $B_{N1} = -B_{N2}$. With these conditions, the expression for phase shift becomes

$$\Delta\phi = \tan^{-1}\left[-\frac{(1 - 0.5B_{N1}^2)}{B_{N1}}\right] - \tan^{-1}\left[\frac{(1 - 0.5B_{N1}^2)}{B_{N1}}\right]$$

$$= \pi - 2\tan^{-1}\left[\frac{1 - 0.5B_{N1}^2}{B_{N1}}\right]$$

$$= 2\tan^{-1}\left[\frac{B_{N1}}{1 - 0.5B_{N1}^2}\right] \qquad (8.56)$$

The bandwidth is governed by the tolerable VSWR and phase error. For a specified VSWR and phase error, the smaller the phase bit, the larger the bandwidth is. For example, within a VSWR of 1.5 and a phase shift error of $\pm2°$, a 22.5° phase bit is reported to offer nearly an octave bandwidth, whereas, for a 45° phase bit, the bandwidth reduces to about 25% [6]. For higher values of phase shifts, the VSWR becomes excessively high.

A narrowband, perfectly matched, loaded line phase shifter can be designed by setting $B_{N2} = 0$ and B_{N1} equal to some positive number. The spacing between the two susceptances can be adjusted to make $|S_{11}| = 0$. Setting $|S_{11}| = 0$ in (8.55), we get

$$\tan\theta = \frac{2}{B_{N1}} \qquad (8.57)$$

Substituting (8.57) in (8.54) gives

$$\phi_1 = \pi - \tan^{-1}\left(\frac{2}{B_{N1}}\right) \tag{8.58}$$

With $B_{N2} = 0$, the phase delay ϕ_2 is equal to the electrical length θ so that

$$\phi_2 = \theta = \tan^{-1}\left(\frac{2}{B_{N1}}\right) \tag{8.59}$$

The phase shift is given by

$$\Delta\phi = \phi_1 - \phi_2 = \pi - 2 \tan^{-1}\left(\frac{2}{B_{N1}}\right) \tag{8.60}$$

From (8.59) and (8.60), we can express B_{N1} and θ in terms of the desired phase shift $\Delta\phi$.

$$B_{N1} = 2 \tan\frac{\Delta\phi}{2} \tag{8.61}$$

$$\theta = \frac{1}{2}(\pi - \Delta\phi) \tag{8.62}$$

The loaded line phase shifter is particularly useful for small phase shifts up to 45°. This is due to the fact that the susceptance magnitude must be kept small for a good input match over the desired frequency band.

8.8 HIGH-PASS LOW-PASS PHASE-SHIFTER CIRCUITS

The phase-shifter circuits described in the preceding sections involve sections of transmission lines. With the appropriate choice of transmission media, these circuits can be fabricated in convenient sizes at microwave and millimeter-wave frequencies. At lower frequencies, because of the need to use long line lengths, the phase shifters become physically large in size. The high-pass low-pass type of phase-shifter circuits shown in Figure 8.13 are recommended for use at lower frequencies, particularly in the UHF band [6]. These structures are also the most promising ones for monolithic realization using either planar Schottky diodes or MESFETs as switching elements [10]. When the switches are connected to the low-pass filter, the signal passing through the circuit undergoes a phase delay, and when the switches are connected to the

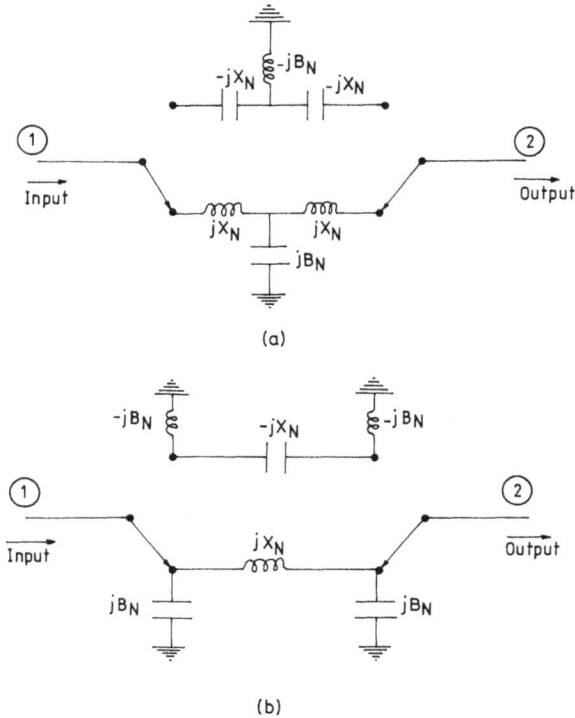

Figure 8.13 High-pass low-pass phase shifters: (a) T-type and (b) π-type.

high-pass filter, the circuit provides a phase advance. Phase shift is obtained by switching between the two filter circuits.

As in the case of loaded line phase shifter, the analysis of this circuit can be easily carried out using the *ABCD* matrices of various elements. For the T-type circuit (Figure 8.13(a)), when the switches are connected to the low-pass circuit, the transmission coefficient S_{21} is given by [6]

$$S_{21} = \frac{2}{2(1 - B_N X_N) + j(B_N + 2X_N - B_N X_N^2)} \tag{8.63}$$

where X_N and B_N represent the normalized reactance and susceptance, respectively. The transmission phase ϕ_1 is given by

$$\phi_1 = \tan^{-1}\left[-\frac{B_N + 2X_N - B_N X_N^2}{2(1 - B_N X_N)} \right] \tag{8.64}$$

When the circuit is switched to the high-pass state, the transmission coefficient S'_{21} and the transmission phase ϕ_2 are obtained by replacing B_N by $-B_N$ and X_N by $-X_N$ in (8.63) and (8.64). The differential phase shift is given by

$$\Delta\phi = \phi_1 - \phi_2 = 2 \tan^{-1}\left[-\frac{B_N + 2X_N - B_N X_N^2}{2(1 - B_N X_N)}\right] \qquad (8.65)$$

Assuming the phase shifter to be lossless and imposing the condition of perfect match, namely $|S_{21}| = 1$, we obtain the relation

$$B_N = \frac{2X_N}{1 + X_N^2} \qquad (8.66)$$

Substituting for B_N from (8.66) in (8.65), we get

$$\Delta\phi = 2 \tan^{-1}\left(\frac{2X_N}{X_N^2 - 1}\right) \qquad (8.67)$$

Alternatively, X_N and B_N can be expressed in terms of phase shift as

$$X_N = \tan\left(\frac{\Delta\phi}{4}\right) \qquad (8.68)$$

$$B_N = \sin\left(\frac{\Delta\phi}{2}\right) \qquad (8.69)$$

Similar analysis for the π-type phase shifter yields the following expressions for X_N and B_N:

$$X_N = \sin\left(\frac{\Delta\phi}{2}\right) \qquad (8.70)$$

$$B_N = \tan\left(\frac{\Delta\phi}{2}\right) \qquad (8.71)$$

The frequency response of the two filter circuits is such that as the frequency is increased, the increase in phase delay in the low-pass state is compensated by the decrease in phase advance in the high-pass state. Thus, constant phase shift with low VSWR is achieved over a fairly large bandwidth. For example, a 90° phase-shifter bit is reported to have a bandwidth of nearly an octave within a phase error of ±2°, and smaller phase bits have even larger bandwidth [6].

The peak and average power capabilities of this phase shifter are the same as those presented in Section 8.6 for the switched line phase shifter.

8.9 OTHER SWITCHED NETWORK PHASE SHIFTERS

In the switched line phase shifter described in Section 8.6, phase shift is achieved by switching between two path lengths, and in the high-pass low-pass phase shifter, phase shift is obtained by switching between two lumped-element filters. Both these circuit forms may be designed to give any phase shift value. Using the same switched network concept but other types of networks in the two paths, broadband fixed phase bit circuits can be realized. We present below two such examples, one for a 180° phase bit and the other for a 90° phase bit.

Broadband 180° Bit Phase Shifter [11]

Boire, Degenford, and Cohn [11] have reported a 180° digital phase bit circuit, which uses switching between a parallel-coupled line and a π-network. The circuit is shown in Figure 8.14. The two networks are equivalent and have identical bandpass filter characteristics [12]. While the magnitude of the transmission coefficient remains the

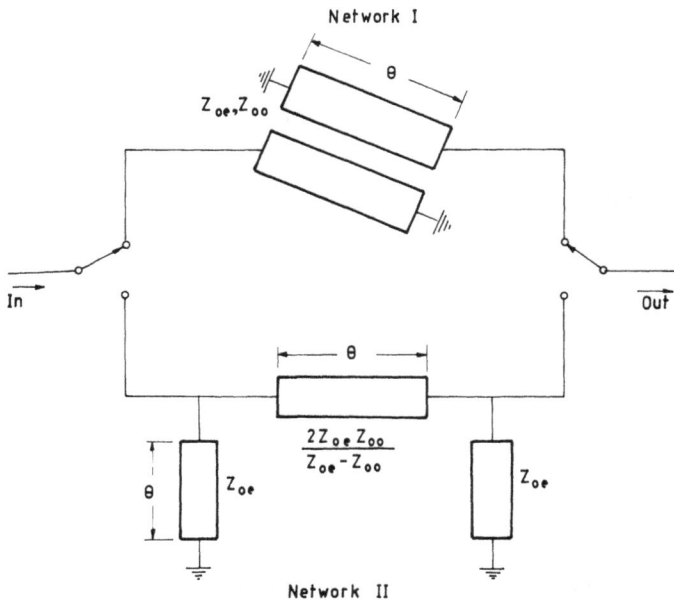

Figure 8.14 Switched network 180° bit phase shifter. (After Boire, Degenford, and Cohn [11].)

same, the transmission phase differs exactly by 180° when switched between the two networks. This property can be seen from the *ABCD* matrices of the two networks. For the parallel-coupled line, the matrix is given by [11]

$$
\begin{bmatrix} A & B \\ C & D \end{bmatrix} = \begin{bmatrix} -\dfrac{(Y_{0o} + Y_{0e})}{(Y_{0o} - Y_{0e})}\cos\theta & -\dfrac{j2\sin\theta}{(Y_{0o} - Y_{0e})} \\ -\dfrac{(Y_{0o} + Y_{0e})^2 \cos^2\theta - (Y_{0o} - Y_{0e})^2}{j2(Y_{0o} - Y_{0e})\sin\theta} & -\dfrac{(Y_{0o} + Y_{0e})\cos\theta}{(Y_{0o} - Y_{0e})} \end{bmatrix}
$$

(8.72)

where Y_{0e} ($= 1/Z_{0e}$) and Y_{0o} ($= 1/Z_{0o}$) are the even- and odd-mode admittances, respectively, of the parallel-coupled line, and θ is its electrical length. Noting A', B', C', and D', as the parameters of the π-network, we have

$$
\begin{bmatrix} A' & B' \\ C' & D' \end{bmatrix} = \begin{bmatrix} -1 & 0 \\ 0 & -1 \end{bmatrix} \begin{bmatrix} A & B \\ C & D \end{bmatrix}
$$

(8.73)

There is a phase reversal in the *ABCD* parameters of the two networks and this phase change is independent of θ. The circuit thus offers a 180° phase shift that is independent of frequency.

Broadband 90° Bit Phase Shifter [13]

Figure 8.15 shows a broadband 90° bit phase shifter based on switching between a straight transmission line and a stub-based network [13]. The characteristics of the

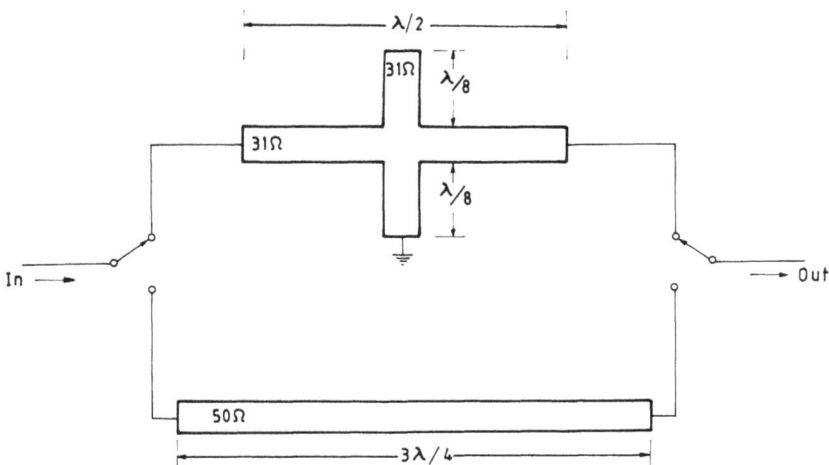

Figure 8.15 Switched network 90° bit phase shifter. (After Wilds [13].)

stub-based network are presented in Chapter 2 of the companion volume under the section on fixed phase shifters. This phase shifter is reported to offer a bandwidth of nearly an octave.

8.10 CONCLUDING REMARKS

Phase-shifter circuits using switching devices are broadly classified into the following categories:

1. Reflection-type circuits
   * Circulator coupled
   * Hybrid coupled
2. Transmission-type circuits
   * Switched line
   * Loaded line
   * High-pass low-pass
   * Switched network

While all the above circuit forms are suited for digital phase shifting, the circulator coupled, hybrid coupled, and loaded line circuits can be used for realizing analog phase shifters also by adopting voltage-variable control devices, such as the varactors. Among the two reflective-type circuits, the hybrid coupled is simpler and easier to realize, and hence is commonly preferred over the circulator type for both digital and analog phase shifting.

All circuit forms can be realized in hybrid microwave integrated circuit (MIC) version. The most popular circuits for multibit *p-i-n* diode phase shifters in hybrid MIC are the switched line, the hybrid coupled, and the loaded line. The loaded line circuit is particularly advantageous for small-size phase bits. Detailed description and comparison of various types of *p-i-n* diode phase shifters are provided in Chapter 9. FET phase shifters in hybrid MIC are covered in Chapter 10.

Except for the circulator coupled circuit, all other phase-shifter types can be easily implemented in monolithic form with GaAs FET as the most natural control device. The high-pass low-pass lumped circuit is particularly suited for monolithic realization. Various types of monolithic phase shifters are discussed in Chapter 12.

REFERENCES

1. R.V. Garver, *Microwave Diode Control Devices,* Artech House, MA, 1978.
2. K.E. Mortenson and J.M. Borrego, *Design, Performance and Applications of Microwave Semiconductor Control Components,* Reprint Volume, Artech House, MA, 1972.
3. J.F. White, *Microwave Semiconductor Engineering,* Van Nostrand, NJ, 1982.
4. M.E. Hines, "Fundamental Limitations in RF Switching and Phase Shifting Using Semiconductor Diodes," *Proc. IEEE,* Vol. 52, June 1964, pp. 697–708.

5. J.H.C. Van Heuven, "P-I-N Switching Diodes in Phase Shifters for Electronically Scanned Aerial Arrays," *Philips Tech. Rev.*, Vol. 32, No. 9/10/11/12, 1971, pp. 405–412.

6. R.V. Garver, "Broad-Band Diode Phase Shifters," *IEEE Trans. on Microwave Theory and Tech.*, Vol. MTT-20, May 1972, pp. 314–323.

7. J.F. White, "Diode Phase Shifters for Array Antennas," *IEEE Trans. on Microwave Theory and Tech.*, Vol. MTT-22, June 1974, pp. 658–674.

8. J.F. White, "High Power, p-i-n Controlled, Microwave Transmission Phase Shifters," *IEEE Trans. on Microwave Theory and Tech.*, Vol. MTT-13, March 1965, pp. 233–243.

9. W.A. Davis, "Design Equations and Bandwidth of Loaded-Line Phase Shifters," *IEEE Trans. on Microwave Theory and Tech.*, Vol. MTT-22, May 1974, pp. 561–563.

10. R.V. Garver, "Microwave Semiconductor Control Devices," *IEEE Trans. on Microwave Theory and Tech.*, Vol. MTT-27, May 1979, pp. 523–529.

11. D.C. Boire, J.E. Degenford, and M. Cohn, "A 4.5 to 18 GHz Phase Shifter," *IEEE MTT Int. Microwave Symp. Digest*, 1985, pp. 601–604.

12. G.L. Matthaei, L. Young, and E.M.T. Jones, *Microwave Filters, Impedance Matching Networks, and Coupling Structures*, McGraw-Hill, New York, 1964.

13. R.B. Wilds, "Try $\lambda/8$ Stubs for Fast Fixed Phase Shifts," *Microwaves*, December 1979, pp. 67–68.

Chapter 9
P-I-N Diode Phase Shifters

9.1 INTRODUCTION

P-I-N diode phase shifters are basically digital in nature. The function of the *p-i-n* diode is to act as an electronic switch when operated between fixed forward and reverse bias states. All the phase shifter configurations discussed in Chapter 8 can be used to build *p-i-n* diode phase shifters by replacing the ideal on-off switches with practical *p-i-n* diodes. This chapter is devoted exclusively to the analysis and design aspects of *p-i-n* diode phase shifters employing practical transmission media. The most commonly used transmission media are the *strip-* and *microstrip*-like transmission lines. In particular, the microstrip line is used the most extensively at microwave frequencies. Other transmission lines, which have found less frequent usage, are the *stripline* and *slotline*. Potential structures for use at millimeter-wave frequencies are the *suspended stripline* and *fin-line*. Among the various circuit forms considered in Chapter 8, the commonly employed ones are the switched line, the hybrid coupled, and the loaded line types. *P-I-N* diode phase shifters in all the configurations are considered, taking into account various practical aspects, such as impedance matching, equalization of insertion loss, and broadbanding techniques.

9.2 *P-I-N* DIODE AND ITS EQUIVALENT CIRCUIT [1]

The *p-i-n* diode consists of an intrinsic semiconductor layer sandwiched between heavily doped *p-* and *n*-type regions. In practice, the intrinsic region is a very weakly doped *p*-type or *n*-type silicon. Figure 9.1 shows the schematic diagram of a *p-i-n* diode and its typical dc voltage-current characteristic. The ac behavior of a *p-i-n* diode at low frequencies (below about 1 KHz) is essentially the same as that of a *p-n* diode. This behavior is governed by the dc voltage-current characteristics, and rectification takes place in both the devices. The behavior of the *p-i-n* diode at microwave frequencies, however, is entirely different from that of a *p-n* diode. The *p-i-n* diode can be used as a switch when operated between fixed forward and reverse bias states for switching microwave signals. When a forward dc bias is applied to

(a)

(b)

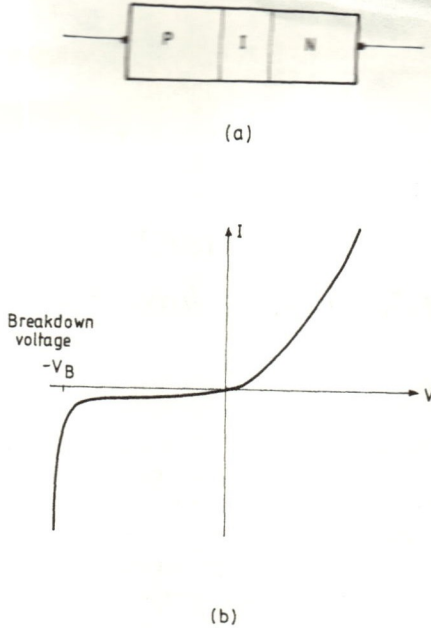

Figure 9.1 (a) Schematic diagram and (b) dc voltage-current characteristic of *p-i-n* diode.

the *p-i-n* diode, free charges from the *p*- and *n*-regions flood the I-region, thus converting it into a conducting medium. The diode behaves essentially like a short circuit, and any microwave signal superimposed on it passes without any rectification. On the other hand, when a reverse bias is applied, the I-region gets completely depleted of charge carriers, and the diode behaves virtually like an open circuit having a large capacitive reactance. For a microwave signal superimposed on the dc voltage, the time period of the positive half cycle would be too short when compared with the life time of the charge carriers, with the result that no conduction can take place through the I-region. Thus, for microwave signals, the *p-i-n* diode acts as a linear device whose impedance is governed by the slope of the dc characteristic at the operating point.

A packaged *p-i-n* diode can be represented by a general equivalent circuit, as shown in Figure 9.2(a), where L_p and C_p are the stray inductance and capacitance, respectively, because of the encapsulation of the diode. Typical values of L_p and C_p are in the range 0.1 to 2 nH, and 0.1 to 0.5 pF, respectively. For a chip diode, these parameters may be omitted. The other parameters, R_j, C_j, and R_b, represent the junction

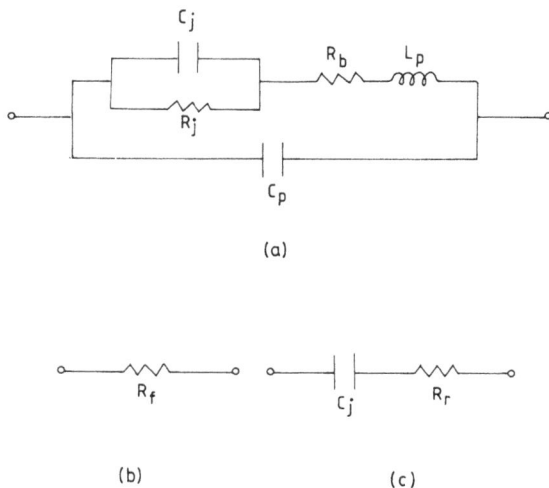

(a)

(b)   (c)

Figure 9.2 Equivalent circuit of (a) packaged *p-i-n* diode, (b) chip diode under forward bias, and (c) chip diode under reverse bias.

resistance, junction capacitance, and bulk semiconductor resistance, respectively, of the diode. These parameters take different values under different bias conditions. For example, the equivalent circuit of a *p-i-n* diode chip under forward dc bias can be represented by a small resistance R_f (Figure 9.2(b)). It is the sum of the resistances of the *p*- and *n*-layers and the contact resistance. The value of R_f is typically in the range 0.2Ω to 5Ω. Under reverse bias, the *p-i-n* diode chip can be represented by a capacitance C_j in series with a small resistance R_r, as shown in Figure 9.2(c). Typical values of C_j and R_r are in the range of 0.01 to 2 pF and 0.2Ω to 5Ω, respectively.

9.3 SWITCHED-LINE STRIPLINE AND MICROSTRIP PHASE SHIFTERS

The basic switched line configurations using (1) series-mounted switches and (2) shunt-mounted switches are discussed in Chapter 8 (Section 8.6). In practice, each switched line phase shifter bit makes use of four identical *p-i-n* diodes to switch between the two transmission line sections. In the following, we present the analysis and practical aspects of both series- and shunt-type circuits in striplines and microstrip lines. Some of the other configurations that use fewer than four diodes are also discussed.

9.3.1 Series-Diode Switched Line Circuit

Figure 9.3(a) shows a typical circuit layout of a series-diode switched line phase shifter in stripline/microstrip-transmission line. The characteristic impedance of the transmission line is Z_0. D_1, D_2, D_3, and D_4 represent four identical p-i-n diodes having a forward-bias impedance of Z_f and a reverse-bias impedance of Z_r. The transmission coefficient of this network can be determined by analyzing it in terms of even- and odd-mode excitations [2, 3]. In the even-mode excitation, the input and output lines are connected to voltage generators having voltages of equal magnitude and phase (for example, $V_0/2$). The plane of symmetry PP' can then be represented

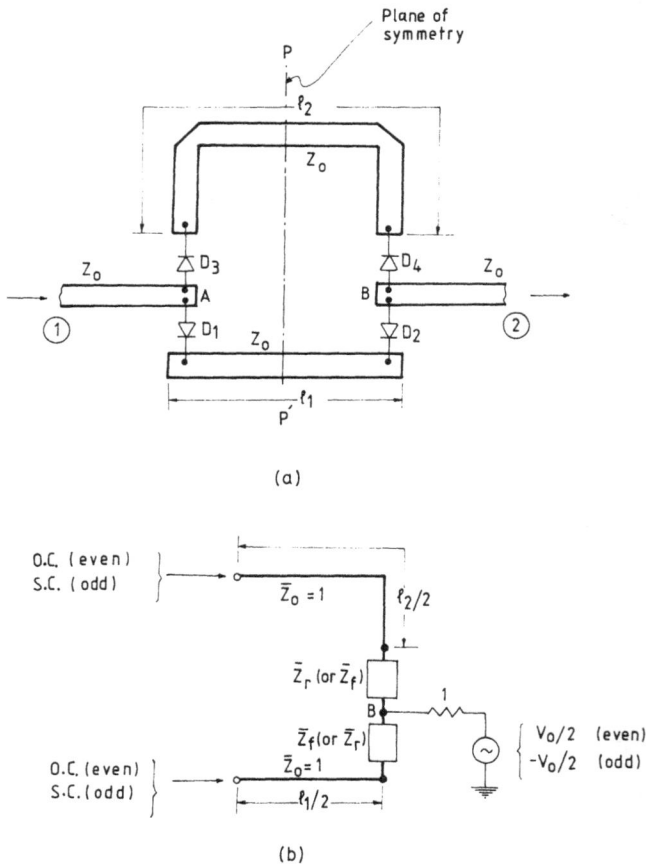

(a)

(b)

Figure 9.3 (a) Series-diode switched line phase shifter; (b) one-half of the circuit for the purpose of analysis.

In the odd-mode excitation, the same two generators are con-
by an open circuit, so that the plane of symmetry PP' represents a short circuit.
nected out of phase. impedance is chosen to be Z_0, which is equal to the characteristic
The generator the transmission line sections. When the two excitations are superimposed, the left generator generates the full voltage V_0, with no voltage in the right
generator.

Figure 9.3(b) shows the symmetric right half of the circuit for the purpose of
determining the transmission coefficient. All impedances are normalized with respect
to Z_0. Thus, $\bar{Z}_f = Z_f/Z_0$ and $\bar{Z}_r = Z_r/Z_0$. Let V_B^e and V_B^o be the voltages appearing
at the output point B for even- and odd-mode excitations, respectively. These volt-
ages can be expressed as

$$V_B^e = \frac{V_0}{2(1 + \bar{Y}_B^e)} \tag{9.1a}$$

$$V_B^o = \frac{-V_0}{2(1 + \bar{Y}_B^o)} \tag{9.1b}$$

where $\bar{Y}_B^e$ and $\bar{Y}_B^o$ are the normalized admittances looking into the transmission line
sections at point B for even mode and odd mode, respectively. When the two excitations are superimposed, the total voltage at B is $(V_B^e + V_B^o)$ and voltage of point
A is $V_0/2$. The transmission coefficient S_{21} is then given by

$$S_{21} = \frac{V_B^e + V_B^o}{V_0/2} = \left[\frac{1}{(1 + \bar{Y}_B^e)} - \frac{1}{(1 + \bar{Y}_B^o)} \right] \tag{9.2}$$

We now consider the two switching states of the circuit.

State 1: In state 1, let diodes D_1 and D_2 be forward biased and diodes D_3 and D_4 be
reverse biased. Using $S_{21}^{(1)}$ as the transmission coefficient, we can write

$$S_{21}^{(1)} = |S_{21}^{(1)}|e^{j\phi_1} = \left[1 + \frac{1}{(\bar{Z}_f - j\cot\beta l_1/2)} + \frac{1}{(\bar{Z}_r - j\cot\beta l_2/2)} \right]^{-1}$$
$$- \left[1 + \frac{1}{(\bar{Z}_f + j\tan\beta l_1/2)} + \frac{1}{(\bar{Z}_r + j\tan\beta l_2/2)} \right]^{-1} \tag{9.3}$$

where $\beta = 2\pi/\lambda$ is the propagation constant of the transmission line.

State 2: In state 2, diodes D_1 and D_2 are reverse biased, and D_3 and D_4 are forward
biased. The expression for the transmission coefficient, expressed as $S_{21}^{(2)} = |S_{21}^{(2)}|e^{j\phi_2}$, is obtained by interchanging $\bar{Z}_f$ and $\bar{Z}_r$ in (9.3).

The differential phase shift $\Delta\phi$ is given by

$$\Delta\phi = (\phi_2 - \phi_1) \tag{9.4}$$

The insertion loss in the two switching states can be obtained using the relation

$$\alpha(dB) = 20 \log_{10}|S_{21}^{(1)}|, \quad \text{state 1}$$

$$= 20 \log_{10}|S_{21}^{(2)}|, \quad \text{state 2} \tag{9.5}$$

If the *p-i-n* diodes are ideal switches, then, as discussed in Chapter 8, the phase is given by $\beta(l_2 - l_1)$ which is simply a linear function of frequency. For practical *p-i-n* diodes, $Z_f \approx R_f$ and $Z_r \approx -jX_r$, where $X_r = 1/\omega C_j$. The forward-bias resistance R_f is generally very small in comparison with the characteristic impedance Z_0 of the transmission line so that in the on-diode path $\bar{Z}_f$ can be assumed to be nearly zero. In the off-diode path, the large but finite reactive impedance of the *p-i-n* diode can be considered as equivalent to adding a small length of line. Thus, even with non-ideal diodes, the phase-frequency characteristic is nearly linear, except at some resonant frequencies where the effective length of the off-diode path (electrical length of the line plus the equivalent lengths of the capacitive off-diodes) is equal to $\lambda/2$ or integral multiples of $\lambda/2$. At resonance, the reflected signal adds up in phase with the incident signal so as to give large phase errors as well as insertion loss spikes. However, for a 180° phase bit the phase error would be zero for all values of l_1, provided all four diodes are assumed identical [3]. In general, in order to ensure minimum phase error and equal insertion loss in both the switching states, the value of βl_1 should be chosen between 20° and 50°.

9.3.2 Shunt-Diode Switched Line Circuit

In the shunt-diode switched line phase shifter, the diodes are shunt mounted at a distance of l_0 from the main junction (Figure 9.4). If the *p-i-n* diodes offer a negligibly small forward-bias resistance in the on-state, then choosing $l_0 = \lambda/4$ ensures high impedance at the main junction and, hence, a large isolation of the on-diode path. In practice, when the lead inductance of the diode needs to be taken into account, the distance l_0 will have to be adjusted so as to achieve maximum impedance at the main junction due to the on-diode. The design of the phase shifter therefore involves the correct choice of l_0, l_1, and l_2 so as to obtain the desired phase shift and minimum insertion loss. This can be easily done by first obtaining the expressions for transmission and reflection coefficients of the circuit, and then computationally optimizing the dimensional parameters. Because the structure is symmetric about PP', the even- and odd-mode technique discussed in Section 9.3.1 can be applied by considering the right half of the structure.

State 1: As in the series-diode configuration, we assume that in state 1 transmission is predominantly through the lower path. For this, diodes D_1 and D_2 are to be reverse biased, while D_3 and D_4 are to be forward biased. Denoting $\bar{Z}_f$ and $\bar{Z}_r$ as the normalized impedances of the diode under forward- and reverse-biased states, respectively, we obtain the expression for $S_{21}^{(1)}$ as

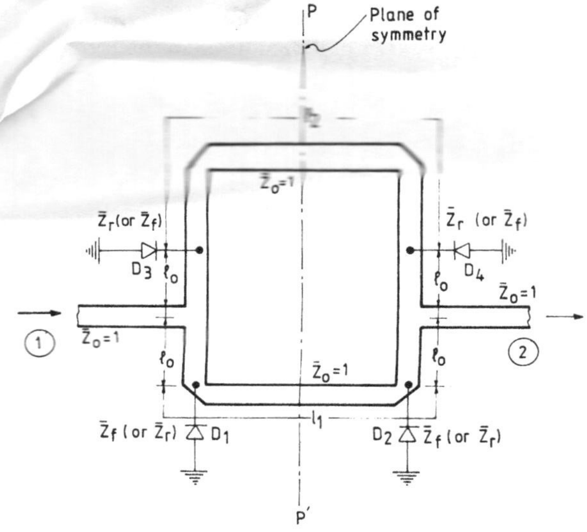

Figure 9.4 Shunt-diode switched line phase shifter.

$$
S_{21}^{(1)} = |S_{21}^{(1)}| e^{j\phi_1} = \left[1 + \frac{\bar{Z}_f \left(\tan\left(\frac{\beta l_2}{2}\right) \cot \beta l_0 + 1 \right) - j \cot \beta l_0}{1 - j\bar{Z}_f \left(\cot \beta l_0 - \tan\left(\frac{\beta l_2}{2}\right) \right)} \right.
$$

$$
\left. + \frac{\bar{Z}_r \left(\tan\left(\frac{\beta l_1}{2}\right) \cot \beta l_0 + 1 \right) - j \cot \beta l_0}{1 - j\bar{Z}_r \left(\cot \beta l_0 - \tan\left(\frac{\beta l_1}{2}\right) \right)} \right]^{-1}
$$

$$
- \left[1 + \frac{\bar{Z}_f \left(-\cot\left(\frac{\beta l_2}{2}\right) \cot \beta l_0 + 1 \right) - j \cot \beta l_0}{1 - j\bar{Z}_f \left(\cot \beta l_0 + \cot\left(\frac{\beta l_2}{2}\right) \right)} \right.
$$

$$
\left. + \frac{\bar{Z}_r \left(-\cot\left(\frac{\beta l_1}{2}\right) \cot \beta l_0 + 1 \right) - j \cot \beta l_0}{1 - j\bar{Z}_r \left(\cot \beta l_0 + \cot\left(\frac{\beta l_1}{2}\right) \right)} \right]^{-1} \tag{9.6}
$$

State 2: In state 2, the expression for the transmission coefficient $S_{21}^{(2)} = |S_{21}^{(2)}|e^{j\phi_2}$ is the same as in Equation (9.6), except that $\bar{Z}_f$ and $\bar{Z}_r$ get interchanged. The differential phase shift is then given by (9.4) and the insertion loss in the two switching states can be obtained using (9.5).

9.3.3 Other Switched Line Configurations

The series- and shunt-diode switched line phase shifters considered above make use of a minimum of four *p-i-n* diodes. A different switched line configuration, which makes use of three *p-i-n* diodes, two in series and one in shunt, has been reported by Dubost and Guero [4]. Figure 9.5 illustrates the scheme in microstrip. The shunt

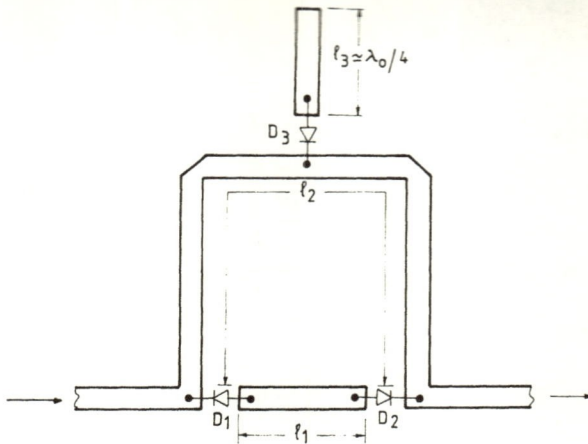

Figure 9.5 Series-shunt switched line phase shifter. (After Dubost and Guero [4].)

diode D_3 is located at the center of the longer transmission line section. The open-ended transmission line section of length l_3 is approximately a quarter-wavelength long in microstrip so as to terminate the diode in an effective short. When all three diodes are forward biased, signal flows through the lower line, and when they are reverse biased, signal passes through the upper line. A single-bit $180°$ phase shifter designed at a center frequency of 15 GHz is reported to offer a phase error of $±3°$, and an insertion loss of 1.5 and 0.8 dB in forward- and reverse-biased states, respectively [4].

shows the schematic of a diode loop binary phase shifter reported by Figure [5]. It is a switched line network with just two p-i-n diodes. The circuit by Co an all-pass C-section (also known as Schiffman section) [6], which is ed through a single diode, with a second diode used to improve the input ch. Assuming the diodes to be ideal, and applying the even- and odd-mode analysis technique, Connerney [5] has derived the following design equations for this phase shifter:

$$\theta_0 = \tan^{-1}[\bar{Y}_{0o} \tan(\Delta\phi_0/2)] \tag{9.7a}$$

$$\Delta\phi = -2 \tan^{-1}[\bar{Y}_{0e} \tan(\omega\theta_0/\omega_0)]$$
$$- \tan^{-1}[\bar{Y}_{0e} \cot(\pi\omega/2\omega_0)] \tag{9.7b}$$

$$|\Gamma_1| = 0 \tag{9.7c}$$

$$|\Gamma_2| = [1 + \bar{Z}_{0e}^2 \tan^2(\pi\omega/2\omega_0)]^{-1/2} \tag{9.7d}$$

In the above equations, θ_0 is the electrical length of the coupled-line section at the design frequency f_0, $\Delta\phi$ is the differential phase shift, and $\Delta\phi_0$ is the differential phase shift at f_0. $\bar{Z}_{0e} = 1/\bar{Y}_{0e}$ and $\bar{Z}_{0o} = 1/\bar{Y}_{0o}$ are the normalized even- and odd-mode impedances of the coupled-line section such that $(\bar{Z}_{0e} \bar{Z}_{0o})^{1/2} = (\bar{Y}_{0e} \bar{Y}_{0o})^{1/2} = 1$. The parameters $|\Gamma_1|$ and $|\Gamma_2|$ are the magnitudes of the reflection coefficients in the two switching states of the phase shifter.

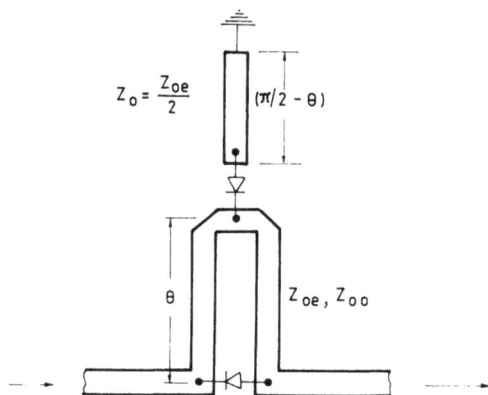

Figure 9.6 Diode loop binary phase shifter. (After Connerney [5].)

9.3.4 Broadbanding Techniques

Stripline Medium

Schiffman [6] observed that by switching between an all-pass C-section and a uniform nondispersive line of suitably chosen length, a nearly constant phase shift can be achieved over a wide bandwidth. The C-section is a section of coupled transmission line with two of its ends connected together. In a homogeneous medium, the Schiffman section has a phase-frequency response as shown in Figure 9.7. Its input impedance, when looking into either of the two free ports (refer to inset in Figure 9.7), is given by

$$Z_{in} = \sqrt{Z_{0e} Z_{0o}} \tag{9.8}$$

where Z_{0e} and Z_{0o} are the even- and odd-mode characteristic impedances, respectively, of the coupled section. The phase-frequency response is governed by the equation

$$\cos\phi = \frac{R - \tan^2\theta}{R + \tan^2\theta} \tag{9.9}$$

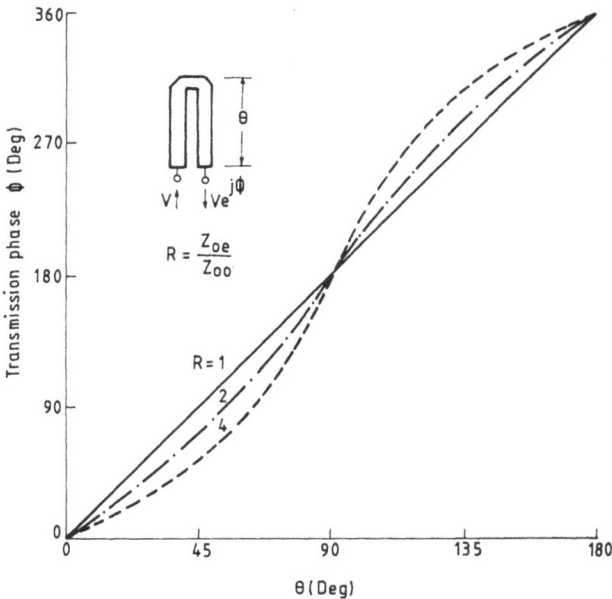

Figure 9.7 Transmission phase-frequency response of homogeneous C-section.

where

$$R = \frac{Z_{0e}}{Z_{0o}} \qquad (9.10)$$

For the special case $R = 1$, the Schiffman section reduces to a straight transmission line of electrical length 2θ.

Of the various switched line configurations considered in the preceding sections, the series and shunt types shown in Figures 9.3 and 9.4 are the most extensively used. However, they are basically time-delay devices; wide bandwidth on the order of an octave can be achieved by introducing a Schiffman section in one of the transmission paths [6, 7]. Figure 9.8 shows a series-type switched line Schiffman phase shifter. The mode of operation is the same as that described in Section 9.3.1 for the series switched line phase shifter. Because the Schiffman section is symmetrically placed in the upper transmission path, the even- and odd-mode analysis applies also to this configuration. The general expression for the transmission coefficient of the phase shifter is given by Equation (9.2). The even- and odd-mode normalized admittances $\bar{Y}_B^e$ and $\bar{Y}_B^o$ appearing in (9.2) are given by the following expressions:

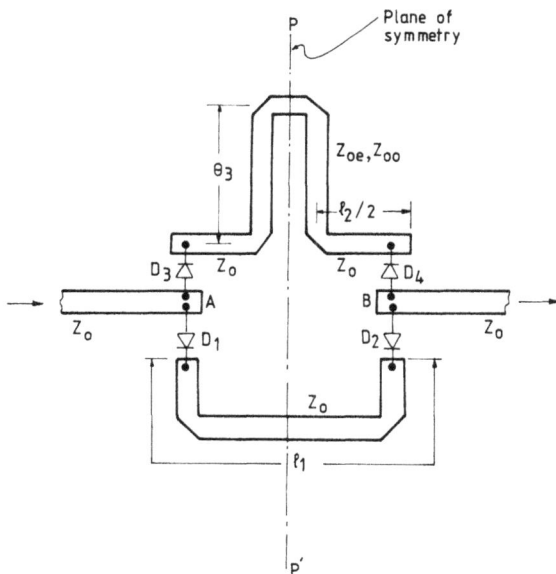

Figure 9.8 Series-type switched line Schiffman phase shifter.

When diodes D_1 and D_2 are forward biased, and D_3 and D_4 reverse biased, we have

$$\bar{Y}_B^e = \frac{1}{[\bar{Z}_f - j\cot(\beta l_1/2)]} + \frac{1}{\left[\bar{Z}_r + j\left\{\dfrac{-\bar{Z}_{0e}\cot\theta_3 + \tan(\beta l_2/2)}{1 + \bar{Z}_{0e}\cot\theta_3\tan(\beta l_2/2)}\right\}\right]}$$

and

$$\bar{Y}_B^o = \frac{1}{[\bar{Z}_f + j\tan(\beta l_1/2)]} + \frac{1}{\left[\bar{Z}_r + j\left\{\dfrac{\bar{Z}_{0o}\tan\theta_3 + \tan(\beta l_2/2)}{1 - \bar{Z}_{0o}\tan\theta_3\cdot\tan(\beta l_2/2)}\right\}\right]} \tag{9.11b}$$

where $\bar{Z}_{0e}$ and $\bar{Z}_{0o}$ are the even- and odd-mode impedances of the coupled-line section normalized with respect to Z_0, and θ_3 is its electrical length. All other notations are the same as those used in Section 9.3.1. When the switching states are reversed, the expressions for $\bar{Y}_B^e$ and $\bar{Y}_B^o$ remain the same as those given by (9.11) except that $\bar{Z}_r$ and $\bar{Z}_f$ get interchanged.

If the diodes are assumed to be ideal, the differential phase shift of the phase shifter is given by

$$\Delta\phi = \beta(l_1 - l_2) - \cos^{-1}\left\{\frac{R - \tan^2\theta_3}{R + \tan^2\theta_3}\right\} \tag{9.12}$$

At the center frequency, the coupled-line section is a quarter-wavelength long ($\theta_3 = \pi/2$), and the transfer phase through it is π. Hence, for a 90° phase shift section, $\beta(l_1 - l_2)$ is chosen equal to $3\pi/2$ at the center frequency. Figure 9.9 illustrates the theoretical bandwidth achievable and the resulting phase error (ε) when a Schiffman section of $R = 4$ is chosen. The phase error is determined by drawing a line (AA') passing through the point $A_0(\theta = \pi/2, \phi = \pi)$ and parallel to the line showing variation of $\beta(l_1 - l_2)$ as a function of θ. It can be seen that with $R = 4$, a phase shift of 90° $\pm$ 12.2° can be achieved over an approximate range of 45° $\le \theta \le$ 135°, which corresponds to a frequency ratio of 3:1. With a smaller value of R for the coupled-line section, both bandwidth and phase error get reduced. For example, with $R = 2.7$, the differential phase shift is 90° $\pm$ 2.5° over a frequency ratio of 1.95:1 [6].

Microstrip Medium

In an inhomogeneous medium, the even- and odd-mode phase velocities in the coupled-line C-section are no longer equal. For example, in a coupled microstrip, the odd

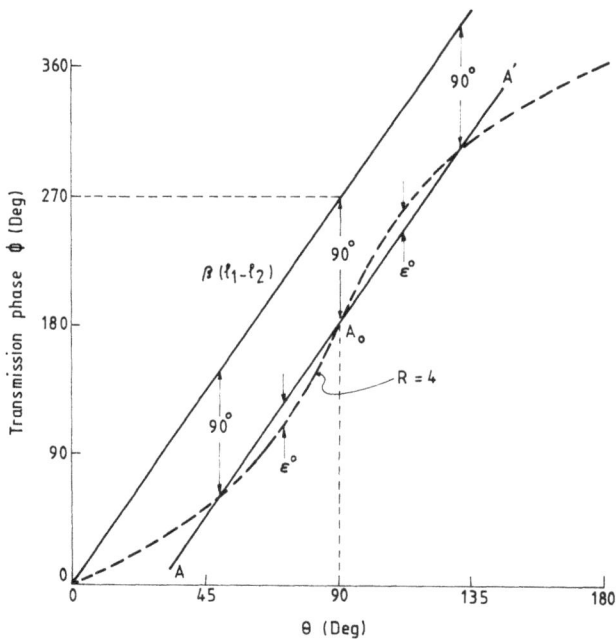

Figure 9.9 Typical bandwidth performance of a Schiffman phase shifter with ideal diodes. Phase shift $\Delta\phi = 90° \pm \varepsilon°$.

mode travels faster than the even mode. Thus, the formulas given in Equations (9.8) and (9.9) are not applicable. Furthermore, the input impedance of the coupled-line C-section will not be independent of frequency, resulting in deterioration of input matching of the coupled line with the rest of the circuit. Several techniques have been reported in the literature for equalizing the two mode velocities [8–11]. Podell [8] has suggested wiggling the input edges of the coupled-line section in order to slow down the odd mode without affecting the even mode. This method is useful when the coupling is less than about 8 dB. Furthermore, wiggling the line is known to introduce losses and phase dispersion [11]. Another method of achieving mode velocity equalization is by means of an interdigitated structure [9, 10]. With this structure, both tight and loose couplings are possible over a wide frequency range.

As compared with the above two methods, a much simpler and convenient technique of achieving broadband matching has been reported by Schiek and Köhler [12]. It makes use of a stepped-impedance C-section (Figure 9.10(a)) in place of the single C-section (Figure 9.10(b)), which serves as a prototype for deriving the design formulas. The stepped-impedance C-section considered here consists of two sections of equal length, with its overall electrical length equal to that of the prototype. The

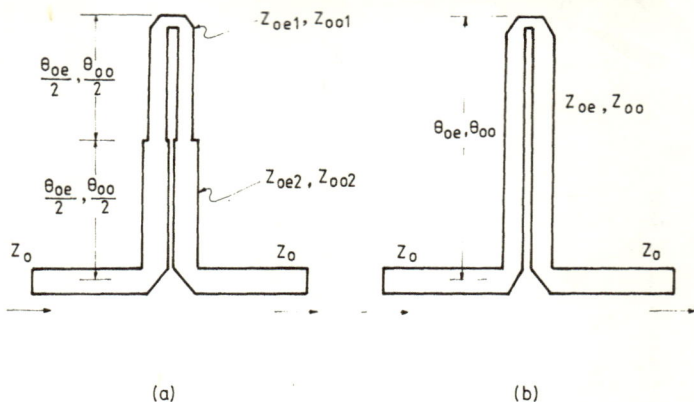

Figure 9.10 (a) A two-section stepped-impedance C-section in microstrip; (b) single C-section in microstrip.

electrical length θ of the prototype is chosen as the arithmetic mean of its even- and odd-mode electrical lengths θ_{0e} and θ_{0o}. At the center frequency $\theta = \pi/2$, specifying $\theta_{0e} = \theta^c_{0e}$ and $\theta_{0o} = \theta^c_{0o}$, we can set

$$\frac{\theta^c_{0e} + \theta^c_{0o}}{2} = \frac{\pi}{2} \tag{9.13}$$

Furthermore, it is assumed that

$$\frac{\theta_{0e}}{\theta_{0o}} = \frac{\theta^c_{0e}}{\theta^c_{0o}} \tag{9.14}$$

For the design of the stepped-impedance C-section, the parameters of the prototype C-section are determined first. That is, for a specified phase shift and tolerable phase error, the ratio $R = (Z_{0e}/Z_{0o})$ of an ideal transverse electromagnetic (TEM) C-section is determined. Z_{0e} and Z_{0o} satisfy the relation $Z_0 = \sqrt{Z_{0e}Z_{0o}}$, where Z_0 is the impedance of the input line with which it is to be matched. Next, corresponding to the ratio R, the even- and odd-mode guide wavelengths (λ_{0e} and λ_{0o}) of the coupled line in microstrip can be determined [13–15]. If λ^c_{0e} and λ^c_{0o} are the values at the center frequency f_0, then the length l of the prototype and, hence, that of the stepped-impedance C-section can be obtained from

$$l = \left[2\left(\frac{1}{\lambda^c_{0e}} + \frac{1}{\lambda^c_{0o}} \right) \right]^{-1} \tag{9.15}$$

The design formulas for determining the even- and odd-mode impedances Z_{0ei} and Z_{0oi} of the two sections ($i = 1, 2$) of the stepped-impedance C-section are given by [12]

$$Z_{0ei} = z_{0ei} Z_0 / \sqrt{R}, \quad i = 1, 2 \qquad (9.16a)$$

$$Z_{0oi} = z_{0oi} Z_0 / \sqrt{R}, \quad i = 1, 2 \qquad (9.16b)$$

where

$$z_{0e1} = \frac{1 + \alpha_{e1}}{\beta_{e1}} \qquad (9.17a)$$

$$z_{0e2} = \alpha_{e1} z_{0e1} \qquad (9.17b)$$

$$z_{0o1} = \frac{\beta_{o1}}{1 + \alpha_{o1}} \qquad (9.17c)$$

$$z_{0o2} = \frac{z_{0o1}}{\alpha_{o1}} \qquad (9.17d)$$

$$\alpha_{e1} = \tan^2\left(\frac{\theta_{0e}^c}{2}\right), \quad \theta_{0e}^c = \frac{2\pi l}{\lambda_{0e}^c} \qquad (9.17e)$$

$$\beta_{e1} = \frac{2\theta_{0e}^c}{\pi} \cos^{-2}\left(\frac{\theta_{0e}^c}{2}\right) \qquad (9.17f)$$

$$\alpha_{o1} = \tan^2\left(\frac{\theta_{0o}^c}{2}\right), \quad \theta_{0o}^c = \frac{2\pi l}{\lambda_{0o}^c} \qquad (9.17g)$$

$$\beta_{o1} = \frac{2\theta_{0o}^c}{\pi} \cos^{-2}\left(\frac{\theta_{0o}^c}{2}\right) \qquad (9.17h)$$

In deriving the above formulas, Schiek and Köhler [12] have assumed that the even- and odd-mode electrical lengths θ_{0e} and θ_{0o} are directly proportional to frequency, and that the phase velocities do not depend on the magnitudes of the even- and odd-mode characteristic impedances. A stepped-impedance C-section designed for a 90° phase shifter (with $R = 2.73$ for the prototype C-section) and matched to a 50Ω system is reported to offer a return loss of above 28 dB across an octave bandwidth

[12]. Matching over a wider frequency range can be achieved by increasing the number of impedance steps in the coupled line. For example, with a stepped imped-ance coupled line with three sections, a 90° phase shifter is reported to offer a the-oretical return loss of better than 30 dB over more than 1:5 bandwidth [12].

While the theoretical bandwidth of Schiffman phase shifters with ideal switches can be extended to as high as a few octaves using the techniques enumerated above, the operational bandwidth is limited in practice because of resonances in the switched-off path. At frequencies where the effective length of the switched-off path becomes equal to half-wavelength or integral multiples of half-wavelength, this path forms a high Q resonant line, causing sharp insertion loss spikes and large phase errors [3]. Coats [7] has compared the off-path insertion loss of switched line Schiffman phase shifters employing series- and shunt-mounted diodes. It was shown that with 20-dB isolation diodes, the series-diode configuration does not offer sufficient separation between minimum isolation points to permit octave band operation between them, whereas the shunt diode configuration offers far superior bandwidth performance. The shunt-type switched line Schiffman phase shifter employing diodes with low parasitic series inductance is reported to yield an octave bandwidth without any res-onance spikes [7].

Suppression of Resonance Spikes

Several techniques have been reported in the literature [16, 17] for reducing the resonance effects in switched line phase shifters. One way is to use two transfer switches instead of two single-pole double-throw (SPDT) switches to terminate the switched off-path in its characteristic impedance [16]. The basic principle is to reduce the unloaded Q of the resonant line so that less power gets coupled to it. Another way of achieving this is to resistively terminate the potentially resonant path in its characteristic impedance by means of an additional switching diode. A circuit con-figuration attributable to Lynes *et al.* [17] illustrating this technique is shown in Figure 9.11. When the transmission is through the lower path, diode D_5 is forward biased, which results in the elimination of any insertion loss spikes. However, when the transmission is through the upper path, the finite isolation of this diode under reverse bias causes a slight increase in the insertion loss of the phase shifter. Both the techniques mentioned above increase the complexity of the driver circuit, and bandwidth limitation occurs because of the RF properties of the diode and dc bias circuit. Because of these limitations, it is preferable to modify the design so as to avoid the occurrence of resonance spikes within the frequency band of interest. As mentioned above, the switched line Schiffman phase shifter with shunt-mounted diodes can be suitably designed to avoid resonances within about an octave bandwidth. Another choice is to use all-pass C-sections of different lengths and couplings in both the transmission paths. Particularly, for a 180° phase shifter, it is possible to achieve large bandwidth by using a coupled exponential line all-pass C-section [18]

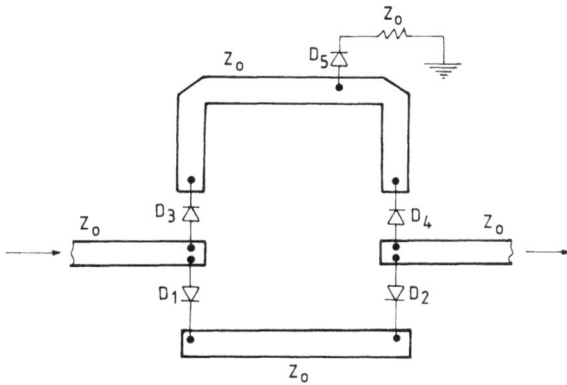

Figure 9.11 Modified series-diode switched line phase shifter for eliminating insertion loss resonance spikes. (After Lynes *et al.* [17].)

in one of the transmission paths, with the other path retained as a straight section or replaced by an all-pass C-section.

A flat phase shift *versus* frequency response over a limited bandwidth of about 12% has been reported by Burns, Holden, and Tang [19] using a rather simple technique. In this technique, the straight reference path of length l_1 in the switched line phase shifter (Figure 9.3) is compensated by short-circuited stubs of length $\lambda_0/4$ where λ_0 is the guide wavelength in the medium of propagation. The modified reference section due to Burns, Holden, and Tang [19] is shown in Figure 9.12. This type of loading equalizes the slopes of the phase-frequency characteristics of the two switched paths about the center frequency.

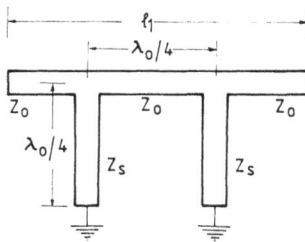

Figure 9.12 Modified reference path of switched line phase shifter shown in Figure 9.3. (After Burns, Holden, and Tang [19].)

9.3.5 Insertion Loss and Peak Power Capacity

Insertion Loss

The even- and odd-mode technique discussed in Section 9.3.1 provides an exact method of analyzing switched line phase shifters possessing symmetry about the input and output ports. If the circuit is not symmetric, as in Figure 9.11, the analysis can be carried out using the conventional node-voltage equations and then modified to incorporate the transmission line equations. The exact insertion loss in the two switching states can then be obtained from the magnitudes of the transmission coefficients using Equation (9.5). In practice, particularly for phased-array applications, it is desirable to optimize the phase shifters for equalized insertion loss in both the switching states. Circuit conditions are suitably imposed in order to satisfy this criterion, and the peak power capacity of the phase shifter is determined accordingly.

Stark, Burns, and Clark [20] have derived approximate closed-form expressions for the insertion loss and peak power capacity of the switched line phase shifter. The derivation is based on the equivalent circuit of the SPDT reverse-mode shunt-diode switch. In the following, we consider a forward-mode SPDT switch and use the same procedure as given by Stark, Burns, and Clark to derive the expression for insertion loss. Figure 9.13(a) shows the circuit representation of an SPDT forward-mode shunt-diode switch. The diodes are mounted at a distance of $\lambda_0/4$ from the T-junction. The forward-mode switch gives series resonance under forward bias and parallel resonance under reverse bias. In Figure 9.13(a), R_f represents the resistance

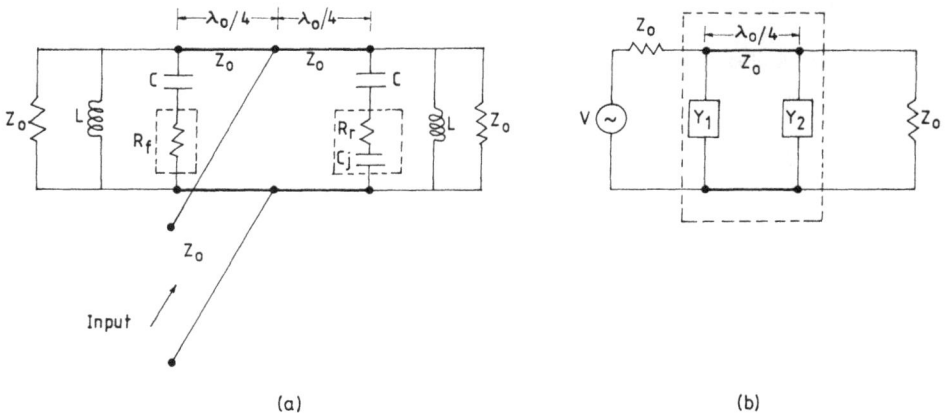

(a)   (b)

Figure 9.13 (a) Schematic of a forward-mode SPDT switch with shunt-mounted diodes and (b) its simplified equivalent circuit.

of the forward-biased diode, and R_r and the series capacitance C_j represent the parameters of the reverse-biased diode. Figure 9.13(b) shows a simplified equivalent circuit where the low resistance R_f is transformed to a high resistance Z_0^2/R_f by the $\lambda_0/4$ line at the T-junction. Thus, in Figure 9.13(b), $Y_1 = R_f/Z_0^2$. The $ABCD$ matrix of the circuit shown within the dotted line is given by

$$\begin{bmatrix} A & B \\ C & D \end{bmatrix} = \begin{bmatrix} jY_2Z_0 & jZ_0 \\ \dfrac{j}{Z_0} + jY_1Y_2Z_0 & jY_1Z_0 \end{bmatrix} \tag{9.18}$$

The insertion loss of the phase shifter can be expressed in terms of the transmission coefficient S_{21} as

$$\alpha(\text{dB}) = 20 \log_{10} \left| \frac{1}{2} \left(A + \frac{B}{Z_0} + CZ_0 + D \right) \right| \tag{9.19}$$

Substituting for A, B, C, and D from (9.18) in (9.19) gives

$$\alpha(\text{dB}) = 20 \log_{10} \left[1 + \frac{1}{2} (g_1 + g_2 + g_1g_2) \right] \tag{9.20}$$

where g_1 and g_2 are normalized conductances shunting the transmission line and are given by

$$g_1 = \frac{R_f}{Z_0} \tag{9.21}$$

$$g_2 = \frac{R_r Z_0}{X_c^2} \tag{9.22}$$

In deriving the expression for g_2, the approximation $R_r \ll X_c$ has been used. Because g_1 and g_2 are both small, we can neglect the product term g_1g_2 in (9.20). Furthermore, if the condition $(g_1 + g_2) \ll 1$ is satisfied, then, applying the mathematical relation $\ln(1 + x) \simeq x$ for $|x| \ll 1$, (9.20) gets simplified to the following:

$$\alpha(\text{dB}) = 4.34 \left(\frac{Z_0 R_r}{X_c^2} + \frac{R_f}{Z_0} \right) \tag{9.23}$$

The condition for minimum insertion loss is obtained by setting the derivative with respect to Z_0 equal to zero. We then obtain

$$Z_0 = X_c \left\{ \frac{R_f}{R_r} \right\}^{1/2} \tag{9.24}$$

Substituting (9.24) into (9.23) yields the expression for minimum loss for an SPDT switch. It is given by

$$\alpha(\text{dB}) = 8.68 \frac{\sqrt{R_r R_f}}{X_c} = 8.68 \frac{f}{f_c} \tag{9.25}$$

where

$$f_c = \frac{1}{2\pi C_j \sqrt{R_r R_f}} \tag{9.26}$$

is the cut-off frequency of the p-i-n diode and f is the operating frequency. Since a single-bit switched line phase shifter employs two SPDT switches, its insertion loss is twice that of one SPDT switch when transmission line losses are neglected. Thus, the expression for minimum insertion loss of a single-bit switched line phase shifter is

$$\alpha(\text{dB}) \simeq 17.36 \frac{\sqrt{R_r R_f}}{X_c} = 17.36 \frac{f}{f_c} \tag{9.27}$$

The characteristic impedance Z_0 must satisfy the relation in (9.24). If the forward- and reverse-bias resistances of the diode are equal, then Z_0 should be equal to the magnitude of diode reactance for minimum insertion loss. It is worthwhile to note that the insertion loss given by (9.27) is the same for both the switching states of a phase bit and also for all bit sizes.

Peak Power Capacity

The peak power that a p-i-n diode can handle, without taking into account the junction heating, is limited by its breakdown voltage under reverse bias. A general expression for the peak power handling capacity of a switched line phase shifter is given by (8.46). The peak power capacity of the phase shifter under the condition of minimum loss has been reported by Stark, Burns, and Clark [20] by considering the model of SPDT switch and by calculating the RF voltage across a reverse-biased diode as a function of the peak power incident on it. If $\hat{P}_i$ is the incident power and

V is the incident voltage, then we have

$$\hat{P}_i = \frac{V^2}{2Z_0}$$

(9.28)

where Z_0 is the characteristic impedance of the junction. The peak RF voltage V_d across the reverse-biased diode (assuming $|X_c| \gg R_r$) is given by

$$V_d = \frac{VX_c}{Z_0}$$

(9.29)

Substituting for V from (9.29) into (9.28), we obtain

$$\hat{P}_i = \frac{V_d^2 Z_0}{2X_c}$$

(9.30)

Now, using the condition on Z_0 for minimum loss as given by (9.24) in (9.30), and setting $R_f = R_r$, we obtain the following expression for the peak power capacity of the switch:

$$\hat{P}_i = \frac{V_d^2}{2Z_0} = \frac{V_B^2}{8Z_0}$$

(9.31)

where V_B is the breakdown voltage of the diode.

9.3.6 Design Examples

As an example, we consider the design of a series-diode switched line phase shifter bit and a series-diode Schiffman phase shifter bit for the same set of specifications, and compare their performance characteristics.

Specifications: Phase shift $\Delta\phi = 90°$, frequency band = 8.8 to 10 GHz, impedance of input/output lines $Z_0 = 50\Omega$; p-i-n diode parameters: R_f (forward bias) = 1Ω, C_f (reverse bias) = 0.06 pf; configuration: microstrip; substrate: alumina having $\varepsilon_r = 9.8$ and thickness $h = 0.635$ mm.

(1) Series-Diode Switched Line Phase Shifter

In order to avoid resonance, it is important to ensure that the effective length of the off-diode path does not become equal to $180°$ or multiples of $180°$ within the operating frequency range. For $\Delta\phi = 90°$, the effective length of the longest path with off-diodes is chosen at less than 180° at the highest operating frequency. Referring to Figure 9.3(a), we require $\beta(l_2 + 2l') < 180°$ where $\beta l'$ is the equivalent electrical length of the off-diode at the highest operating frequency. The value of $\beta l'$ is calculated by equating the reactance of the off-diode with the input impedance of a section of an open-circuited transmission line having the same characteristic impedance Z_0 and propagation constant β as those of the switched path. Thus,

$$-jZ_0 \cot \beta l' = -\frac{j}{\omega C_j} \tag{9.32}$$

or

$$\beta l' = \cot^{-1}\left(\frac{1}{\omega C_j Z_0}\right) \tag{9.33}$$

Substituting $Z_0 = 50\Omega$, $C_j = 0.06$ pf and $f = 10$ GHz, we obtain the equivalent electrical length of the off-diode as

$$\beta l' = 10.68° \text{ (say } 11°) \tag{9.34}$$

In order to satisfy the condition $\beta(l_2 + 2l') < 180°$, we require $\beta l_2 < 158°$. Correspondingly, $\beta l_1 < 68°$ for $\Delta\phi = 90°$ at $f = 10$ GHz.

As a starting value, we choose $\beta l_1 = 40°$, $\beta l_2 = 130°$ at the center frequency $f = 9.4$ GHz. We have $\bar{Z}_f = R_f/Z_0 = 0.02$, and $\bar{Z}_r = j/(\omega C_j Z_0) = -j\,5.643$. Using these parameters, the transmission coefficient $|S_{21}^{(1)}|e^{j\phi_1}$ for state 1 (with l_1 as the on-diode path) is calculated using (9.3). Similarly the value of $|S_{21}^{(2)}|e^{j\phi_2}$ for state 2 (with l_2 as the on-diode path) is calculated by interchanging $\bar{Z}_f$ and $\bar{Z}_r$ in (9.3). The differential phase shift is obtained as $\Delta\phi = \phi_2 - \phi_1 = 89.55°$. The deviation in $\Delta\phi$ from 90° can be rectified with a small iteration in βl_2. In this case, the actual value of βl_2 is 129.24°.

The actual path lengths l_1 and l_2 can now be calculated from $\beta l_1 = 40°$ and $\beta l_2 = 129.24°$, obtained at 9.4 GHz. For a 50Ω microstrip on alumina substrate having $\varepsilon_r = 9.8$, $h = 0.635$ mm, the effective dielectric constant ε_{eff} is 6.83 so that $\beta = 2\pi/\lambda_0 = 0.5145$ rad [13]. The path lengths l_1 and l_2, and the width w of a 50Ω microstrip are obtained as $l_1 = 1.357$ mm, $l_2 = 4.384$ mm, $w = 0.637$ mm.

All the parameters of the switched line phase shifter circuit (Figure 9.3(a)) are now known. Equations (9.3) to (9.5) can be used to compute the phase shift $\Delta\phi$ and insertion loss as a function of frequency. Figure 9.14(a) shows the computed plot of $\Delta\phi$, and Figure 9.14(b) shows the plots of insertion loss in the two states expressed as α_1(dB) and α_2(dB). As expected, the plot of $\Delta\phi$ versus frequency is nearly linear. Over the frequency range 8.8 to 10 GHz, the maximum phase error is ±5.7°.

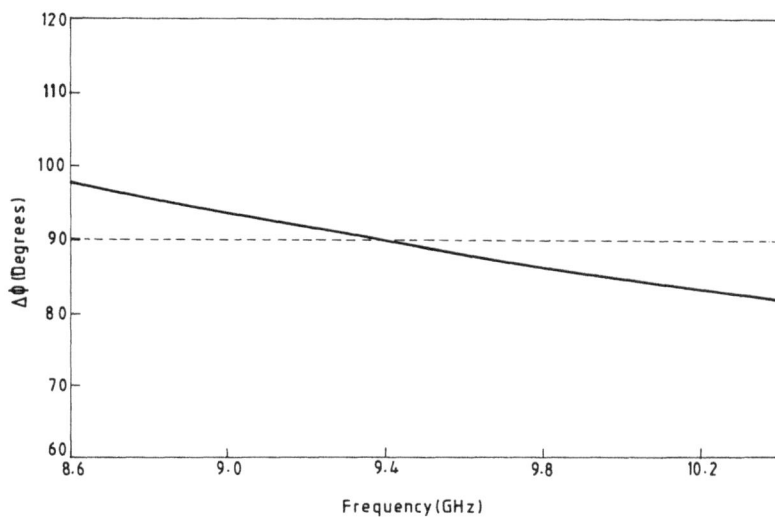

(a)

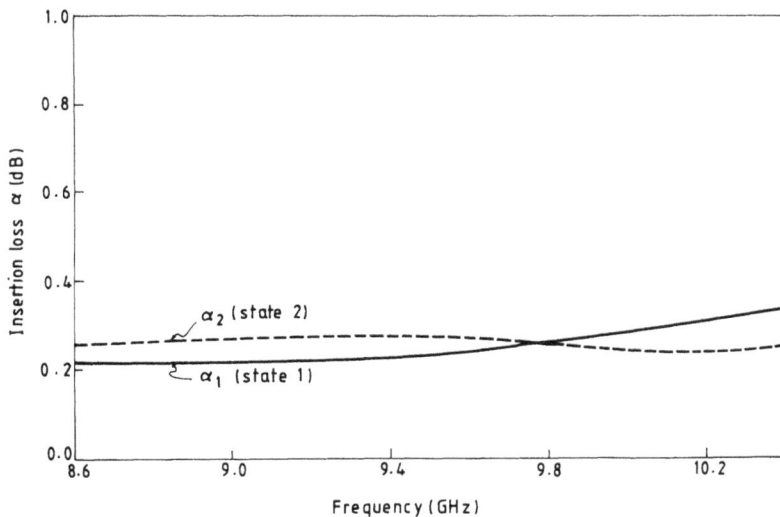

(b)

Figure 9.14 Theoretical plots of (a) phase shift and (b) insertion loss in the two states as a function of frequency for a series-diode switched line 90° phase shifter (Figure 9.3(a)). $\beta l_1 = 40°$, $\beta l_2 = 129.24°$ at $f = 9.4$ GHz, $Z_0 = 50\Omega$, p-i-n diode: $R_f = 1\Omega$, $C_j = 0.06$ pf.

---

(2) Series-Diode Schiffman Phase Shifter

We now evaluate the increase in bandwidth by introducing a Schiffman section in the series-diode switched line phase shifter considered at (1) above (see Figure 9.8).

Referring to Figure 9.8 and Equation (9.12), we note that at the center frequency, 9.4 GHz, $\theta_3 = 90°$, and, hence, for $\Delta\phi = 90°$, we have

$$\beta(l_1 - l_2) = 270° \tag{9.35}$$

This condition is valid in the case of ideal diodes, and serves as a useful guideline for design. The diode parameters are $R_f = 1\Omega$ and $C_j = 0.06$ pf, so that with $Z_0 = 50\Omega$,

$$\bar{Z}_f = 0.02, \quad \bar{Z}_r = -j\,5.643 \quad \text{at} \quad 9.4\,\text{GHz} \tag{9.36}$$

For the coupled-line section, the choice of R depends on the bandwidth and the tolerable phase error. In this example, since the bandwidth is small, we choose a small value of R; that is, $R = 2$. We have $Z_0 = \sqrt{Z_{0e}Z_{0o}} = 50\Omega$. The normalized even- and odd-mode impedances of the coupled line are given by

$$\bar{Z}_{0e} = \frac{Z_{0e}}{Z_0} = 1.4142, \quad Z_{0e} = 70.71\Omega \tag{9.37}$$

$$\bar{Z}_{0o} = \frac{Z_{0o}}{Z_0} = 0.7071, \quad Z_{0o} = 35.36\Omega \tag{9.38}$$

Using the formulas for the coupled microstrip [13], the width w_0 of the strip conductors and the spacing s between them can be obtained. For a 50Ω microstrip line, the width w is 0.637 (as calculated in example (1) above).

The transmission coefficient of the Schiffman phase shifter is determined by first calculating $\bar{Y}_B^e$ and $\bar{Y}_B^o$ from (9.11) and then substituting the values in (9.2). Let $|S_{21}^{(1)}|e^{j\phi_1}$ denote the transmission coefficient in state 1. For determining the transmission coefficient $|S_{21}^{(2)}|e^{j\phi_2}$ in state 2, the values of $\bar{Z}_f$ and $\bar{Z}_r$ in (9.11) are interchanged. The phase shift $\Delta\phi$ and insertion loss in the two states are then calculated from (9.4) and (9.5), respectively. For the calculation of $\bar{Y}_B^e$ and $\bar{Y}_B^o$, we need to specify βl_1 and βl_2.

To start with, choose a sample value of βl_2 in the range of 20° to 45°, $\beta l_1 = (\beta l_2 + 270°)$ at $f = 9.4$ GHz, and plot $\Delta\phi$ as a function of frequency. If resonances occur, iterate βl_2 until resonances are removed from the operating band. With this value of βl_2 at 9.4 GHz, iterate βl_1 around the value given by $(\beta l_2 + 270°)$ until $\Delta\phi$ is 90° at 9.4 GHz. For the set of parameters considered in this example ($R_f = 1\Omega$,

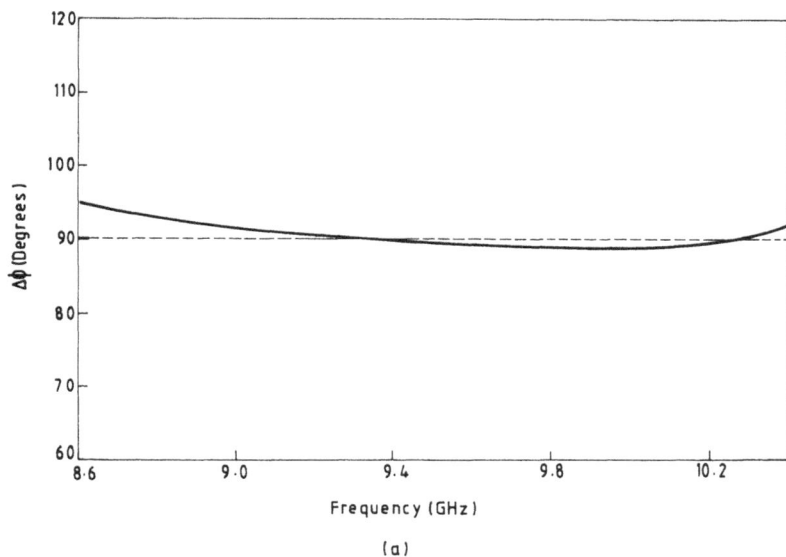

(a)

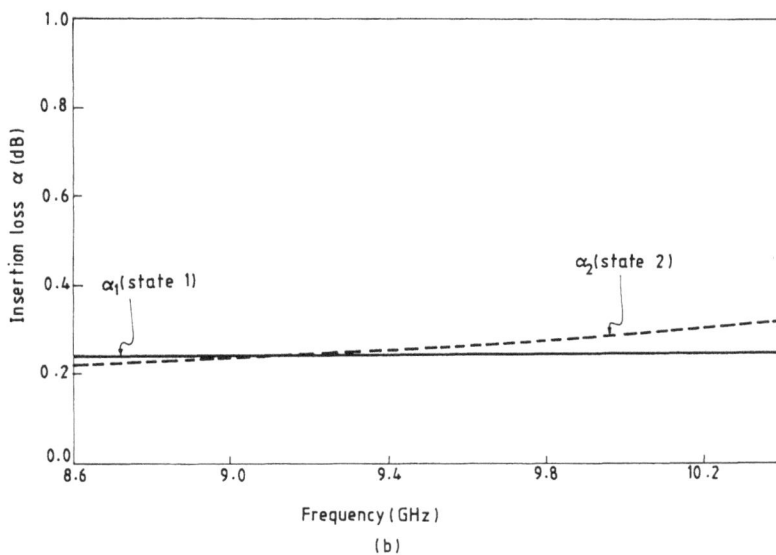

(b)

Figure 9.15 Theoretical plots of (a) phase shift and (b) insertion loss in the two states as a function of frequency for a series-diode Schiffman 90° phase shifter (Figure 9.8). $\beta l_2 = 25°$, $\beta l_1 = 294.46°$ at $f = 9.4$ GHz, $Z_0 = 50\Omega$, p-i-n diode: $R_f = 1\Omega$, $C_r = 0.06$ pf.

$C_j = 0.06$ pf, $R = 2$, $Z_0 = 50\Omega$, $Z_{0e} = 70.71\Omega$, $Z_{0o} = 35.36\Omega$), we obtain

$$\beta l_2 = 25°, \quad \beta l_1 = 294.46° \quad \text{at} \quad f = 9.4 \text{ GHz} \tag{9.39}$$

or

$$l_2 = 0.848 \text{ mm}, \quad l_1 = 9.99 \text{ mm} \tag{9.40}$$

Figure 9.15 shows the theoretical plots of $\Delta\phi$ and insertion loss in the two states as a function of frequency. Comparing it with Figure 9.14, it can be seen that for a tolerable phase error of $\pm 5°$, the operating band of Schiffman phase shifter is 8.6 to 10.4 GHz, whereas that of a switched line phase shifter is 8.9 to 9.9 GHz.

9.3.7 Typical Circuit Layout

Practical switched line phase shifters must make provision for impedance match at the two ports, isolation between the two switching paths, and also isolation between RF and driver circuit [7, 17, 20, 21]. Figure 9.16 shows a typical input/output circuit

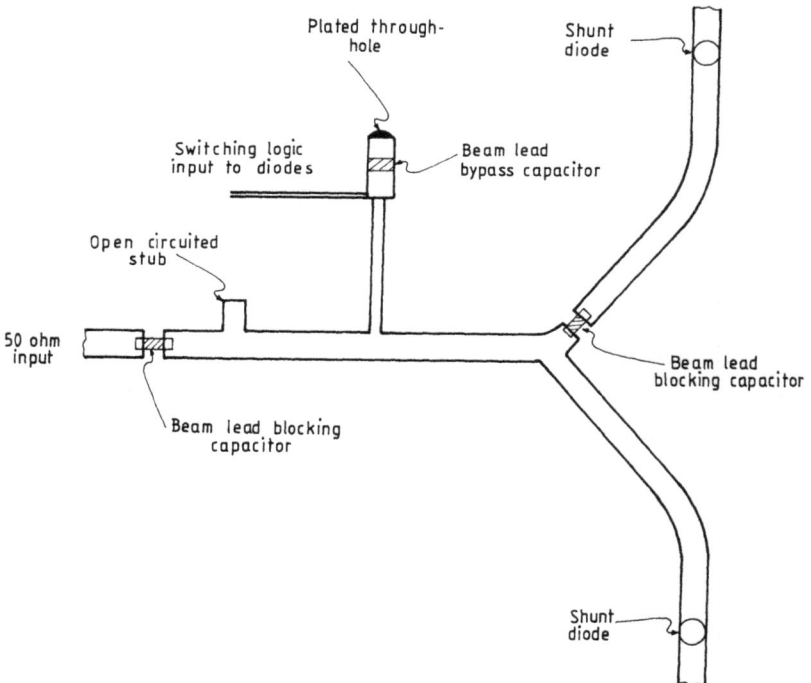

Figure 9.16 Typical input/output circuit configuration for switched line phase shifter. (From Coats [7], copyright © 1973 IEEE, reprinted with permission.)

of a switched line phase shifter reported by Coats [7]. Impedance match at the input port is achieved by means of a shorted stub. As shown in Figure 9.16, the stub is shorted by means of a plated through-hole via a beam lead bypass capacitor. The driver output is connected to the stub end of this capacitor. General driver circuit considerations are discussed in Section 9.8. Isolation between the two switching paths is achieved by means of a blocking capacitor near the Y-junction. The beam lead capacitor on the input line provides isolation between the RF and diode-switching logic circuits. Using this type of input/output circuit, Coats [7] has reported an octave band 3-bit Schiffman-type switched line phase shifter in microstrip.

9.4 HYBRID COUPLED STRIPLINE-LIKE AND MICROSTRIP-LINE PHASE SHIFTERS

The principle of operation of the hybrid coupled phase shifter is discussed in Chapter 8 (Section 8.5). Of the three circuit schematics shown in Figure 8.9, the circuit employing a 3-dB, 90° branchline hybrid is the most commonly used in practice. In the following, we discuss the analysis and performance characteristics of this type of phase shifter in detail. Important features of phase shifters using other types of hybrid couplers are also covered.

9.4.1 Branchline Hybrid Coupled Phase Shifter

Most of the studies reporting on the branchline hybrid coupled phase shifter assume the coupler to be ideal and concentrate on the analysis and design of a reflective phase-shifting network [22–28]. Some studies have also been reported [29] that combine the performance of the coupler with that of the reflective network.

Analysis

Figure 9.17(a) shows the schematic of a branchline hybrid coupled phase shifter. The reflective network incorporates a shunt-mounted p-i-n diode backed by a section of a short-circuited transmission line. Some of the other forms of reflective terminations are shown in Figures 9.17(b) to (e).

Since the analysis of the phase shifter includes the performance of the coupler, we first present the design equations governing the coupler performance. The analysis of the four port symmetric coupler using the standard even- and odd-mode technique is available in the literature [30]. The design of the branchline coupler, which takes into account the junction parasitics, has also been reported [31, 32]. The scattering parameters of the 3-dB, 90° branchline coupler shown in Figure 9.17(a) are given by

$$S_{11} = S_{22} = S_{33} = S_{44} \tag{9.41a}$$

$$S_{12} = S_{21} = S_{34} = S_{43} \tag{9.41b}$$

$$S_{13} = S_{31} = S_{24} = S_{42} \tag{9.41c}$$

$$S_{14} = S_{41} = S_{23} = S_{32} \tag{9.41d}$$

$$S_{11} = \frac{1}{2}(S_{11e} + S_{11o}) \tag{9.41e}$$

$$S_{12} = \frac{1}{2}(S_{11e} - S_{11o}) \tag{9.41f}$$

$$S_{13} = \frac{1}{2}(S_{14e} - S_{14o}) \tag{9.41g}$$

$$S_{14} = \frac{1}{2}(S_{14e} + S_{14o}) \tag{9.41h}$$

where

$$S_{11e} = -\frac{j}{\Delta_e}\left[2p\cos\theta - \frac{p^2}{\sqrt{2}}\sin\theta + \frac{\sin\theta}{\sqrt{2}}\right] \tag{9.42a}$$

$$S_{11o} = -\frac{j}{\Delta_o}\left[-\frac{2\cos\theta}{p} - \frac{\sin\theta}{\sqrt{2}\,p^2} + \frac{\sin\theta}{\sqrt{2}}\right] \tag{9.42b}$$

$$S_{14e} = \frac{2}{\Delta_e} \tag{9.42c}$$

$$S_{14o} = \frac{2}{\Delta_o} \tag{9.42d}$$

$$\Delta_e = (2\cos\theta - \sqrt{2}\,p\sin\theta) + j\left(2p\cos\theta - \frac{p^2}{\sqrt{2}}\sin\theta + \frac{3}{\sqrt{2}}\sin\theta\right) \tag{9.42e}$$

$$\Delta_o = \left(2\cos\theta + \frac{\sqrt{2}\sin\theta}{p}\right) + j\left(-\frac{2\cos\theta}{p} - \frac{\sin\theta}{\sqrt{2}\,p^2} + \frac{3}{\sqrt{2}}\sin\theta\right) \tag{9.42f}$$

$$p = \tan \frac{\theta}{2} \tag{9.42g}$$

$$\theta = \frac{2\pi}{\lambda} l = \frac{\pi}{2} \frac{\lambda_0}{\lambda} \tag{9.42h}$$

In (9.42), λ is the guide wavelength in the propagating medium at frequency f, and λ_0 is the guide wavelength at the center frequency f_0. At $f = f_0$, we have $\theta = 90°$, $S_{11} = S_{12} = 0$, $S_{13} = -1/\sqrt{2}$ and $S_{14} = -j/\sqrt{2}$.

We now consider the overall performance of the phase shifter (Figure 9.17(a)) by considering the coupler along with the reflective load. Let $\Gamma = |\Gamma| e^{j\phi_0}$ be the reflection coefficient of the reflective load. For a unit signal fed to port 1, the signals

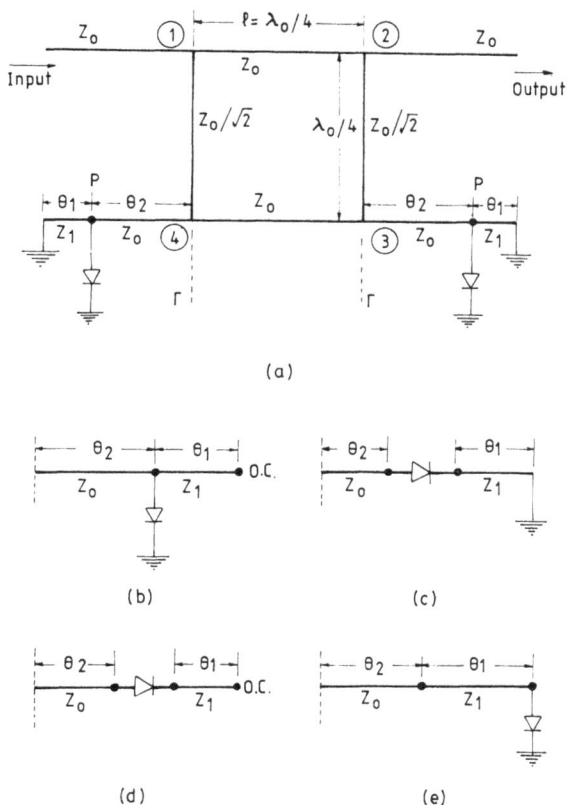

(a)

(b)

(c)

(d)

(e)

Figure 9.17 (a) Branchline hybrid coupled phase shifter; (b) to (e) different forms of reflective terminations.

S_{13} and S_{14} emerging out of ports 3 and 4 are reflected back. The reflected signals ΓS_{13} and ΓS_{14} form inputs to ports 3 and 4, respectively. These signals are then distributed between the four ports. Signals emerging from ports 1 and 2 are absorbed as they are match terminated, but the signals emerging from ports 3 and 4 are reflected again by the terminations. Considering multiple reflections such as these from ports 3 and 4, Kori and Mahapatra [29] have derived expressions for the overall transmission coefficient T_{21} from port 1 to 2 and the overall reflection coefficient Γ_{in} at port 1. For symmetric terminations as considered in Figure 9.17(a), the expressions are given by

$$T_{21} = S_{12} + \Gamma[S_{13}\,S_{14}] \begin{bmatrix} 1 - \Gamma S_{11} & -\Gamma S_{12} \\ -\Gamma S_{12} & 1 - \Gamma S_{11} \end{bmatrix}^{-1} \begin{bmatrix} S_{14} \\ S_{13} \end{bmatrix} \qquad (9.43)$$

$$\Gamma_{in} = S_{11} + \Gamma[S_{14}\,S_{13}] \begin{bmatrix} 1 - \Gamma S_{11} & -\Gamma S_{12} \\ -\Gamma S_{12} & 1 - \Gamma S_{11} \end{bmatrix}^{-1} \begin{bmatrix} S_{14} \\ S_{13} \end{bmatrix} \qquad (9.44)$$

In deriving (9.43) and (9.44), it is assumed that the magnitudes of all the eigen values of

$$\begin{bmatrix} \Gamma S_{11} & \Gamma S_{12} \\ \Gamma S_{12} & \Gamma S_{11} \end{bmatrix}$$

are less than 1. The expressions given by (9.43) and (9.44) are applicable for any hybrid coupled phase shifter using symmetric four-port hybrid and symmetric reflective terminations. From these, the performance of the phase shifter in terms of voltage standing wave ratio (VSWR), phase shift, and insertion loss can be easily evaluated.

Design of 3-dB Branchline Coupler

The design of a 3-dB branchline coupler in microstrip is straightforward when the junction discontinuities are neglected. As illustrated in Figure 9.17(a), the impedances involved are Z_0, $Z_0/\sqrt{2}$, and the branch lengths are each $\lambda_0/4$, where Z_0 is the characteristic impedance of the input and output lines, and λ_0 is the guide wavelength in the medium at the center frequency f_0. In practical circuits, the junction discontinuities play a major role in affecting the coupler performance, particularly beyond the C-band. It is important to find the exact location of the reference planes at the junctions with respect to which the quarter-wavelengths are to be taken. Figure 9.18 shows the layout of a T-junction illustrating the shift in the locations of the reference planes [32]. Furthermore, a load located at some point in the shunt arm will appear transformed by a ratio n^2. The parameters that are required for evaluating

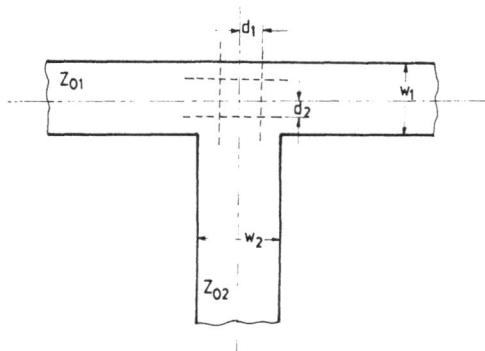

Figure 9.18 Microstrip T-junction showing typical shifts in the reference planes.

the shifts d_1 and d_2 are the effective widths of the main and shunt arms denoted as w_{eff1} and w_{eff2}, respectively, and the transformer ratio n. Closed-form formulas for calculating d_1 and d_2 are documented in the literature [32]. Referring to Figure 9.18, the formulas are as follows:

$$w_{eff1} = \frac{120 \pi h}{Z_{01}\sqrt{\varepsilon_{eff1}}} \tag{9.45a}$$

$$w_{eff2} = \frac{120 \pi h}{Z_{02}\sqrt{\varepsilon_{eff2}}} \tag{9.45b}$$

where h is the height of the substrate; Z_{01} and Z_{02} are the characteristic impedances of the main and shunt arms, respectively; and ε_{eff1} and ε_{eff2} are the effective dielectric constants of the microstrip corresponding to Z_{01} and Z_{02}, respectively. The transformer ratio is given by

$$n^2 = \left[\frac{\sin\{\pi(w_{eff1}/\lambda_{01})(Z_{01}/Z_{02})\}}{\pi(w_{eff1}/\lambda_{01})(Z_{01}/Z_{02})}\right]^2$$

$$\cdot [1 - \{\pi(w_{eff1}/\lambda_{01})(d_2/w_{eff1})\}^2] \tag{9.46}$$

$$\lambda_{01} = \frac{\lambda^a}{\sqrt{\varepsilon_{eff1}}} \tag{9.47}$$

where λ^a is the wavelength in air. The shift in the reference plane for the main line is obtained from

$$\frac{d_1}{w_{eff2}} = 0.05 \, n^2 \frac{Z_{01}}{Z_{02}} \tag{9.48}$$

The shift in the reference plane for the shunt arm is obtained from

$$\frac{d_2}{w_{eff1}} = \left[0.5 - \left\{ 0.076 + 0.2\left(\frac{2w_{eff1}}{\lambda_{01}}\right)^2 + 0.663 \exp\left(-1.71\frac{Z_{01}}{Z_{02}}\right) \right. \right.$$
$$\left. \left. - 0.172 \ln\left(\frac{Z_{01}}{Z_{02}}\right) \right\} \frac{Z_{01}}{Z_{02}} \right] \tag{9.49}$$

As an example, we consider the design of a 3-dB branchline coupler in microstrip at 10 GHz. The substrate is alumina having $\varepsilon_r = 9.8$ and $h = 0.635$ mm.

In this design, we make use of the microstrip data on ε_{eff} and Z generated using the rigorous spectral domain method. The two impedances involved are $Z_{01} = 50\Omega$ for the main line, and $Z_{02} = (50/\sqrt{2})\Omega$ for the shunt arm at $f = 10$ GHz. The spectral domain results at 10 GHz are as follows: For $w_1 = 0.639$ mm, $\varepsilon_{eff1} = 6.8661$, $Z_{01} = 50\Omega$. For $w_2 = 1.225$ mm, $\varepsilon_{eff2} = 7.4533$, $Z_{02} = 35.35\Omega$. From (9.45), we obtain the effective widths as

$$w_{eff1} = \frac{376.7 \times 0.635}{50\sqrt{6.8661}} = 1.8257 \text{ mm} \tag{9.50}$$

$$w_{eff2} = \frac{376.7 \times 0.635}{35.35 \times \sqrt{7.4533}} = 2.4785 \text{ mm} \tag{9.51}$$

The guide wavelength λ_{01} is obtained from (9.47) as

$$\lambda_{01} = \frac{30}{\sqrt{6.8661}} = 11.4489 \text{ mm} \tag{9.52}$$

Similarly, $\lambda_{02} = 30/\sqrt{7.4533} = 10.9887$ mm. Substituting the values for w_{eff1}, λ_{01}, Z_{01}, and Z_{02} in (9.49), the shift d_2 is calculated.

$$\frac{d_2}{w_{eff1}} = 0.3643; \quad d_2 = 0.665 \text{ mm} \tag{9.53}$$

Using (9.46), the transformer ratio is obtained as

$$n^2 = 0.8157 \tag{9.54}$$

Using (9.48), the shift d_1 is obtained as

$$\frac{d_1}{w_{eff2}} = 0.05762; \quad d_2 = 0.1427 \text{ mm} \tag{9.55}$$

The final layout of the coupler with all the relevant dimensions is shown in Figure 9.19.

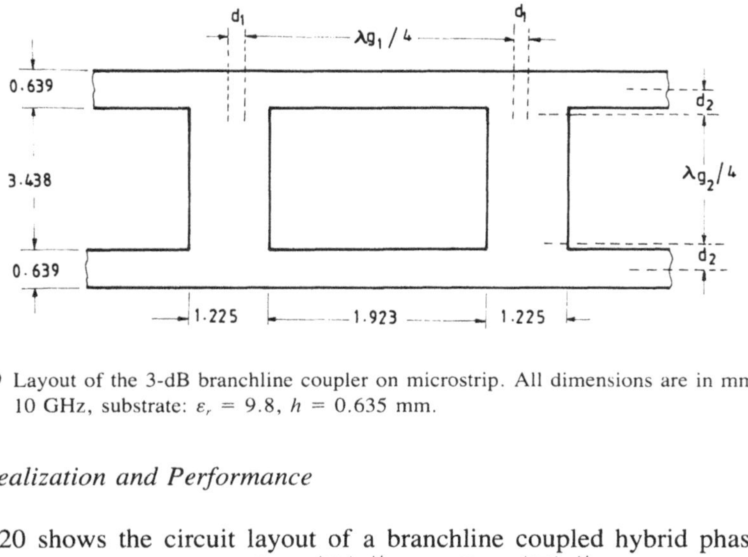

Figure 9.19 Layout of the 3-dB branchline coupler on microstrip. All dimensions are in mm. Freq. = 10 GHz, substrate: $\varepsilon_r = 9.8$, $h = 0.635$ mm.

Circuit Realization and Performance

Figure 9.20 shows the circuit layout of a branchline coupled hybrid phase shifter with biasing arrangement. Let $\Gamma_1 = |\Gamma_1|e^{j\phi_1}$ and $\Gamma_2 = |\Gamma_2|e^{j\phi_2}$ denote the reflection coefficients of the reflective circuit corresponding to the two switching states of the p-i-n diode. With $\phi_1 = (\phi_0 - \Delta\phi/2)$ and $\phi_2 = (\phi_0 + \Delta\phi/2)$, the differential phase shift obtained at the reflective circuit is $\Delta\phi$. If the coupler is ideal, then the differential phase shift at the output of the coupler at the center frequency is also $\Delta\phi$. At any other frequency f, let $\Delta\phi'$ be the differential phase shift at the output of the coupler. Then the phase shift error $(\Delta\phi' - \Delta\phi)$ as well as the input VSWR depend not only on the frequency but also on the value of ϕ_0. Thus, for achieving maximum bandwidth performance for a specified phase bit, the value of ϕ_0 must be chosen carefully. The desired value of ϕ_0 can be obtained by appropriately choosing the electrical length θ_2 (see Figure 9.20).

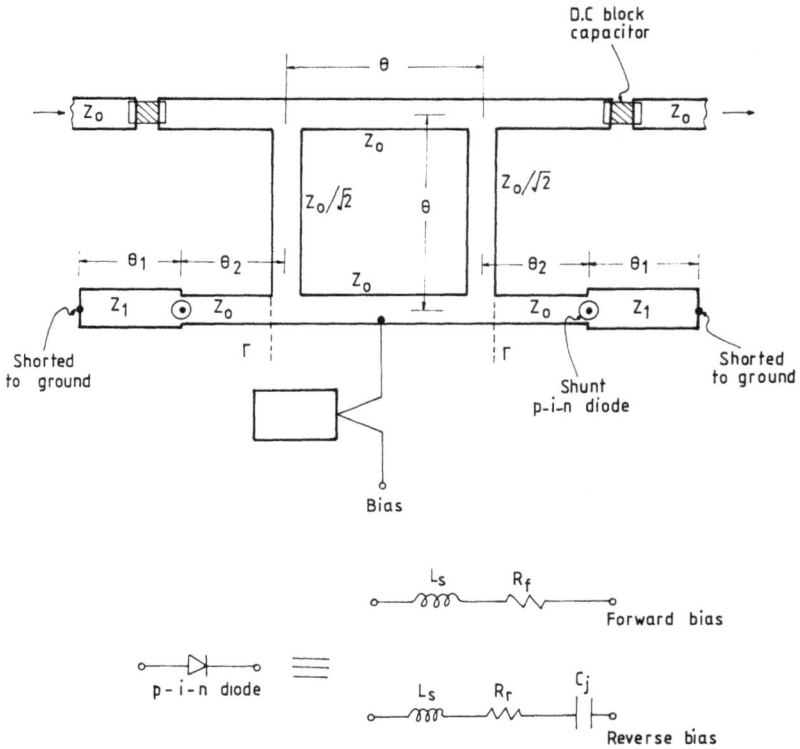

Figure 9.20 Layout of branchline hybrid coupled phase shifter bit in microstrip.

Figure 9.21 shows the theoretical phase shift *versus* frequency response of a 90° phase shifter for different values of ϕ_0 [29]. The circuit configuration corresponds to Figure 9.16. For a fixed set of diode parameters and for specified Z_1, θ_1 is varied to give $\Delta\phi = 90°$ at $f_0 = 10$ GHz. This calculation is made considering only the reflective circuit, and the coupler is assumed to be ideal. For the diode parameters considered in Figure 9.21 and with $Z_1 = Z_0 = 50\Omega$, the value of θ_1 is 41.54° for $\Delta\phi = 90°$. It can be seen from Figure 9.21 that when only the reflective circuit is considered, the phase shift increases with an increase in frequency. This characteristic changes considerably when the coupler response is included, and in particular with a change in the electrical length θ_2 or, equivalently, ϕ_0. For the parameters considered in Figure 9.21, $\theta_2 = 92.3°$ ($\phi_0 = 270°$) gives the least phase error over the largest bandwidth.

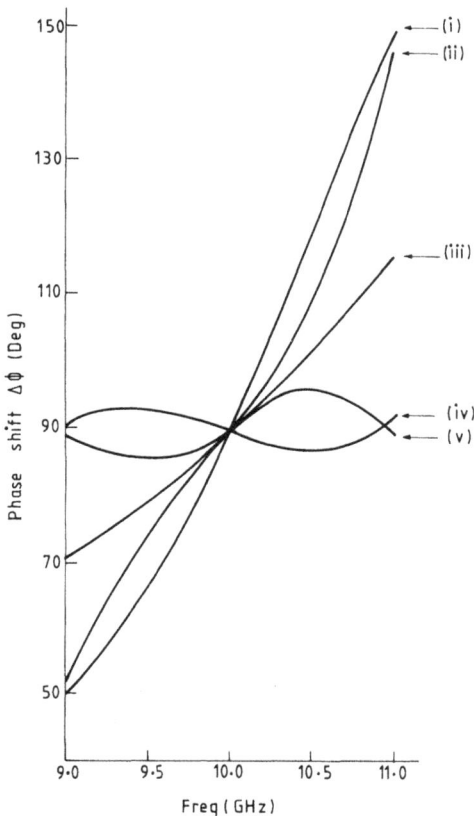

Figure 9.21 Phase shift against frequency for different ϕ_0 of a hybrid coupled phase shifter ($\Delta\phi = 90°$). Refer to Figure 9.16 for circuit. $Z_1 = Z_0 = 50$ ohms; *P-I-N* diode: $L_s = 0.05$ nH, $R_f = 0.6\Omega$, $C_j = 0.15$ pF, $R_r \approx 0$, (i) $\theta_2 = 2.3°$, $\phi_0 = 90°$, (ii) $\theta_2 = 137.3°$, $\phi_0 = 180°$, (iii) Reflection circuit without coupler, (iv) $\theta_2 = 92.3°$, $\phi_0 = 270°$, (v) $\theta_2 = 47.3°$, $\phi_0 = 0°$. (From Kori and Mahapatra [29], IEE Proc., Pt.H, 1987 Vol. 134, pp. 156–162, reprinted with permission of IEE.)

9.4.2 Backward-Wave Hybrid Coupled Phase Shifter

Hybrid coupled phase shifters that use parallel-line backward-wave couplers in stripline are reported in the literature [28]. The main advantage of the backward-wave coupler over the branchline coupler is its inherent broad bandwidth. For example, the bandwidth of a 3-dB, 90° branchline coupler is approximately 10%, whereas that of a 3-dB, 90° backward-wave coupler is around 35%. The schematic of a hybrid coupled phase shifter incorporating a 3-dB, 90° backward-wave coupler is shown in Figure 8.9(c). While this is a useful hybrid for incorporation in phase-shifter circuits, its realization in microstrip configuration poses practical problems. This is because for 3-dB coupling, the spacing between the coupled lines becomes so small that it is difficult to etch the pattern with reasonable dimensional accuracy. This problem is overcome in suspended stripline geometry. Figure 9.22 shows the layout of a phase shifter in suspended stripline. The backward-wave coupler can be easily realized by using edge-coupled strip conductors. The coupling length is $\lambda_0/4$, where λ_0 is the guide wavelength in suspended stripline at the center frequency f_0 (assuming the even- and odd-mode phase velocities in the coupled line to be equal). This geometry also permits dc block to be etched directly on the substrate by using overlapping strip conductors, as shown in Figure 9.22. The low-pass filter for introducing bias is realized in the form of a distributed circuit consisting of an open-circuited low-impedance line of electrical length 90°, and a high-impedance line also of electrical length 90°.

The general expressions given by Equations (9.43) and (9.44) can be directly applied to the backward-wave coupled hybrid phase shifter to determine its overall

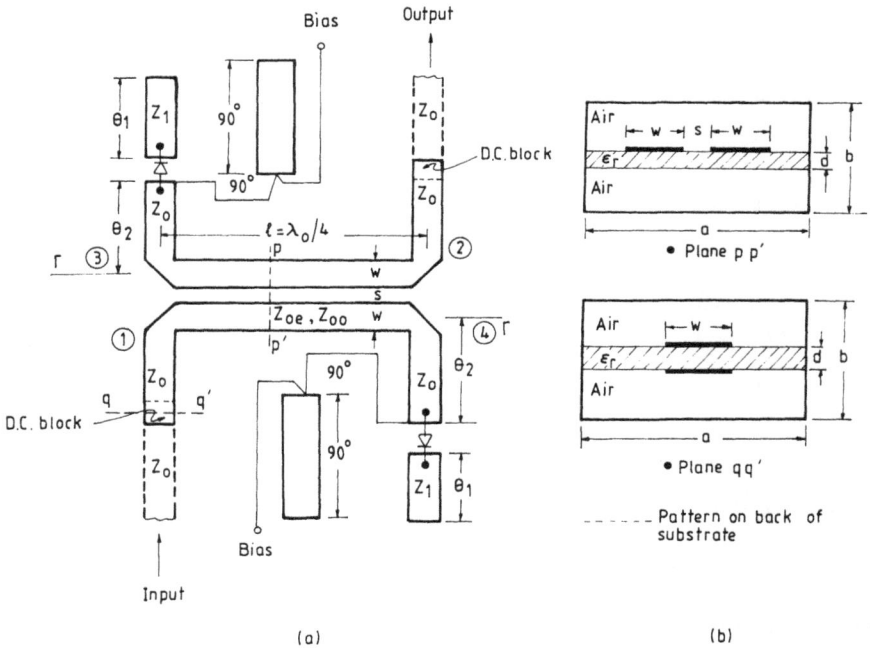

Figure 9.22 (a) Backward-wave hybrid coupled phase shifter in suspended stripline. Dotted line shows layout on the lower side of the substrate. (b) Cross-sectional view at planes pp' and qq'.

input reflection coefficient Γ_{in} and transmission coefficient T_{21}. The scattering parameters of the basic backward-wave coupler are given by [33]

$$S_{11} = 0 \tag{9.56a}$$

$$S_{12} = 0 \tag{9.56b}$$

$$S_{13} = \frac{jC_0 \sin\theta}{[\sqrt{1 - C_0^2} \cos\theta + j \sin\theta]} \tag{9.56c}$$

$$S_{14} = \frac{\sqrt{1 - C_0^2}}{[\sqrt{1 - C_0^2} \cos\theta + j \sin\theta]} \tag{9.56d}$$

where

$$C_0 = \frac{Z_{0e} - Z_{0o}}{Z_{0e} + Z_{0o}} \tag{9.57a}$$

$$\theta = \frac{2\pi}{\lambda} l = \frac{\pi}{2} \frac{\lambda_0}{\lambda} \tag{9.57b}$$

C_0 is the midband voltage coupling factor, Z_{0e} and Z_{0o} are the even- and odd-mode impedances, respectively, of the coupled line, and λ and λ_0 are the guide wavelengths in the propagating medium at frequency f and center frequency f_0, respectively.

If we set $S_{11} = S_{12} = 0$ in (9.43) and (9.44), the expressions for Γ_{in} and T_{21} simplify to the following:

$$\Gamma_{in} = \Gamma(S_{13}^2 + S_{14}^2) \tag{9.58}$$

$$T_{21} = 2\Gamma S_{13} S_{14} \tag{9.59}$$

The reflection coefficient Γ, due to the reflective termination, is to be determined with reference to port 3 (or port 4).

9.4.3 Rat-Race Hybrid Coupled Phase Shifter

In a rat-race coupler the two output arms that provide equal power division are out of phase by 180°. The phase difference of 90° desired for implementation in phase-shifter circuits is achieved by extending the reference plane of one of the output arms by 90°. The practical bandwidth of a rat-race coupler in terms of isolation, power split, and VSWR is approximately 20%, which is more than that of a 3-dB, 90° branchline coupler. Another advantage of the rat-race coupler is that the discontinuity effects are considerably less, because the ring is of uniform width.

General expressions for the overall reflection and transmission coefficients of a rat-race hybrid coupled phase shifter can be derived by considering the schematic shown in Figure 9.23. These expressions must include the frequency-dependent scattering parameters of the rat-race coupler. The scattering parameters of the rat-race coupler can be easily derived using the even- and odd-mode technique [30]. Referring to the port terminology considered in Figure 9.23, the scattering parameters of the rat-race coupler are given by

$$S_{11} = S_{44} \tag{9.60a}$$

$$S_{22} = S_{33} \tag{9.60b}$$

$$S_{12} = S_{21} = S_{34} = S_{43} \tag{9.60c}$$

$$S_{13} = S_{31} = S_{24} = S_{42} \tag{9.60d}$$

$$S_{14} = S_{41} \tag{9.60e}$$

$$S_{23} = S_{32} \tag{9.60f}$$

where

$$S_{11} = \frac{1}{2}(S_{11e} + S_{11o}) \tag{9.61a}$$

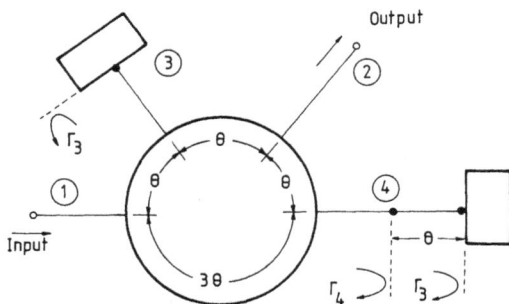

Figure 9.23 Rat-race coupled hybrid terminated in reflective loads, $\theta = \pi/2$ at $f = f_0$.

$$S_{22} = \frac{1}{2}(S_{33e} + S_{33o}) \qquad (9.61b)$$

$$S_{12} = \frac{1}{2}(S_{13e} - S_{13o}) \qquad (9.61c)$$

$$S_{13} = \frac{1}{2}(S_{13e} + S_{13o}) \qquad (9.61d)$$

$$S_{14} = \frac{1}{2}(S_{11e} - S_{11o}) \qquad (9.61e)$$

$$S_{23} = \frac{1}{2}(S_{33e} - S_{33o}) \qquad (9.61f)$$

$$S_{11e} = \frac{1}{\Delta_e}\left[(p_1 - p_2)\sin\theta \right.$$
$$\left. + \frac{j}{\sqrt{2}}\{(1 + p_1 p_2)\sin\theta - (p_1 + p_2)\cos\theta\} \right] \qquad (9.62a)$$

$$S_{11o} = \frac{1}{\Delta_o}\left[\left(-\frac{1}{p_1} + \frac{1}{p_2}\right)\sin\theta + \frac{j}{\sqrt{2}}\left\{\left(1 + \frac{1}{p_1 p_2}\right)\sin\theta \right.\right.$$
$$\left.\left. + \left(\frac{1}{p_1} + \frac{1}{p_2}\right)\cos\theta\right\} \right] \qquad (9.62b)$$

$$S_{33e} = \frac{1}{\Delta_e}\left[(p_2 - p_1)\sin\theta \right.$$
$$\left. + \frac{j}{\sqrt{2}}\{(1 + p_1 p_2)\sin\theta - (p_1 + p_2)\cos\theta\} \right] \qquad (9.62c)$$

$$S_{33o} = \frac{1}{\Delta_o}\left[\left(-\frac{1}{p_2} + \frac{1}{p_1}\right)\sin\theta + \frac{j}{\sqrt{2}}\left\{\left(1 + \frac{1}{p_1 p_2}\right)\sin\theta \right.\right.$$
$$\left.\left. + \left(\frac{1}{p_1} + \frac{1}{p_2}\right)\cos\theta\right\} \right] \qquad (9.62d)$$

$$S_{13e} = \frac{2}{\Delta_e} \tag{9.62e}$$

$$S_{13o} = \frac{2}{\Delta_o} \tag{9.62f}$$

$$\Delta_e = [2\cos\theta - (p_1 + p_2)\sin\theta]$$
$$+ \frac{j}{\sqrt{2}}[(3 - p_1 p_2)\sin\theta + (p_1 + p_2)\cos\theta] \tag{9.62g}$$

$$\Delta_o = \left[2\cos\theta + \left(\frac{1}{p_1} + \frac{1}{p_2}\right)\sin\theta\right]$$
$$+ \frac{j}{\sqrt{2}}\left[\left(3 - \frac{1}{p_1 p_2}\right)\sin\theta - \left(\frac{1}{p_1} + \frac{1}{p_2}\right)\cos\theta\right] \tag{9.62h}$$

$$p_1 = \tan\frac{3\theta}{2} \tag{9.62i}$$

$$p_2 = \tan\frac{\theta}{2} \tag{9.62j}$$

$$\theta = \frac{\pi}{2}\frac{\lambda_0}{\lambda} \tag{9.62k}$$

At the center frequency $f = f_0$, $\theta = \pi/2$, the scattering parameters reduce to $S_{11} = S_{22} = S_{12} = 0$, $S_{13} = S_{23} = -j/\sqrt{2}$ and $S_{14} = j/\sqrt{2}$.

Referring to the phase shifter schematic shown in Figure 9.23, and considering port 1 as the input port, the expression for the overall input reflection coefficient is given by

$$\Gamma_{in} = S_{11} + (S_{13}\Gamma_3 S_{13} + S_{14}\Gamma_4 S_{14}) + (S_{13}\Gamma_3 S_{22} + S_{14}\Gamma_4 S_{12})\Gamma_3 S_{13}$$
$$+ (S_{13}\Gamma_3 S_{12} + S_{14}\Gamma_4 S_{11})\Gamma_4 S_{14} + \ldots \tag{9.63}$$

The overall transmission coefficient T_{21} from port 1 to port 2 is given by

$$T_{21} = S_{12} + (S_{13}\Gamma_3 S_{23} + S_{14}\Gamma_4 S_{13}) + (S_{13}\Gamma_3 S_{22} + S_{14}\Gamma_4 S_{12})\Gamma_3 S_{23}$$
$$+ (S_{13}\Gamma_3 S_{12} + S_{14}\Gamma_4 S_{11})\Gamma_4 S_{13} + \ldots \tag{9.64}$$

The termination at port 4 has an additional section of transmission line of electrical length θ, which is 90° at the center frequency. Thus, $\Gamma_4 = \Gamma_3 e^{-j2\theta}$. Neglecting higher order reflection terms in (9.63) and (9.64), we can write

$$\Gamma_{in} = S_{11} + \Gamma_3(S_{13}^2 + S_{14}^2 e^{-j2\theta}) \tag{9.65}$$

$$T_{21} = S_{12} + \Gamma_3 S_{13}(S_{23} + S_{14} e^{-j2\theta}) \tag{9.66}$$

Figure 9.24 shows the layout of a rat-race hybrid coupled phase shifter. As in the case of the branchline hybrid coupled phase shifter, the transmission line section of electrical length θ_2 helps in adjusting the phase-frequency response of the phase shifter to maximize the bandwidth.

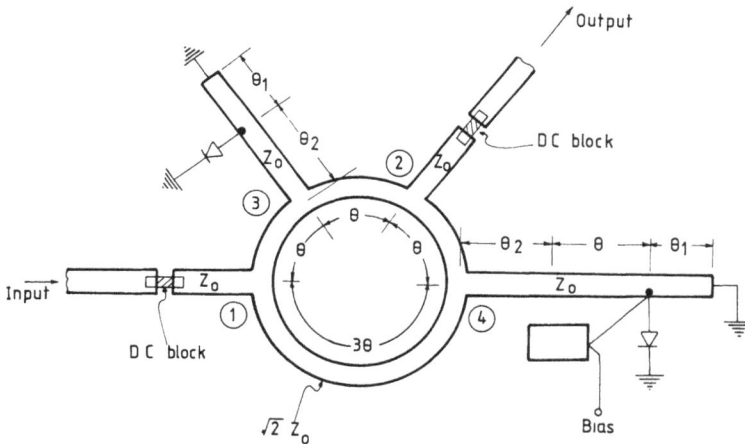

Figure 9.24 Layout of rat-race hybrid coupled phase shifter, $\theta = \pi/2$ at $f = f_0$.

9.4.4 Optimization of Bandwidth

As discussed in Section 9.4.1, the bandwidth performance of a hybrid coupled phase shifter can be improved by suitably designing the reflective circuit to compensate for the frequency-dependent phase shift error due to the coupler. Another approach is to design the coupler and the reflective circuit independently so that each offers broadband performance.

In a single-section branchline coupler, an improved bandwidth of approximately 30% has been reported by using a quarter-wavelength transformer shunted by a half-wavelength open-circuited stub in each of the output arms of the coupler

[34]. By using suitable matching techniques, an increase in bandwidth, even up to an octave, has been reported [35]. In the case of the backward-wave coupler, bandwidth of more than an octave can be achieved by using asymmetric coupled lines [36, 37]. The bandwidth of the rat-race coupler can be increased from 20% to about an octave by replacing the three-quarter-wavelength-long section of the ring with a shorted quarter-wavelength-long parallel-coupled line [38].

When the coupler has a flat VSWR and phase-frequency response within the desired frequency band, the design of the reflective network can be carried out independently. The method for optimizing reflective network for the hybrid coupled phase shifter has been reported by Starski [24]. In order to illustrate the procedure, we consider the example of a reflective termination shown in Figure 9.25. Let Z_{1f} and Z_{2f} denote the input impedances at planes PP' and QQ', respectively, when the diode is forward biased, and let Z_{1r} and Z_{2r} denote the corresponding impedances when the diode is reverse biased. These impedances can be expressed as

$$Z_{2f} = R_{2f} + jX_{2f} = Z_2 \left[\frac{Z_{1f} + jZ_2 \tan\theta_2}{Z_2 + jZ_{1f} \tan\theta_2} \right] \tag{9.67}$$

$$Z_{2r} = R_{2r} + jX_{2r} = Z_2 \left[\frac{Z_{1r} + jZ_2 \tan\theta_2}{Z_2 + jZ_{1r} \tan\theta_2} \right] \tag{9.68}$$

$$Z_{1f} = R_{1f} + jX'_{1f}, \quad X'_{1f} = X_{1f} + Z_1 \tan\theta_1 \tag{9.69}$$

$$Z_{1r} = R_{1r} + jX'_{1r}, \quad X'_{1r} = X_{1r} + Z_1 \tan\theta_1 \tag{9.70}$$

where R_{1f} and X_{1f} are the resistance and reactance, respectively, of the diode under forward bias, and R_{1r} and X_{1r} are the corresponding quantities under reverse bias. For a given set of diode parameters, we choose the characteristic impedance Z_1 and electrical length θ_1 of the shorted transmission line section such that $R_{1f} << |X'_{1f}|$ and $R_{1r} << |X'_{1r}|$. Equations (9.67) and (9.68) then reduce to

$$\bar{Z}_{2f} = j\bar{Z}_2 \left[\frac{\bar{X}'_{1f} + \bar{Z}_2 \tan\theta_2}{\bar{Z}_2 - \bar{X}_{1f} \tan\theta_2} \right] = j\bar{X}_{2f} \tag{9.71}$$

$$\bar{Z}_{2r} = j\bar{Z}_2 \left[\frac{\bar{X}'_{1r} + \bar{Z}_2 \tan\theta_2}{\bar{Z}_2 - \bar{X}_{1r} \tan\theta_2} \right] = j\bar{X}_{2r} \tag{9.72}$$

The bar notation indicates normalization with respect to the characteristic impedance Z_0 of the coupled port of the hybrid coupler. Using Γ_f and Γ_r as the reflection coefficients at plane QQ' under forward- and reverse-bias states, respectively, we can write

$$\Gamma_f = \frac{(j\bar{X}_{2f} - 1)}{(j\bar{X}_{2f} + 1)} \tag{9.73}$$

$$\Gamma_r = \frac{(j\bar{X}_{2r} - 1)}{(j\bar{X}_{2r} + 1)} \tag{9.74}$$

The phase shift $\Delta\phi$ is then given by

$$\Delta\phi = 2[\tan^{-1} \bar{X}_{2f} - \tan^{-1} \bar{X}_{2r}] \tag{9.75}$$

A condition for minimum phase shift error with respect to frequency can be obtained by setting $d(\Delta\phi)/df = 0$. From (9.75), we can write

$$\frac{1}{(1 + \bar{X}_{2f}^2)} \frac{d\bar{X}_{2f}}{df} + \frac{1}{(1 + \bar{X}_{2r}^2)} \frac{d\bar{X}_{2r}}{df} = 0 \tag{9.76}$$

The condition for equal insertion loss is obtained by equating the magnitudes of Γ_f and Γ_r. Including the forward- and reverse-bias resistances of the diode in the expressions for Γ_f and Γ_r, and setting $|\Gamma_f| = |\Gamma_r|$, we have

$$|\Gamma_f| = \left[\frac{(\bar{R}_{2f} - 1)^2 + \bar{X}_{2f}^2}{(\bar{R}_{2f} + 1)^2 + \bar{X}_{2f}^2}\right]^{1/2} = \left[\frac{(\bar{R}_{2r} - 1)^2 + \bar{X}_{2r}^2}{(\bar{R}_{2r} + 1)^2 + \bar{X}_{2r}^2}\right]^{1/2} \tag{9.77}$$

Neglecting the terms $\bar{R}_{2f}^2$ and $\bar{R}_{2r}^2$ in comparison with $\bar{X}_{2f}^2$ and $\bar{X}_{2r}^2$ in (9.77), we obtain

$$\frac{\bar{R}_{2f}}{\bar{R}_{2r}} = \frac{(1 + \bar{X}_{2f}^2)}{(1 + \bar{X}_{2r}^2)} \tag{9.78}$$

With this condition, the insertion loss of the phase shifter can be calculated using

$$\alpha(\text{dB}) = -20 \log_{10} |\Gamma_f| \tag{9.79}$$

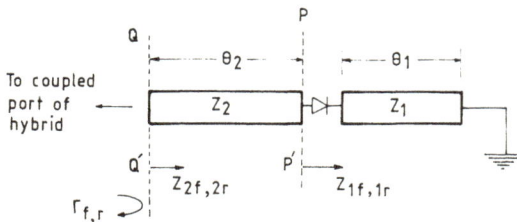

Figure 9.25 Layout of a reflective termination of hybrid coupled phase shifter.

Equations (9.75) and (9.79) in conjunction with (9.77) and (9.78) yield optimum phase shift *versus* frequency response and equalized insertion loss in the two switching states of the diode.

9.4.5 Insertion Loss and Power-Handling Capability—Approximate Formulas

Insertion Loss

Approximate closed-form expression for the insertion loss of a hybrid coupled phase shifter can be derived based on the assumption that the *p-i-n* diode sustains maximum rated RF current (I_{m1}) when it is forward biased and maximum rated RF voltage (V_{m0}) when it is reverse biased. If P_{df} and P_{dr} represent the power dissipated in the diode under forward- and reverse-bias states, respectively, we can write

$$P_{df} = I_{m1}^2 R_f \tag{9.80}$$

$$P_{dr} = \frac{V_{m0}^2 R_r}{R_r^2 + X_r^2} \tag{9.81}$$

where R_f is the forward-bias resistance, and R_r and X_r ($= 1/2\pi f C_j$) are the reverse-bias resistance and reactance, respectively, of the diode. Under the assumption $R_r \ll X_r$, which is valid for practical diodes, (9.81) can be approximated to

$$P_{dr} = \frac{V_{m0}^2 R_r}{X_r^2} \tag{9.82}$$

Multiplying (9.80) and (9.82) and setting $P_{df} = P_{dr} = P_d$, we can write

$$P_d = V_{m0} I_{m1} \left(\frac{f}{f_c} \right) \tag{9.83}$$

where f_c is the cut-off frequency of the diode and is given by

$$f_c = \frac{1}{2\pi C_j \sqrt{R_f R_r}} \tag{9.84}$$

According to the maximum power-handling theorem from Hines [39], we have (see Section (8.2))

$$V_{m0} I_{m1} = 4 P_{max} \sin \frac{\Delta\phi}{2} \tag{9.85}$$

where P_{max} is the maximum RF power incident on the diode and $\Delta\phi$ is the phase shift in the reflective network. Using (9.85) in (9.83), we get

$$P_d = 4P_{max} \frac{f}{f_c} \sin \frac{\Delta\phi}{2} \qquad (9.86)$$

The insertion loss is given by

$$\alpha(\text{dB}) = -10 \log_{10}\left(1 - \frac{P_d}{P_{max}}\right)$$

$$= -10 \log_{10}\left(1 - 4\frac{f}{f_c} \sin \frac{\Delta\phi}{2}\right)$$

$$\approx 17.4 \frac{f}{f_c} \sin\left(\frac{\Delta\phi}{2}\right) \qquad (9.87)$$

Peak Power Capability

Expressions for the average and peak power-handling capability of a reflective circuit are derived in Section 8.4. The maximum average power that a hybrid coupled phase shifter can handle is the same as that of a single reflective circuit and is given by (8.46). Its peak power-handling capability is twice that of the single reflective circuit and is given by (8.50).

9.5 LOADED-LINE STRIPLINE AND MICROSTRIP PHASE SHIFTERS

Loaded line phase shifters have been studied extensively by several investigators [40–48]. The basic circuit and operating principle of a two-element loaded line phase shifter are discussed in Section 8.7. In this section, we consider some practical circuits, their design, insertion loss, and operational bandwidth. Salient features of this phase shifter are compared with those of the switched line and hybrid coupled phase shifters.

9.5.1 General Analysis

Figure 9.26 shows a general equivalent circuit representation of a single-bit loaded line phase shifter, where Y_{si} represents the complex admittance of the shunt stub loading the main line. The analysis of this circuit can be easily carried out in terms

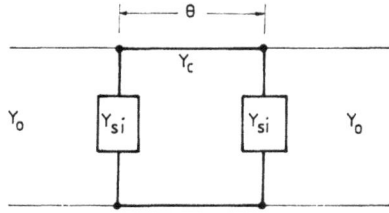

Figure 9.26 General equivalent circuit of a loaded line phase shifter.

of its ABCD matrix [3, 40, 41]. For the equivalent circuit shown in Figure 9.26, the elements of the ABCD matrix are given by

$$A = D = (\cos\theta - B_{si} Z_c \sin\theta) + jG_{si} Z_c \sin\theta \tag{9.88a}$$

$$B = jZ_c \sin\theta \tag{9.88b}$$

$$C = 2G_{si}(\cos\theta - B_{si}Z_c \sin\theta) + jZ_c$$

$$\times [2B_{si}Y_c \cos\theta + (Y_c^2 + G_{si}^2 - B_{si}^2) \sin\theta] \tag{9.88c}$$

where

$$Y_{si} = G_{si} + jB_{si}, \quad i = 1 \quad \text{or} \quad 2 \tag{9.89}$$

The subscript $i = 1$ or 2 corresponds to either of the two switching states. In deriving (9.88), the transmission line is assumed to be lossless. G_{si} accounts for loss due to the finite resistance of the diode. The scattering coefficients of the circuit can be obtained from its ABCD parameters by using the formulas given by (8.24) and (8.25). They are

$$S_{11} = S_{22} = \frac{BY_0 - CZ_0}{2A + BY_0 + CZ_0} \tag{9.90a}$$

$$S_{12} = S_{21} = \frac{2}{2A + BY_0 + CZ_0} \tag{9.90b}$$

where Z_0 is the characteristic impedance of the input and output arms of the phase shifter.

Lossless Case ($G_{si} = 0$)

We assume the *p-i-n* diode to be lossless. Setting $G_{si} = 0$ in (9.88), we get

$$A = D = (\cos\theta - B_{si}Z_c \sin\theta) \tag{9.91a}$$

$$B = jZ_c \sin\theta \tag{9.91b}$$

$$C = jZ_c[2B_{si}Y_c \cos\theta + (Y_c^2 - B_{si}^2) \sin\theta] \tag{9.91c}$$

Substituting for A, B, C, and D from (9.91) in (9.90a), the magnitude of the input reflection coefficient is obtained as

$$|S_{11}| = \frac{\left| B_{si}Z_0 \cos\theta + \dfrac{Z_cZ_0}{2} (Y_c^2 - Y_0^2 - B_{si}^2) \sin\theta \right|}{\left[(\cos\theta - B_{si}Z_c \sin\theta)^2 + \left\{ B_{si}Z_0 \cos\theta + \dfrac{Z_cZ_0}{2} (Y_0^2 + Y_c^2 - B_{si}^2) \sin\theta \right\}^2 \right]^{1/2}} \tag{9.92}$$

The transmission coefficient is obtained from (9.90b). The expression is given by

$$S_{21} = \frac{1}{\left[(\cos\theta - B_{si}Z_c \sin\theta) + j\left\{ B_{si}Z_0 \cos\theta + \dfrac{Z_cZ_0}{2} (Y_0^2 + Y_c^2 - B_{si}^2) \sin\theta \right\} \right]} \tag{9.93}$$

The insertion loss of the phase shifter is given by

$$\alpha(\text{dB}) = -20 \log_{10}|S_{21}|$$

$$= -10 \log_{10}\left[(\cos\theta - B_{si}Z_c \sin\theta)^2 + \left\{ B_{si}Z_0 \cos\theta \right.\right.$$

$$\left.\left. + \frac{Z_cZ_0}{2} (Y_0^2 + Y_c^2 - B_{si}^2) \sin\theta \right\}^2 \right] \tag{9.94}$$

We now impose the condition that the input port must be perfectly matched. Thus, for S_{11} to be equal to zero, we require

$$BY_0 = CZ_0 \tag{9.95}$$

Substituting for B and C from (9.91), we obtain the condition for input match as

$$(Y_0^2 - Y_c^2 + B_{si}^2) \sin\theta = 2B_{si}Y_c \cos\theta \tag{9.96}$$

With this condition, the expression for S_{21} becomes

$$S_{21} = |S_{21}|e^{j\phi} = \frac{1}{(A + BY_0)} \tag{9.97}$$

Substituting for A and B from (9.91), we obtain

$$S_{21} = \frac{1}{[(\cos\theta - B_{si}Z_c \sin\theta) + jZ_cY_0 \sin\theta]} \tag{9.98}$$

Under matched conditions, $|S_{21}| = 1$ and the phase angle ϕ can be expressed in the form

$$\cos\phi = \cos\theta - B_{si}Z_c \sin\theta \tag{9.99a}$$

$$\sin\phi = -Z_cY_0 \sin\theta \tag{9.99b}$$

The expression for the differential phase shift is given by

$$\Delta\phi = \cos^{-1}(\cos\theta - B_{s1}Z_c \sin\theta) - \cos^{-1}(\cos\theta - B_{s2}Z_c \sin\theta) \tag{9.100}$$

According to (9.99), $\cos\phi$ assumes two values for the two switching states $i = 1,2$, whereas $\sin\phi$ remains unaltered. As pointed out by Atwater [42], this condition can be satisfied if ϕ is switched symmetrically about $90°$ by increments of $\pm\Delta\phi/2$. Using $\phi = (90° \pm \Delta\phi/2)$ in (9.99) gives

$$Y_c = Y_0 \sin\theta \sec\left(\frac{\Delta\phi}{2}\right) \tag{9.101}$$

$$B_{si} = Y_0\left[\cos\theta \sec\left(\frac{\Delta\phi}{2}\right) \pm \tan\left(\frac{\Delta\phi}{2}\right)\right], \quad i = 1,2 \tag{9.102}$$

These form the basic design equations for a lossless loaded line phase shifter.

Opp and Hoffman [43] have defined three classes of loading modes. In class I, $B_{s1} \neq B_{s2} \neq 0$. In class II, $B_{s1} = 0$ and $B_{s2} \neq 0$. When $B_{s1} = 0$, there is no load on the main line, and the phase of S_{21} is $\phi = \theta$. In class III, $B_{s1} = -B_{s2}$. In general, when the load Y_{si} is complex, switching is done between complex conjugate loads, that is, $Y_{s2} = Y_{s1}^*$. In class III mode, the main line length θ should be 90°, and when the susceptance is switched between B_{s1} and $-B_{s1}$, the phase changes by $\pm\Delta\phi/2$, about 90°. In both class II and III modes, there is an additional constraint that the input VSWR be equal to unity in both diode states. It can be seen that the design relations (9.101) and (9.102) reduce to a class III solution when $\theta = 90°$. The choice of $\theta = 90°$ is also reported to give maximum bandwidth [3, 44]. For example, for a 22.5° phase bit, the percentage bandwidth over which VSWR is <2 and phase shift error is <2% is reported to be about 43% when $\theta = 90°$, and 12% when $\theta = 75°$ [44].

Effect of Diode Loss [42]

The *ABCD* parameters of the equivalent circuit with complex Y_{si} are given by (9.88). The condition for input match yields the following two equations corresponding to the real and imaginary parts of (9.95):

$$G_{si}(\cos\theta - 2B_{si}Z_c \sin\theta) = 0 \tag{9.103a}$$

$$(Y_0^2 - Y_c^2 + B_{si}^2 - G_{si}^2)\sin\theta - 2B_{si}Y_c \cos\theta = 0 \tag{9.103b}$$

As pointed out by Atwater [42], for the lossless case, condition (9.103a) is automatically satisfied. Condition (9.103b) reduces to (9.96), which is still a quadratic in B_{si} and, therefore, can be satisfied for two values of switching susceptances. When $G_{si} \neq 0$, condition (9.103a) cannot be satisfied for two values of B_{si}. However, for small values of G_{si}, (9.103b) approximates the lossfree case, yielding a quasi-input matched condition. Since the equation is quadratic in B_{si}, it can be satisfied by two values of B_{si}. With (9.103b), the expression for the transmission coefficient can be written as [42]

$$S_{21} = \frac{(\cos\phi + j \sin\phi)}{(1 + G_{si}Z_0)(1 + Z_c^2 G_{si}^2 \sin^2\theta)^{1/2}} \tag{9.104}$$

where

$$\cos\phi = \frac{\cos\theta - B_{si}Z_c \sin\theta}{(1 + Z_c^2 G_{si}^2 \sin^2\theta)^{1/2}} \tag{9.105a}$$

$$\sin\phi = \frac{-Z_c Y_0 \sin\theta}{(1 + Z_c^2 G_{si}^2 \sin^2\theta)^{1/2}} \tag{9.105b}$$

If ϕ is assumed to switch between the two values $(90° \pm \Delta\phi/2)$ as in the lossless case, then (9.105) can be recast in the form

$$Y_c = Y_0\left[1 - G_{si}^2 Z_c^2 \cos^2\left(\frac{\Delta\phi}{2}\right)\right]^{1/2} \sin\theta \sec\left(\frac{\Delta\phi}{2}\right) \qquad (9.106)$$

$$B_{si} = Y_0\left[\left\{1 - G_{si}^2 Z_c^2 \cos^2\left(\frac{\Delta\phi}{2}\right)\right\}^{1/2} \cos\theta \sec\left(\frac{\Delta\phi}{2}\right)\right.$$
$$\left. \pm \tan\left(\frac{\Delta\phi}{2}\right)\right] \qquad (9.107)$$

It may be noted that with $G_{si} = 0$, (9.106) and (9.107) reduce to the loss-free case given by (9.101) and (9.102), respectively. The insertion loss of the phase shifter under the quasi-input-matched condition may be obtained from (9.104).

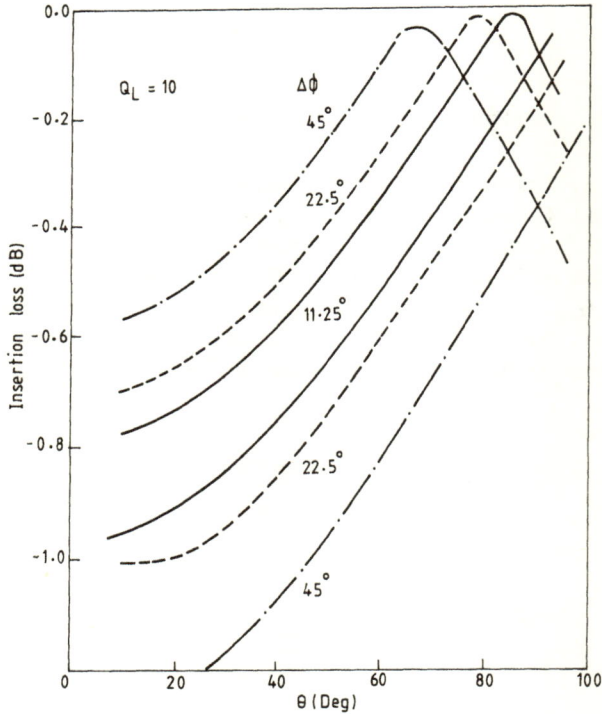

Figure 9.27 Insertion loss *versus* electrical length θ of the main line section for $Q_L = 10$. (From Atwater [42], copyright © 1985 IEEE, reprinted with permission.)

$$\alpha(\text{dB}) = -20\left[\log_{10}(1 + B_{si}Z_0/Q_L) + \frac{1}{2}\log_{10}\{1 + (B_{si}Z_c\sin\theta/Q_L)^2\}\right] \quad (9.108)$$

where

$$Q_L = \frac{|B_{si}|}{G_{si}} \quad (9.109)$$

The variation of insertion loss as a function of the main line electrical length θ for $Q_L = 10$ and for three phase bits of size $\Delta\phi = 11.25°$, $22.5°$, and $45°$ is shown in Figure 9.27 [42]. For each phase bit, two insertion loss curves are shown. These correspond to the two switching states of the phase shifter. It can be seen that the mean value of insertion loss for each phase bit decreases with an increase in θ and reaches a minimum for $\theta = 90°$. Furthermore, at $\theta = 90°$, the insertion loss is the same for the two switching states.

9.5.2 Loaded Line Circuits

Figure 9.28 shows two examples of loaded line circuits employing (1) shunt stub mounting and (2) tandem stub mounting. The design of these two circuits is illustrated below.

Shunt-Stub-Mounted Circuit [45]

In this circuit (Figure 9.28(a)), the *p-i-n* diode is mounted at the end of the stub of characteristic impedance Z_1 and electrical length θ_1. The admittance Y_{si} presented at

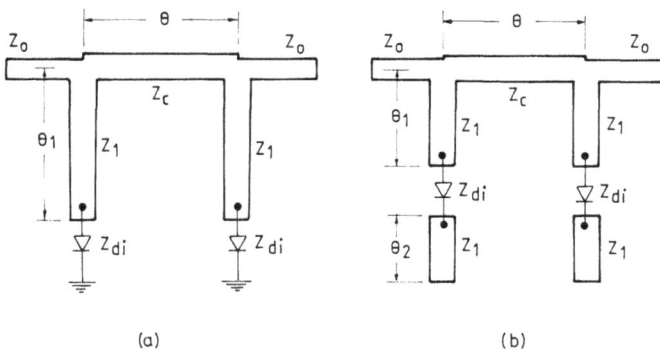

(a) (b)

Figure 9.28 Examples of loaded line phase shifters with (a) shunt stubs and (b) tandem stubs.

the main line due to the stub is given by

$$Y_{si} = \frac{(Z_1 + jZ_{di} \tan\theta_1)}{Z_1(Z_{di} + jZ_1 \tan\theta_1)} = G_{si} + jB_{si}, \quad i = 1,2 \tag{9.110}$$

where $Z_{d1} = R_f + jX_f$ and $Z_{d2} = R_r + jX_r$ are the forward- and reverse-bias impedances of the p-i-n diode, respectively. For low-loss diodes, we may neglect R_f and R_r and use $Z_{d1} = jX_f$ and $Z_{d2} = jX_r$ for deriving the design equations. With this approximation, the susceptances presented at the main line due to the stub are given by

$$B_{s1} = \frac{(X_f \tan\theta_1 - Z_1)}{Z_1(X_f + Z_1 \tan\theta_1)} \tag{9.111}$$

$$B_{s2} = \frac{(X_r \tan\theta_1 - Z_1)}{Z_1(X_r + Z_1 \tan\theta_1)} \tag{9.112}$$

Solving for Z_1 and θ_1, we obtain

$$Z_1 = \left[\frac{(X_f - X_r) - X_f X_r (B_{s1} - B_{s2})}{B_{s1} - B_{s2} - B_{s1} B_{s2}(X_f - X_r)}\right]^{1/2} \tag{9.113}$$

$$\tan\theta_1 = \frac{Z_1(1 + X_f B_{s1})}{(X_f - B_{s1}Z_1^2)} = \frac{Z_1(1 + X_r B_{s2})}{(X_r - B_{s2}Z_1^2)} \tag{9.114}$$

Equations (9.113) and (9.114) together with (9.101) and (9.102) serve as design equations. For the class III mode of operation, we set $\theta = 90°$, $B_{s1} = -B_{s2} = B_s$ in the design relations. This design gives equal VSWR in the two bias states, and also maximizes the bandwidth. The insertion loss can be calculated using (9.108) along with (9.109) and (9.110).

Bahl and Gupta [45] have shown that for this type of phase shifter, the insertion loss decreases with an increase in X_f and $|X_r|$. For a typical set of parameters, $f = 3$ GHz, $Z_0 = 50\Omega$, $\theta = 90°$, $R_f = R_r = 1\Omega$, $X_f = 2\Omega$, and $X_r = -500\Omega$, the insertion loss of a 22.5° phase bit is reported to be about 0.02 dB, and that of a 45° phase bit is about 0.07 dB. With all other parameters remaining the same, if X_r is reduced to -100Ω, then the insertion loss of a 22.5° bit increases to about 0.04 dB, and that of a 45° bit increases to about 0.12 dB. For the same set of parameters, the bandwidth of a 22.5° bit is about 19%, and that of a 45° bit is about 10% [45]. This bandwidth is defined as the frequency range over which VSWR is less than 1.2 and phase error is less than $\pm 2\%$.

Tandem-Stub-Mounted Circuit [42]

Figure 9.28(b) shows a tandem-stub-mounted circuit. As in the case of the shunt-stub-mounted circuit, for the purpose of deriving design expressions, we neglect the values of R_f and R_r and assume $Z_{d1} = jX_f$ and $Z_{d2} = jX_r$ for the *p-i-n* diode. Then the susceptances presented by the stub at the main line in the forward- and reverse-bias states are given by

$$B_{s1} = Y_1 \frac{[(\tan\theta_1 + \tan\theta_2) - X_fY_1 \tan\theta_1 \tan\theta_2]}{(1 - \tan\theta_1 \tan\theta_2 - X_fY_1 \tan\theta_2)} \tag{9.115}$$

$$B_{s2} = Y_1 \frac{[(\tan\theta_1 + \tan\theta_2) - X_rY_1 \tan\theta_1 \tan\theta_2]}{(1 - \tan\theta_1 \tan\theta_2 - X_rY_1 \tan\theta_2)} \tag{9.116}$$

For a specified value of Y_1, the stub lengths θ_1 and θ_2 are obtained from (9.115) and (9.116) by using B_{s1} and B_{s2} as calculated from (9.102). In the following, we present explicit design formulas for two special cases:

(1) Diode reactance under forward bias is zero ($X_f = 0$).
(2) Characteristic impedance of the stub sections is equal to Z_0 ($Y_1 = Y_0$).

Case (1): $X_f = 0$ (Figure 9.28(b)).
    Explicit expressions for θ_1 and θ_2 have been reported by Atwater [42] by considering $X_f = 0$ for the forward-biased diode. With $X_f = 0$, (9.110) simplifies to

$$B_{s1} = Y_1 \tan(\theta_1 + \theta_2) \tag{9.117}$$

Using trigonometric identities, (9.116) and (9.117) are combined to yield the following expressions for θ_1 and θ_2:

$$\theta_1 = \tan^{-1}\left[\frac{1}{M} (N + \sqrt{N^2 + MP}) \right] \tag{9.118}$$

$$\theta_2 = \tan^{-1}(B_{s1}/Y_1) - \theta_1 \tag{9.119}$$

where

$$M = \left[1 + \frac{(B_{s1} - B_{s2})}{X_rY_1^2} \right] \tag{9.120a}$$

$$N = \frac{(B_{s1} + B_{s2})}{2Y_1} \tag{9.120b}$$

$$P = \frac{1}{Y_1^2} \left[\frac{(B_{s2} - B_{s1})}{X_r} - B_{s1}B_{s2} \right] \qquad (9.120c)$$

Equations (9.118) to (9.120) together with (9.101) and (9.102) serve as design formulas. It may be noted that the characteristic admittance Y_1 of the stub sections is arbitrary and may be iterated within the realizable range to provide the most suitable stub design.

Case (2): $Y_1 = Y_0$ (Figure 9.28(b))

With $Y_1 = Y_0 = 1/Z_0$, (9.115) and (9.116) can be recast in the form

$$B_{s1} = Y_0 \frac{[(X_p + X_f) \tan\theta_1 - Z_0]}{[(X_p + X_f) + Z_0 \tan\theta_1]} \qquad (9.121)$$

$$B_{s2} = Y_0 \frac{[(X_p + X_r) \tan\theta_1 - Z_0]}{[(X_p + X_r) + Z_0 \tan\theta_1]} \qquad (9.122)$$

where

$$X_p = -Z_0 \cot\theta_2 \qquad (9.123)$$

It may be noted that jX_p is the input impedance of the open-circuited stub of length θ_2 at the diode plane. Eliminating $\tan\theta_1$ from (9.121) and (9.122), we obtain the following quadratic equation in X_p:

$$X_p^2 + X_p(X_f + X_r) + \frac{(X_f - X_r)(B_{s1}B_{s2}Z_0^2 + 1)}{(B_{s2} - B_{s1})} + Z_0^2 + X_f X_r = 0 \qquad (9.124)$$

Using (9.121), stub length θ_1 can be expressed in terms of B_{s1} and X_p:

$$\theta_1 = \tan^{-1} \left[\frac{\{1 + B_{s1}(X_p + X_f)\}Z_0}{(X_f + X_r) - B_{s1}Z_0^2} \right] \qquad (9.125)$$

Alternatively, from (9.122), we can write

$$\theta_1 = \tan^{-1} \left[\frac{\{1 + B_{s2}(X_p + X_r)\}Z_0}{(X_f + X_r) - B_{s2}Z_0^2} \right] \qquad (9.126)$$

Equations (9.123) to (9.126) together with (9.101) and (9.102) serve as design equations. Considering the class III mode of operation (for which $\theta = 90°$, $B_{s2} = -B_{s1}$), for a given phase shift $\Delta\phi$, first Y_c and B_{s1} are calculated from (9.101) and (9.102).

X_r and X_f are known from the diode parameters. With these values, the quadratic equation (9.124) is solved to determine X_p. The stub lengths θ_1 and θ_2 are then calculated from (9.125) and (9.123), respectively.

In addition to the two examples considered above, several other forms of stub loads can be used to build loaded line phase shifters. Figure 9.29 shows some of the useful stub-loading circuits. The design formulas for evaluating the stub lengths in these cases can be easily derived following the same procedure as presented above. It may be noted that for a loaded line circuit with stub loading corresponding to Figure 9.29(b), the design formulas are given by (9.101), (9.102), and (9.124) to (9.126) with $X_p = Z_0 \tan\theta_2$.

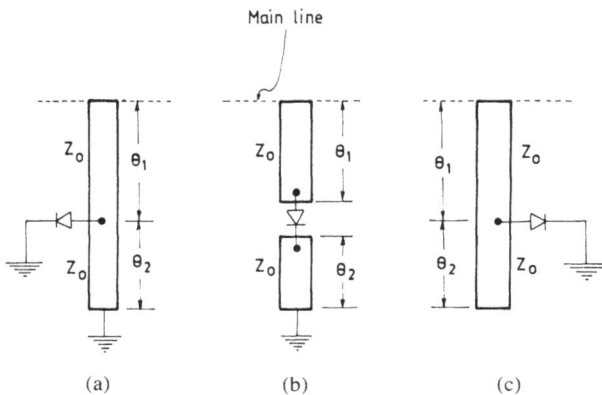

Figure 9.29 Examples of shunt stubs for loaded line phase shifter.

9.5.3 Loaded Line Circuits for Large Phase Bits

The loaded line circuits considered above are commonly used for small phase bits up to 45°. For larger phase bits of 90° and 180° in size, the normalized load susceptances required to perturb the main line are high. This leads to higher VSWR and smaller bandwidth. For matched transmission in either switching state, the phase shift per bit must be small. Larger phase shifts on the order of 90° or more can be achieved by cascading 22.5° or 45° sections. This approach, however, leads to using a large number of diodes. For example, 30 diodes are required for a 4-bit phase shifter to cover 360° in steps of 22.5°. This is because the circuit does not have binary phase bits up to 180°, as in the case of switched line or hybrid coupled phase shifter, but uses 15 bits, each 22.5° in size.

One method of reducing the perturbation in a loaded line phase shifter with a large bit size is to use more than two elements [47, 48]. Figure 9.30 shows a three-element loaded line phase shifter. The transmission line section between successive

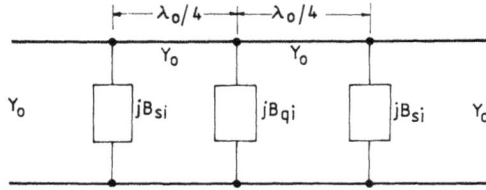

Figure 9.30 Schematic of three-element loaded line phase shifter.

loading elements is a quarter-wavelength long at the center frequency. The *ABCD* matrix of the phase shifter is given by

$$
\begin{bmatrix} A & B \\ C & D \end{bmatrix} = \begin{bmatrix} B_{si}B_{qi}Z_0^2 - 1 & -jB_{qi}Z_0^2 \\ jB_{si}(B_{si}B_{qi}Z_0^2 - 2) & B_{si}B_{qi}Z_0^2 - 1 \end{bmatrix}, \quad i = 1, 2 \quad (9.127)
$$

Imposing the condition for input match given by (9.95), we obtain

$$
B_{qi} = \frac{2B_{si}}{(B_{si}^2 Z_0^2 + 1)} \quad (9.128)
$$

The transmission coefficient under the matched condition can be obtained by using (9.127) in (9.97)

$$
S_{21} = |S_{21}|e^{j\phi} = \frac{1}{[(B_{si}B_{qi}Z_0^2 - 1) - jB_{qi}Z_0]} \quad (9.129)
$$

Substituting for B_{qi} from (9.128) in (9.129), the phase factor is obtained as

$$
\phi = \tan^{-1}\left(\frac{2B_{si}Z_0}{1 - B_{si}^2 Z_0^2}\right) \quad (9.130)
$$

The difference in the two values of ϕ corresponding to B_{si}, $i = 1, 2$ yields the differential phase shift $\Delta\phi$. Further, by setting $B_{s1} = -B_{s2}$, the expression for $\Delta\phi$ becomes

$$
\Delta\phi = 2 \tan^{-1}\left(\frac{2B_{s1}Z_0}{1 - B_{s1}^2 Z_0^2}\right) \quad (9.131)
$$

It can be seen that for $\Delta\phi = 180°$ we require $B_{s1} = -B_{s2} = Y_0$, and from (9.128) it follows that $B_{q1} = -B_{q2} = Y_0$. Thus, switching all susceptances between $\pm Y_0$ gives a phase shift of 180° at the center frequency. Similarly, for $\Delta\phi = 90°$, we require $B_{s1} = -B_{s2} = 0.414 \, Y_0$ and $B_{q1} = -B_{q2} = 0.705 \, Y_0$. The characteristics of a three-element loaded line phase shifter with these susceptance parameters have been reported [48]. It is shown that the phase shift error in both 90° and 180° phase bits is considerably less compared to that in a hybrid coupled phase shifter. However, on the basis of VSWR, the bandwidth of a 180° phase bit is too narrow to be of practical use, whereas the 90° phase bit yields a bandwidth of about 10% for a VSWR within 1.2 [48].

The problem of narrow bandwidth encountered in the three-element phase shifter is overcome in the four-element version shown in Figure 9.31. The overall *ABCD* matrix of this four-element phase bit is given by

$$
\begin{bmatrix} A & B \\ C & D \end{bmatrix} = \begin{bmatrix} Z_0(B_{si} + B_{qi} - B_{si}B_{qi}^2 Z_0^2) & jZ_0(B_{qi}^2 Z_0^2 - 1) \\ jZ_0(B_{si}^2 - B_{si}^2 B_{qi}^2 Z_0^2 + 2B_{si}B_{qi} - Y_0^2) & Z_0(B_{si} + B_{qi} - B_{si}B_{qi}^2 Z_0^2) \end{bmatrix},
$$

$$i = 1, 2 \quad (9.132)$$

The condition for input match gives

$$B_{qi}^2(B_{si}^2 Z_0^2 + 1) = 2B_{si}B_{qi} + B_{si}^2 \quad (9.133)$$

With this condition, the expression for the transmission coefficient is obtained as

$$S_{21} = \frac{1}{Z_0(B_{si} + B_{qi} - B_{si}B_{qi}^2 Z_0^2) + j(B_{qi}^2 Z_0^2 - 1)} \quad (9.134)$$

The phase factor of S_{21} is given by

$$\phi = \tan^{-1} \left| \frac{B_{qi}^2 Z_0^2 - 1}{Z_0(B_{si} + B_{qi} - B_{si}B_{qi}^2 Z_0^2)} \right| \quad (9.135)$$

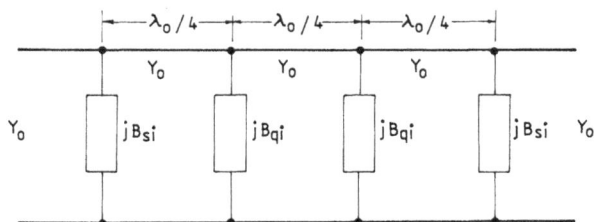

Figure 9.31 Schematic of four-element loaded line phase shifter.

If we set $B_{s1} = -B_{s2}$, then from (9.122) it follows that $B_{q1} = -B_{q2}$. With this condition, the phase shift is given by

$$\Delta\phi = 2 \tan^{-1} \left| \frac{B_{q1}^2 Z_0^2 - 1}{Z_0(B_{s1} + B_{q1} - B_{s1}B_{q1}^2 Z_0^2)} \right| \tag{9.136}$$

For the 180° phase bit, this four-element model is reported to offer the best performance (in terms of bandwidth and insertion loss) compared to the three-element and five-element models.

9.5.4 Design Example

As an example, we present the design of a loaded line phase shifter bit reported in a technical report from the Advanced Centre for Research in Electronics in Bombay [49]. The specifications are $\Delta\phi = 22.5°$, frequency = 9.375 GHz, p-i-n diode: lead inductance $L = 0.2$ nH and capacitance $C_j = 0.08$ pf, characteristic impedance of input/output line $Z_0 = 50\Omega$.

The loaded line phase shifter employing a tandem stub is chosen for the design (Figure 9.28(b) with $Y_1 = Y_0$). For a class III mode of operation, the design formulas are given by (9.101), (9.102), and (9.123) to (9.126). Setting $\theta = 90°$ in (9.101) and (9.102), we have

$$Y_c = Y_0 \sec \frac{\Delta\phi}{2} \tag{9.137}$$

$$B_{si} = \pm Y_0 \tan \frac{\Delta\phi}{2}, \quad i = 1, 2 \tag{9.138}$$

Setting $Y_0 = 1/Z_0 = 0.02\mho$, $\Delta\phi = 22.5°$, we obtain

$$Y_c = 2.039 \times 10^{-2}\mho \quad \text{or} \quad Z_c = 49.05\Omega \tag{9.139}$$

$$B_{s1} = -B_{s2} = 3.9783 \times 10^{-3}\mho \tag{9.140}$$

The diode reactances under forward- and reverse-bias conditions are given by

$$jX_f = j2\pi \times 9.375 \times 0.2 = j\,11.78\Omega \tag{9.141}$$

$$jX_r = j\left(11.78 - \frac{10^3}{2\pi \times 9.375 \times 0.08}\right) = -j\,200.43\Omega \tag{9.142}$$

Substituting for Z_0, B_{s1}, B_{s2}, X_f, and X_r in (9.124), we obtain the following quadratic equation in X_p:

$$X_p^2 - 188.65 \, X_p - 25476.81 = 0 \qquad (9.143)$$

The solution of this equation gives two values for X_p:

$$X_p = 279.72\Omega, \quad -91.08\Omega \qquad (9.144)$$

The stub lengths θ_1 and θ_2 are obtained from (9.126) and (9.123):

$$\theta_1 = \tan^{-1} \left| \frac{\{1 + B_{s1}(X_p + X_f)\}Z_0}{(X_f + X_r)^2 - B_{s1}Z_0^2} \right|$$

$$= 20.98°, \quad 159.02° \qquad (9.145)$$

$$\theta_2 = \cot^{-1}(-X_p/Z_0)$$

$$= 169.87°, \quad 28.77° \qquad (9.146)$$

For a 50Ω microstrip on alumina substrate ($\varepsilon_r = 9.8$, $h = 0.635$ mm), the guide wavelength λ_0 at $f_0 = 9.375$ GHz is 12.245 mm. Thus,

$$\theta_1 = \frac{2\pi}{\lambda_0} l_1 = 20.98°, \quad 159.02° \text{ corresponds to } l_1 = 0.714 \text{ mm}, \quad 5.409 \text{ mm} \qquad (9.147)$$

$$\theta_2 = \frac{2\pi}{\lambda_0} l_2 = 169.87°, \quad 28.77° \text{ corresponds to } l_2 = 5.78 \text{ mm}, \quad 0.978 \text{ mm} \qquad (9.148)$$

Figure 9.32 shows the layout of the phase shifter bit with all the relevant parameters. The shift in reference planes d_1 and d_2 are computed using (9.48) and (9.49).

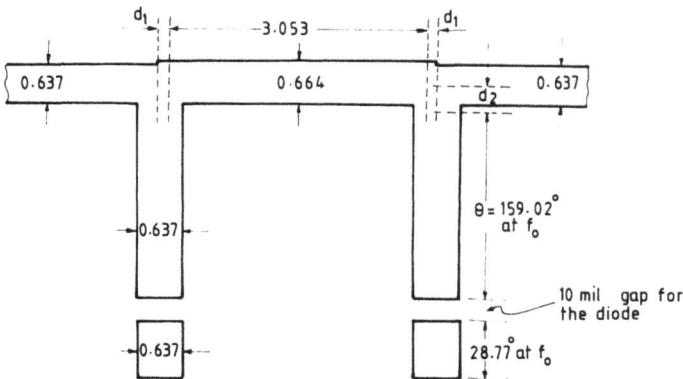

Figure 9.32 Layout of 22.5° loaded line phase shifter bit (all dimensions are in mm).

9.6 SLOT-MICROSTRIP COUPLED PHASE SHIFTER

A 180° phase bit employing a slot-microstrip coupled configuration has been realized by Davis [50]. The circuit layout is shown in Figure 9.33(a). It consists of a hybrid ring of mean path length $3\lambda_0/2$ in microstrip. Two p-i-n diodes (D_1 and D_2) are shunt mounted at a distance of $\lambda_0/4$, one on either side of the microstrip T-junction. The dotted line represents a slot resonator etched from the ground plane side of the microstrip. The total electrical length of the slot resonator is λ_s, where λ_s is the guide wavelength in a slot transmission line having the same slot width as that of the

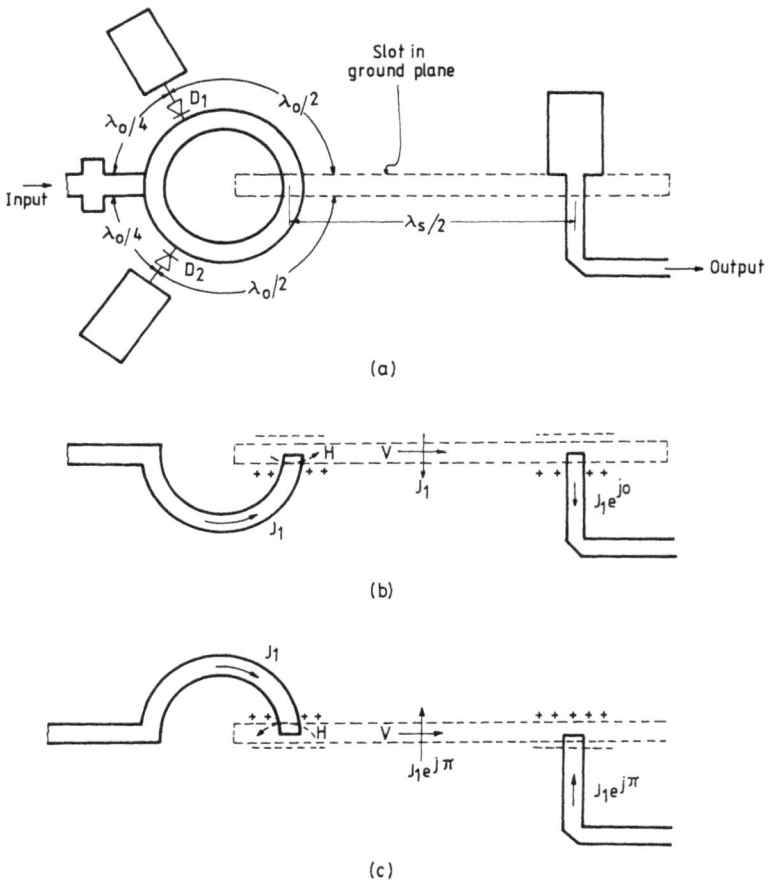

(a)

(b)

(c)

Figure 9.33 (a) Slot-microstrip coupled 180° phase shifter; (b) circuit operation with D_1 short and D_2 open; (c) circuit operation with D_2 short and D_1 open. (From Davis [50], copyright © 1975 IEEE, reprinted with permission.)

resonator. The physical length of the resonator is equal to $(\lambda_s - 2\Delta l)$, where Δl accounts for the inductive end effect at the shorted end of the slot resonator [51].

When diode D_1 is short circuited and D_2 is reverse biased, the input signal travels through the lower half of the hybrid ring, as illustrated in Figure 9.33(b). The short produced by the shunt diode D_1 presents an effective open circuit at the microstrip T-junction, and an effective short circuit with respect to microstrip ground at the slot-microstrip junction, which is located at a distance of $\lambda_0/2$. With the slot oriented perpendicular to the microstrip and shorted at a distance of $(\lambda_s/4 - \Delta l)$ from the junction, maximum coupling takes place between the microstrip and the slot. The magnetic field lines encircling the microstrip conductor at the junction induces the desired E-field across the slot, and the slot mode propagates towards the right along the slot. The second slot-microstrip transition, located at a distance of $\lambda_s/2$ from the first one, enables coupling of slot energy back to the microstrip output port. In the second switching state diode D_2 is short circuited and D_1 is reverse biased, the signal travels clockwise through the upper half of the ring. The electric field induced in the slot at the slot-microstrip junction is out of phase with the E-field induced during the first switching state. This is illustrated in Figure 9.33(c). The device thus produces an exact 180° phase shift between the two switching states. Because the two signal paths corresponding to the two switching states are of the same length, the frequency dependence of the differential phase shift is zero to a first order [50].

9.7 FIN-LINE PHASE SHIFTER

P-I-N diode phase shifter in fin-line geometry can be easily realized using a parallel slot coupled fin-line coupler [52]. The printed pattern of such a fin-line coupler is shown in Figure 9.34. The coupler makes use of an edge-coupled fin-line for the coupled region and unilateral fin-line beyond the coupled region. The layout also shows transition from unilateral fin-line to standard rectangular waveguide at each of the four ports of the coupler. In this type of coupler, the coupling is in the forward direction. Power fed to port 1 gets coupled to ports 2 and 3, and port 4 forms the isolated port. The even and odd modes propagate with different phase velocities. The scattering parameters of this coupler can be determined by using the even- and odd-mode type of analysis [30]. The expressions for the scattering parameters S_{21} and S_{31} are given by [30]

$$S_{21} = \frac{1}{2}(S_{21e} + S_{21o}) \qquad (9.149)$$

$$S_{31} = \frac{1}{2}(S_{21e} - S_{21o}) \qquad (9.150)$$

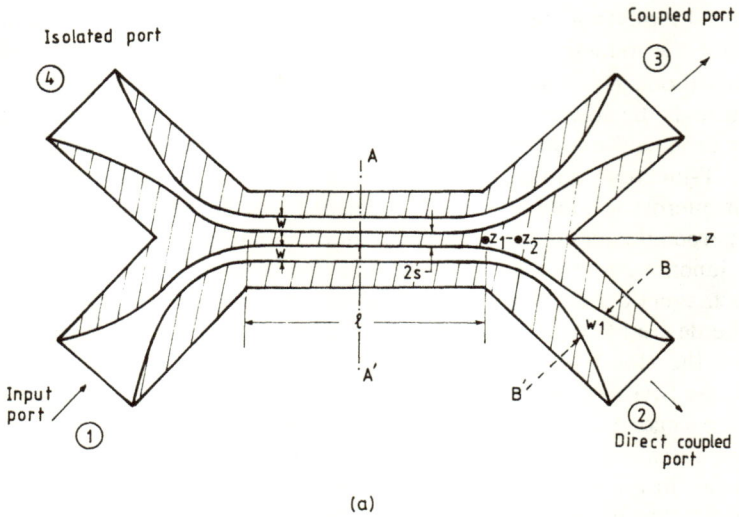

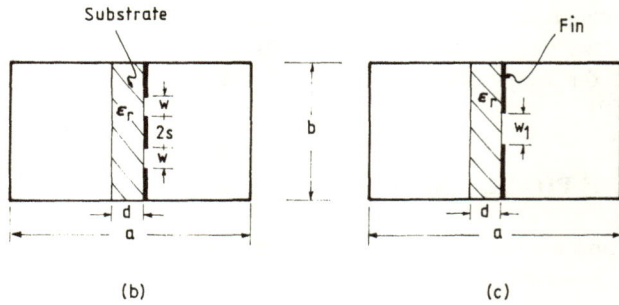

Figure 9.34 (a) Parallel-slot coupled fin-line coupler; (b) edge-coupled fin-line—cross section at AA′; (c) unilateral fin-line—cross section at BB′.

where

$$S_{21e} = \cfrac{2}{\left[2 \cos\theta_{0e} + j\left(\cfrac{Z_{0e}}{Z_0} + \cfrac{Z_0}{Z_{0e}}\right) \sin\theta_{0e} \right]} \qquad (9.151a)$$

$$S_{21o} = \cfrac{2}{\left[2 \cos\theta_{0o} + j\left(\cfrac{Z_{0o}}{Z_0} + \cfrac{Z_0}{Z_{0o}}\right) \sin\theta_{0o} \right]} \qquad (9.151b)$$

$$\theta_{0e} = \beta_{0e}l \tag{9.151c}$$

$$\theta_{0o} = \beta_{0o}l \tag{9.151d}$$

Z_{0e} and Z_{0o} are the even- and odd-mode impedances, and β_{0e} and β_{0o} are the even- and odd-mode propagation constants of the edge-coupled fin-line; Z_0 is the characteristic impedance of the unilateral fin-line at the input and output ports, and l is the length of the uniformly coupled section. For a 3-dB, 90° hybrid coupler, the outputs at ports 2 and 3 must satisfy the relation

$$S_{31} = -jS_{21} \tag{9.152}$$

Using (9.149) through (9.151) in (9.152) yields the following transcendental equation:

$$4 + \left(\frac{Z_{0e}}{Z_0} + \frac{Z_0}{Z_{0e}}\right)\left(\frac{Z_{0o}}{Z_0} + \frac{Z_0}{Z_{0o}}\right)\tan\theta_{0e}\tan\theta_{0o} = 0 \tag{9.153}$$

The solution of (9.153) gives the length l of the coupled section. For good input match (at port 1) and isolation (between ports 1 and 4), Z_{0e} and Z_{0o} should be close to Z_0 [53]. If we assume $Z_{0e} \simeq Z_{0o} = Z_0$, the solution of (9.153) simplifies to

$$l = \frac{\pi}{2(\beta_{0e} - \beta_{0o})} \tag{9.154}$$

The above formulas do not take into account the nonuniform coupling that exists at the bends on either side of the uniformly coupled section. The effective coupling length is slightly longer than l and is given by

$$l_{eff} = l + 2\Delta l \tag{9.155}$$

where

$$\Delta l = \frac{L}{\pi}\int_{z_1}^{z_2}[\beta_{0e}(z) - \beta_{0o}(z)]dz \tag{9.156a}$$

$$L = \frac{\pi}{(\beta_{0e} - \beta_{0o})} \tag{9.156b}$$

It may be noted that L is the coupling length (with uniform coupling) required for complete transfer of power from port 1 to port 3. The relation given by (9.156b) is obtained by setting $|S_{31}| = 1$, with the assumption $Z_{0e} \simeq Z_{0o} = Z_0$. The integration in (9.156a) is carried out in the axial (z) direction of the coupler. As shown in Figure 9.34, the points z_1 and z_2 correspond to the two junctions at the ends of the non-uniformly coupled region. Beyond the junction plane $z = z_2$, the coupling between the two adjacent slots is negligible and the value of β_{0e} is practically equal to β_{0o}. In a practical coupler, the junction effects due to the bends are minimized by keeping the bends as short as possible, while ensuring that the reflection coefficient at the port is within a tolerable limit.

For determining the dimensional parameters of the edge-coupled fin-line and unilateral fin-line, approximate formulas as well as rigorous spectral domain formulas are available in the literature [54]. The procedure for designing the transition from unilateral fin-line to standard rectangular waveguide is also well documented [54, 55].

A fin-line hybrid coupled phase shifter can be realized by terminating ports 2 and 3 of the 3-dB, 90° coupler in symmetric reflective circuits, as shown in Figure 9.35. The reflective circuit consists of a short-circuited section of unilateral fin-line with a *p-i-n* diode mounted across the slot at a distance of l_0 from the shorted end. Narrow half-wavelength-long slits cut on the fins serve as dc blocks. The principle of operation is the same as that of a microstrip hybrid coupled phase shifter. A signal fed to port 1 gets divided equally between ports 2 and 3, but with a phase difference of 90°. When both the diodes are forward biased, the signal gets reflected at the diodes, and when they are reverse biased, reflection takes place at the shorted ends of the fin-line at ports 2 and 3. The reflected signals in the two switching states add

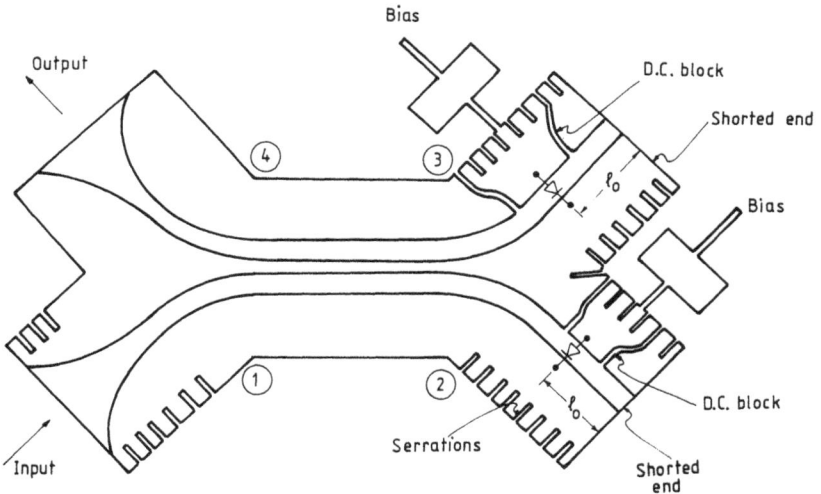

Figure 9.35 Fin-line hybrid coupled phase shifter.

up at port 4 and cancel at port 1. If the switches are ideal, the differential phase shift is $\Delta\phi = 2\beta l_0$, where β is the propagation constant in the unilateral fin-line.

9.8 DRIVER CIRCUIT CONSIDERATIONS

TTL(transistor-transistor logic)-compatible binary state transistor drivers are commonly employed for biasing the *p-i-n* diodes to achieve the two switching states. A comprehensive treatment of such driver circuits for *p-i-n* diode phase shifters is available in the literature [28, 47].

The driver circuit of the phase shifter must deliver to each *p-i-n* diode either a forward-bias current, which can be in the range of 10 to 250 mA at about 1V, or a reverse-bias voltage in a typical range of -10V to -250V at a negligible reverse-bias current (less than 1 mA). The actual bias requirements depend on the power that the phase shifter is required to handle and, therefore, the size of the diode. For example, *p-i-n* diodes in a typical X-band phase shifter operate with a 25-mA forward-bias current and -25V reverse-bias voltage in the two switching states. On the other hand, a high-power L-band phase shifter required to switch several kW of RF pulsed power in ground-based radars might use per diode as much as 200 mA of forward-bias current and -200V of reverse-bias voltage [28]. High-power switching diodes must necessarily have a large breakdown voltage on the order of -1500V or more in order to allow for a large RF voltage swing. When a large RF pulse is applied under reverse bias, a certain amount of pulse leakage current (on the order of a few mA) would be generated, and the driver must be capable of supplying this current pulse without allowing the applied reverse-bias voltage to drop appreciably.

Another aspect of the driver circuit is the switching speed. The switching time of a *p-i-n* diode is determined by the time required for transition from the forward to the reverse-bias state. Under the forward bias, the intrinsic region of the *p-i-n* diode gets flooded with holes and electrons. The total charge Q stored in the I-region is given by $Q = I_0\tau$, where I_0 is the forward bias current and τ is the average life time of the charge carriers. This charge must be withdrawn from the I-region by a momentary current surge provided by the reverse-bias voltage. The time required to withdraw this charge is called the switching time and is given by [28]

$$T_s \simeq \frac{I_0\tau}{I_r} \qquad (9.157)$$

where I_r is the average value of the surge current drawn from the reverse-bias power supply. This expression neglects the charge stored in the p^+ and n^+ regions and also the carrier recombination that occurs during switching. However, the approximate formula given by (9.157) is useful for most practical applications.

The commonly used input control signal to the driver is TTL compatible. TTL compatibility is required for interfacing the phased array with the beam steering computer. Where the beam steering logic must be transmitted over short distances, an unbalanced TTL is used. For transmission over long transmission lines, a higher impedance-balanced signal source is used. Figure 9.36 shows a two-transistor (unbalanced) TTL-compatible driver reported by White [28]. TTL logic refers to the input control signal being in the "zero" state or "one" state. In the logic "zero" state, the TTL signal source must be capable of holding a voltage not more than 0.8V at the input terminal while sinking up to 16 mA of current. The logic "one" state corresponds to high impedance at the input terminal, with a dc voltage between 2V and 5V. The current drawn from the TTL signal source must not exceed 100 μA.

The driver circuit shown in Figure 9.36 switches the *p-i-n* diode load between a reverse bias of $-V_R$, and a forward current whose magnitude can be adjusted by appropriately choosing the value of the resistor R_2. In order to illustrate the operation of the circuit, we first assume that TTL logic "one" is applied to the input terminal *AB* so that V_{AB} is about 5V. With $V_F = 5V$, hardly any current flows through R_3.

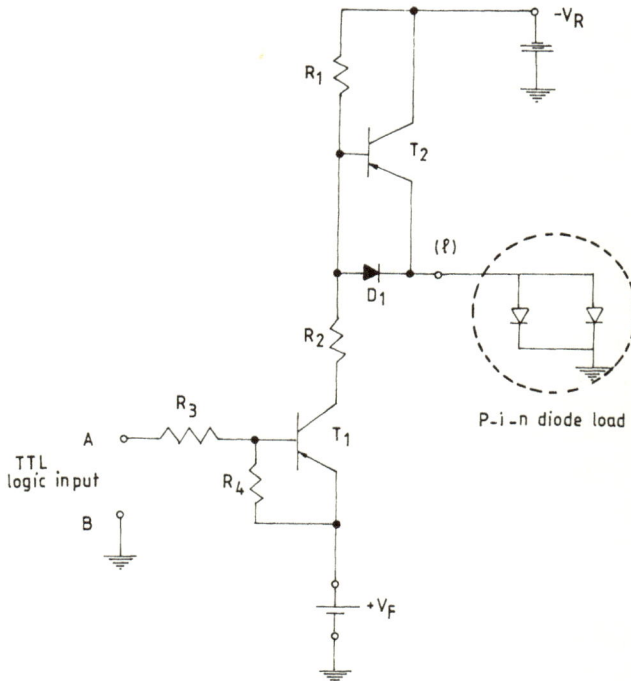

Figure 9.36 Two-transistor (unbalanced) TTL-compatible driver. (From White [28], copyright © 1974 IEEE, reprinted with permission.)

The value of R_4 is chosen such that there is not enough voltage at the emitter-base of T_1 to turn this transistor on. Any leakage current between the input terminal and the voltage supply at the emitter of T_1 will flow through R_4. With T_1 off, the p-i-n diode load is directly connected to the reverse-bias supply $-V_R$. Next, when a TTL logic "zero" is applied at the input terminal, the voltage V_{AB} is between 0 and 0.8V. Since this voltage is much less than V_F, current flows through the emitter base of T_1 and R_3 to the logic signal source. The transistor T_1 is turned on, thereby connecting V_F to the p-i-n diode load through R_2 and diode D_1. In this forward-bias state, there is also a current drain from the $-V_R$ supply through R_1, R_2, transistor T_1, and bias supply V_F. This current is undesirable for establishing the forward-bias current of the p-i-n diode load, and can be minimized by making R_1 as large as possible. On the other hand, this reverse current on amplification by the β-factor of transistor T_2

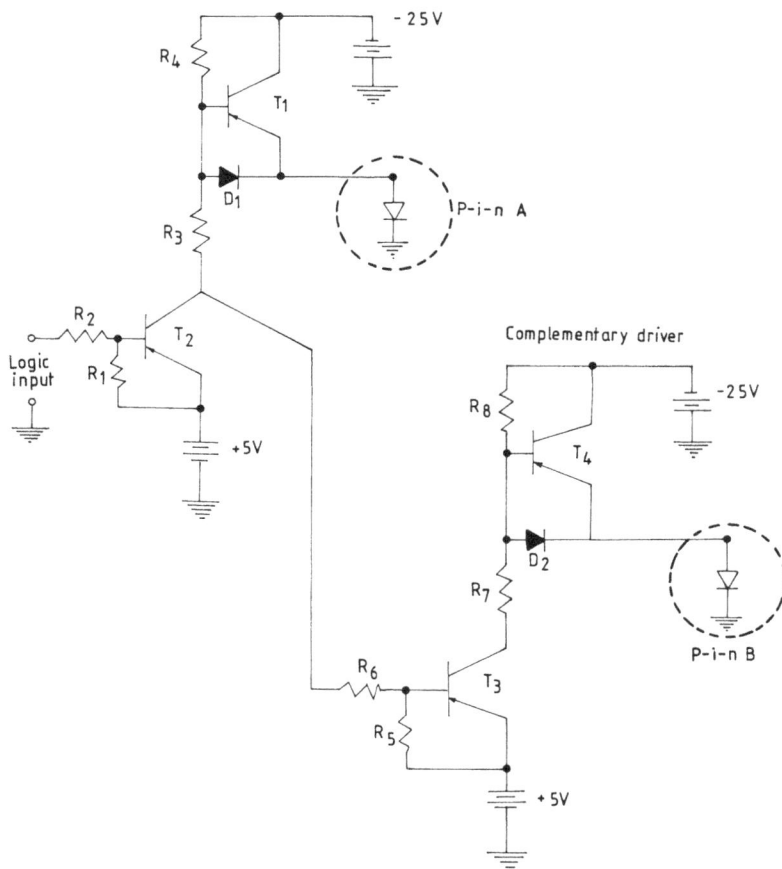

Figure 9.37 Complementary driver pair. (After White [47].)

results in the switching current I_r, which enables faster removal of charge from the I-region during the transition from forward to reverse bias. It thus helps in improving the switching speed. This transient reverse current is also needed to provide the pulse leakage current drawn by the reverse-biased diodes when a high-power RF pulse is incident on them. Therefore, a compromise is to be made between power dissipation and switching speed in choosing the value of R_1. It has been reported by White [28] that typical switching speeds of 2 μs for low-power phase shifters and 5 μs for high-power phase shifters are accomplished with no more than 1 to 5 mA of current through R_1.

Figure 9.37 shows a complementary driver reported by White [47] for operating pairs of *p-i-n* diodes in opposite bias states. This type of driver is suitable for switching the *p-i-n* diodes in a switched line phase shifter. The driver adopts a common logic input. When TTL logic "one" is applied to the input terminal, *p-i-n* diode *A* receives the reverse bias from the master driver. The same logic turns the transistor T_3 on by the emitter-base current through R_6, R_3, and R_4 path to the -25V supply. This current, amplified by the β of transistor T_3, provides the forward bias to *p-i-n* diode *B*. The magnitude of the current can be adjusted by appropriately choosing the value of R_4. With TTL logic "zero" at the input terminal, T_3 gets turned off, with the result that *p-i-n* diode *B* receives the reverse bias of -25V, and *p-i-n* diode *A* gets forward biased.

9.9 COMPARISON OF VARIOUS PHASE SHIFTERS

Practical *p-i-n* diode phase shifters are commonly built in microstrip geometry using hybrid MIC technology. Multibit phase shifters are realized using one or more of the following circuit techniques—switched line, hybrid coupled, and loaded line.

The choice between switched line, hybrid coupled, and loaded line circuits for a particular phase bit depends on factors such as the power handling capability, number of diodes, insertion loss, circuit complexity, and size.

For the purpose of comparing the peak power capability ($\hat{P}_i$) and insertion loss (α) of the three circuit forms, we use the following reported closed-form expressions [2, 20]:

Switched Line Bit (Figure 9.3(a)):

$$\hat{P}_i = \frac{V_d^2}{2Z_0} \tag{9.158}$$

$$\alpha(\text{dB}) = 17.4 \frac{f}{f_c} \tag{9.159}$$

Hybrid Coupled Bit (Figure 9.17(a)):

$$\hat{P}_i = \frac{V_d^2}{4Z_0 \sin^2 \dfrac{\Delta\phi}{2}} \qquad (9.160)$$

$$\alpha(\text{dB}) = 17.4 \frac{f}{f_c} \sin \frac{\Delta\phi}{2} \qquad (9.161)$$

Loaded Line Bit (Figure 9.26):

$$\hat{P}_i = \frac{V_d^2}{8Z_0 \sin^2\left(\dfrac{\Delta\phi}{2}\right)} \qquad (9.162)$$

$$\alpha(\text{dB}) = 17.4 \frac{f}{f_c} \tan\left(\frac{\Delta\phi}{2}\right) \qquad (9.163)$$

where V_d is the peak RF voltage rating and f_c is the cut-off frequency of the *p-i-n* diode. The insertion loss formulas assume equal loss in the two switching states. For the switched line phase shifter, both the peak power capability and the insertion loss are independent of the phase shift. On the other hand, in the hybrid coupled and loaded line phase shifter bits, the smaller the phase shift is, the higher the peak power capability is and the smaller the insertion loss is. If we consider a 180° phase bit, the switched line circuit offers the maximum peak power capability. The insertion loss is the same as that for the hybrid coupled bit. Another advantage of the switched line phase shifter is that diode contribution to insertion loss is practically the same in both the switching states, and any loss variation is due to the path difference between the two switched paths. The circuit is simple because no special matching circuit is required to equalize the insertion loss. The disadvantages are that it requires four diodes per phase bit (for equal diode loss in the two bias states), whereas the hybrid coupled phase bit uses two diodes. Secondly, the switched line phase bit requires complementary bias voltages for the on- and off-paths of the circuit. Because the advantages outweigh the disadvantages, the switched line is the most preferred configuration for the 180° bit. Where peak power is not a serious consideration, the hybrid coupled bit is also used.

For the 90° phase bit, the hybrid coupled configuration offers the best choice. This is because, while it has the same peak power capability as the switched line phase bit, its insertion loss is less by a factor of $1/\sqrt{2}$. In the case of a 45° phase bit also, the hybrid coupled configuration is commonly used, although the loaded

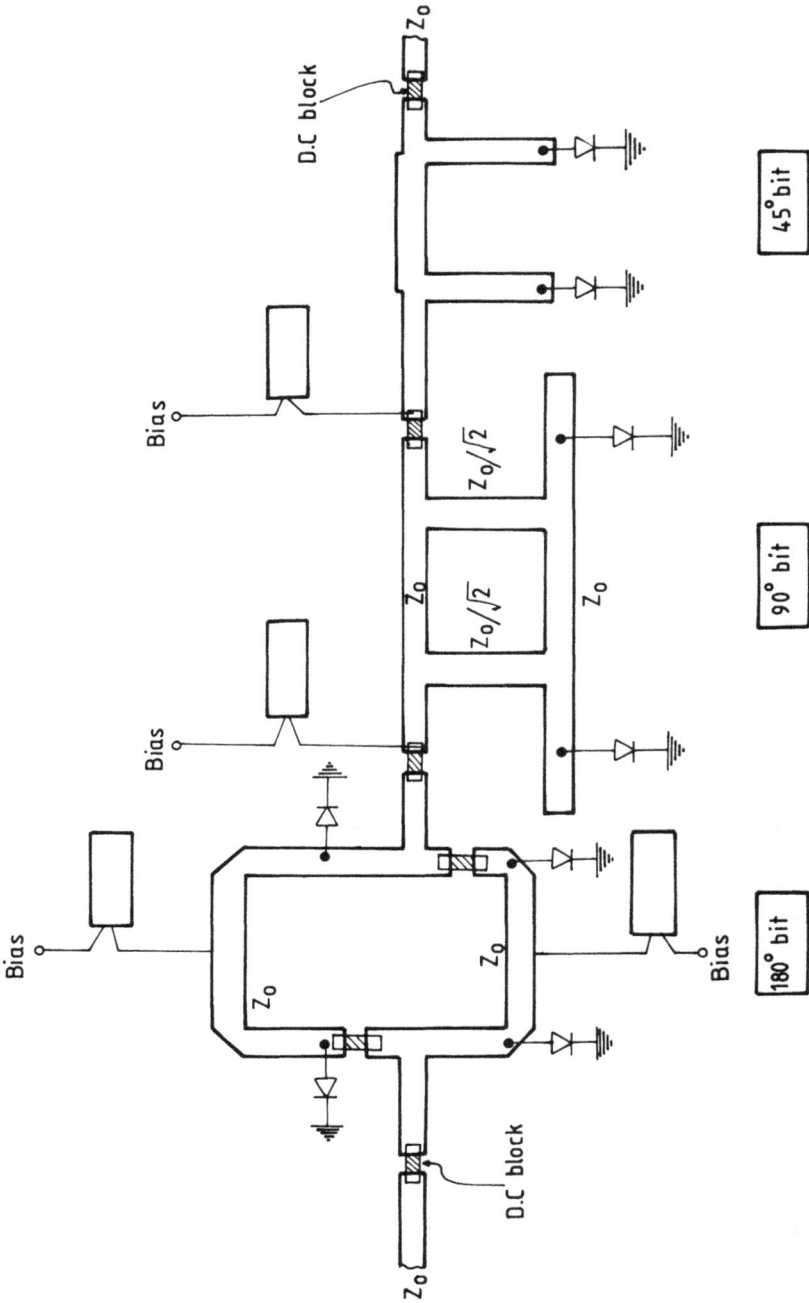

Figure 9.38 Layout of a 3-bit phase shifter.

line circuit offers a good alternative. The loaded line circuit is particularly suited for small phase bits 22.5° and 11.25° in size. The circuit is much simpler. As it dispenses with the need to use any hybrid, the discontinuity effects are also reduced. Loaded line circuits with a small bit size are also best for maximum power handling. For 90° and 180° phase bits, although a three-element and a four-element model, respectively, are reported, from the point of view of performance, they do not score over the hybrid coupled and switched line circuits. The bandwidth of three-element and four-element models is small, and, furthermore, there is no saving in terms of number of diodes used when compared with a hybrid coupled 90° phase bit and a switched line 180° phase bit.

All three basic circuit forms discussed above are useful in a multibit phase shifter required for phased-array applications. Figure 9.38 shows a typical layout of a simple 3-bit phase shifter employing a switched line circuit for the 180° bit, hybrid coupled circuit for the 90° bit, and loaded line circuit for the 45° bit.

Table 9.1
Typical Performance Characteristics of *P-I-N* Diode Phase Shifters

	White [28]	*Burns et al.* [19]	*Terrio et al.* [55]	*Glance* [56]
Frequency	L-band 1.2–1.4 GHZ	S-band 3.1–3.5 GHz	X-band 8.5–10 GHz	Ku-band 14–14.5 GHz
Number of bits	3	5	3	4
Phase bit type	Stripline hybrid coupled— 180°, 90°, 45°	Microstrip switched line— 180°, 90°, 45° Loaded line— 22.5°, 11.25°	Stripline hybrid coupled—180°, 90°, 45°	Microstrip hybrid coupled— 180°, 90°, 45°, 22.5°
Number of diodes	6	16	6	8
Insertion loss	0.6 ± 0.2 dB	1.75 dB (av)	1.6 dB (av)	1.4 ± 0.1 dB
Return loss or VSWR	Ret. loss 14 dB (min)	VSWR 1.5 (max)	VSWR 1.5 (max)	—
Phase shift error	$< \pm 10°$	—	$< \pm 10°$	—
Power handling capacity	1-kW peak (2-ms pulse, 0.06 duty cycle)	—	130-W peak (50 μs pulse, 0.1 duty cycle)	—
Size or weight	—	Size 1 × 2 in^2	Weight 1/3 oz	Size 1.677 × 0.67 in^2

Table 9.1 lists the performance parameters of some of the *p-i-n* diode phase shifters reported in the literature [19, 28, 56, 57]. Over the frequency range covering L- to Ku-bands, the insertion loss is in the range of approximately 0.7 to 1.8 dB. Their utility beyond Ku-band is only at the expense of accepting increased insertion loss. The switching speed of *p-i-n* diode phase shifters is less than 1 μs. The power limitation is determined mainly by the breakdown voltage of the diode, and the average power it can handle is limited by the thermal characteristics of the diode and the effectiveness of heat sinking. Peak power levels of a few tens of kilowatts (depending on the pulse width) and average power of a few hundred watts have been reported [40, 58].

REFERENCES

1. H.A. Watson (ed), *Microwave Semiconductor Devices and Their Circuit Applications*, McGraw-Hill, New York, 1969.

2. R.V. Garver, *Microwave Diode Control Devices*, Artech House, MA, 1976.

3. R.V. Garver, "Broadband Diode Phase Shifters," *IEEE Trans. on Microwave Theory and Tech.*, Vol. MTT-20, May 1972, pp. 314–323.

4. G. Dubost and S. Guero, "A 3 Bit Digital Phase Shifter in Ku-Band for Microstrip Phased-Array," *Proc. 8th Coll. on Microwave Communication Digest*, Budapest, Hungary, August 25–29, 1986, pp. 291–292.

5. W.R. Connerney, "Diode Loop Binary Phase Shifter," *IEEE Trans. on Microwave Theory and Tech.*, Vol. MTT-16, February 1968, pp. 134–135.

6. B.M. Schiffman, "A New Class of Broadband Microwave 90-Degree Phase Shifters," *IRE Trans. on Microwave Theory and Tech.*, Vol. MTT-6, April 1958, pp. 232–237.

7. R.P. Coats, "An Octave Band Switched Line Microstrip 3B Diode Phase Shifter," *IEEE Trans. on Microwave Theory and Tech.*, Vol. MTT-21, July 1973, pp. 444–449.

8. A. Podell, "A High Directivity Microstrip Coupler Technique," *IEEE G-MTT Int. Microwave Symp. Digest*, 1970, pp. 33–36.

9. J. Lange, "Interdigitated Strip Quadrature Hybrid," *IEEE G-MTT Int. Microwave Symp. Digest*, 1969, pp. 1–13.

10. M. Waugh and D. La Combe, "Unfolding the Lange Coupler," *IEEE Trans. on Microwave Theory and Tech.*, Vol. MTT-20, November 1972, pp. 777–792.

11. A. Affandi and M. Sheikh, "Modified Phase-Shifters in Microstrip," *Proc. 8th Coll. on Microwave Communication Digest*, Budapest, Hungary, August 25–29, 1986, pp. 309–311.

12. B. Schiek and J. Köhler, "A Method for Broadband Matching of Microstrip Differential Phase Shifters," *IEEE Trans. on Microwave Theory and Tech.*, Vol. MTT-25, August 1977, pp. 666–671.

13. B. Bhat and S.K. Koul, "Unified Approach to Solve a Class of Strip- and Microstrip-Like Transmission Lines," *IEEE Trans. on Microwave Theory and Tech.*, Vol. MTT-30, May 1982, pp. 679–685. See also correction *IEEE Trans. on Microwave Theory and Tech.*, Vol. MTT-30, November 1982, p. 2067.

14. S.K. Koul and B. Bhat, "Propagation Parameters of Coupled Microstrip-Like Transmission Lines for Millimeter Wave Applications," *IEEE Trans. on Microwave Theory and Tech.*, Vol. MTT-29, December 1981, pp. 1364–1370.

15. T.G. Bryant and J.A. Weiss, "Parameters of Microstrip Transmission Lines and of Coupled Pairs of Microstrip Lines," *IEEE Trans. on Microwave Theory and Tech.*, Vol. MTT-16, December 1968, pp. 1021–1027.

16. G.D. Lynes, "Ultra Broadband Phase Shifters," *IEEE G-MTT Int. Microwave Symp. Digest,* June 1973, pp. 104–106.

17. G.D. Lynes, G.E. Johnson, B.E. Huckleberry, and N.H. Forrest, "Design of Broadband 4-Bit Loaded Switched-Line Phase Shifters," *IEEE Trans. on Microwave Theory and Tech.*, Vol. MTT-22, June 1974, pp. 693–697.

18. S. Yamamoto, T. Azakami, and K. Itakura, "Coupled Nonuniform Transmission Line and Its Applications," *IEEE Trans. on Microwave Theory and Tech.*, Vol. MTT-15, April 1967, pp. 220–231.

19. R.W. Burns, R.L. Holden, and R. Tang, "Low Cost Design Techniques for Semiconductor Phase Shifters," *IEEE Trans. on Microwave Theory and Tech.*, Vol. MTT-22, June 1974, pp. 675–688.

20. L. Stark, R.W. Burns, and W.P. Clark, "Phase Shifters for Arrays," Ch. 12 of *Radar Handbook* by M.I. Skolnik (ed), McGraw-Hill, New York, 1970.

21. C.H. Grauling and B.D. Geller, "A Broadband Frequency Translator with 30 dB Suppression of Spurious Sidebands," *IEEE Trans. on Microwave Theory and Tech.*, Vol. MTT-18, September 1970, pp. 651–652.

22. R.W. Burns and L. Stark, "PIN Diodes Advance High Power Phase Shifting," *Microwaves,* Vol. 4, No. 11, November 1968, pp. 38–48.

23. P. Wahi and K.C. Gupta, "Effect of Diode Parameters on Reflection Type Phase Shifters," *IEEE Trans. on Microwave Theory and Tech.*, Vol. MTT-24, September 1976, pp. 619–621.

24. P.J. Starski, "Optimization of Matching Network for a Hybrid Coupled Phase Shifter," *IEEE Trans. on Microwave Theory and Tech.*, Vol. MTT-25, August 1977, pp. 662–666.

25. S. Mahapatra and Q.H. Bakir, "Computer Aided Design and Evaluation of MIC Hybrid Coupled PIN Diode Phase Shifter," *IEE-IERE Proc.* (India), Vol. 16, January 1978, pp. 19–28.

26. H.A. Atwater, "Reflection Coefficient Transformations for Phase Shifting Circuits, *IEEE Trans. on Microwave Theory and Tech.*, Vol. MTT-28, June 1980, pp. 563–568.

27. K. Watanabe, M. Arima, and T. Yamamoto, "Graph Design of p-i-n Diode Phase Shifter," *IEEE Trans. on Microwave Theory and Tech.*, Vol. MTT-29, August 1981, pp. 829–831.

28. J.F. White, "Diode Phase Shifters for Array Antennas," *IEEE Trans. on Microwave Theory and Tech.*, Vol. MTT-22, June 1974, pp. 658–674.

29. M.H. Kori and S. Mahapatra, "Integral Analysis of Hybrid-Coupled Semiconductor Phase Shifters," *IEE Proc.*, Vol. 134, Pt. H, No. 2, April 1987, pp. 156–162.

30. J. Reed and G.J. Wheeler, "A Method of Analysis of Symmetric Four-Port Networks," *IRE Trans. on Microwave Theory and Tech.*, October 1956, pp. 246–252.

31. M. Cuhaci and G.J.P. Lo, "High Frequency Microstrip Branchline Coupler Design with T-Junction Discontinuity Compensation," *Electronics Letters,* Vol. 17, 1981, pp. 87–89.

32. T.C. Edwards, *Foundations for Microstrip Circuit Design*, John Wiley and Sons, New York, 1981.

33. E.M.T. Jones and J.T. Bolljahn, "Coupled Strip Transmission Line Filters and Directional Couplers," *IRE Trans. on Microwave Theory and Tech.*, Vol. MTT-4, April 1956, pp. 75–81.

34. G.P. Riblet, "A Directional Coupler with Very Flat Coupling," *IEEE Trans. on Microwave Theory and Tech.*, Vol. MTT-26, February 1978, pp. 70–74.

35. F.C. DeRonde, "Octave-Wide Matched Symmetrical Reciprocal 4- and 5-Ports," *IEEE MTT-S Int. Microwave Symp. Digest,* 1982, pp. 521–523.

36. R. Levy, "General Synthesis of Asymmetric Multi-Element Coupled Transmission Line Directional Couplers," *IRE Trans. on Microwave Theory and Tech.*, Vol. MTT-11, July 1963, pp. 226–237.

37. R. Levy, "Tables for Asymmetric Multi-Element Coupled Transmission Line Directional Couplers," *IEEE Trans. on Microwave Theory and Tech.*, Vol. MTT-12, May 1964, pp. 275–279.

38. S. March, "A Wideband Stripline Hybrid Ring," *IEEE Trans. on Microwave Theory and Tech.*, Vol. MTT-16, June 1968, p. 361.

39. M.E. Hines, "Fundamental Limitations in RF Switching and Phase Shifting Using Semiconductor Diodes," *Proc. IEEE*, Vol. 52, June 1964, pp. 697–708.

40. J.F. White, "High-Power, p-i-n Diode Controlled Microwave Transmission Line Phase Shifters," *IEEE Trans. on Microwave Theory and Tech.*, Vol. MTT-13, March 1965, pp. 233–242.

41. T. Yahara, "A Note on Designing Digital Loaded Line Phase Shifters," *IEEE Trans. on Microwave Theory and Tech.*, Vol. MTT-20, October 1972, pp. 703–704.

42. H.A. Atwater, "Circuit Design of the Loaded Line Phase Shifter," *IEEE Trans. on Microwave Theory and Tech.*, Vol. MTT-33, July 1985, pp. 626–634.

43. F.L. Opp and W.F. Hoffman, "Design of Digital Loaded Line Phase Shift Networks for Microwave Thin Film Applications," *IEEE Trans. on Microwave Theory and Tech.*, Vol. MTT-16, July 1968, pp. 462–468.

44. W.A. Davis, "Design Equations and Bandwidth of Loaded Line Phase Shifters," *IEEE Trans. on Microwave Theory and Tech.*, Vol. MTT-22, May 1974, pp. 561–563.

45. I.J. Bahl and K.C. Gupta, "Design of Loaded Line p-i-n Diode Phase Shifter Circuits," *IEEE Trans. on Microwave Theory and Tech.*, Vol. MTT-28, March 1980, pp. 219–224.

46. T. Yahara, Y. Kadowaki, and K. Shirahata, "Optimum Design of Digital-Loaded Line Phase Shifters," *Electronics Commun.* (Japan), Vol. 57B, December 1974, pp. 43–52.

47. J.F. White, *Microwave Semiconductor Engineering*, Van Nostrand, New Jersey, 1982.

48. S. Nayagam and S. Mahapatra, "On the Design of High-Power Multiple-Element Loaded Line Phase Shifters," *Int. J. Electronics*, Vol. 59, No. 2, 1985, pp. 175–186.

49. Technical Report, "Development of Computer Aided Analysis and Design of Microwave Integrated Circuits," Advanced Centre for Research in Electronics, IIT Bombay, 1990.

50. M.E. Davis, "Integrated Diode Phase-Shifter Elements for an X-Band Phased-Array Antenna," *IEEE Trans. on Microwave Theory and Tech.*, Vol. MTT-23, December 1975, pp. 1080–1084.

51. A.K. Agrawal and B. Bhat, "Hybrid Mode Analysis of Slot Resonators with Side Walls," *Arch. fur. Electronik und Ubertragung*, AEÜ-38, July/August 1984, pp. 261–264.

52. G.B. Gajda and C.J. Verver, "Millimeter-Wave QPSK Modulator in Fin-Line," *IEEE MTT-Int. Microwave Symp. Digest*, 1986, pp. 233–236.

53. J. Siegel, "Design and Optimization of Planar Directional Couplers," *15th Eur. Microwave Conf. Digest*, 1985, pp. 853–858.

54. B. Bhat and S.K. Koul, *Analysis, Design and Applications of Fin-Lines*, Artech House, MA, 1987.

55. P. Pramanick and P. Bhartia, "Design Tapered Fin-Lines Using a Calculator," *Microwaves and RF*, Vol. 26, June 1987, pp. 111–114.

56. F.G. Terrio, R.J. Stockton, and W.D. Sato, "A Low Cost PIN Diode Phase Shifter for Airborne Phased Arrays," *IEEE Trans. on Microwave Theory and Tech.*, Vol. MTT-22, June 1974, pp. 688–692.

57. B. Glance, "A Fast Low-Loss Low-Drive 14 GHz Microstrip PIN Diode Phase Shifter," *IEEE Trans. on Microwave Theory and Tech.*, Vol. MTT-28, June 1980, pp. 669–671.

58. P. Onno and A. Pitkins, "Miniature Multi-Kilowatt p-i-n Diode MIC Digital Phase Shifters," *IEEE G-MTT Int. Microwave Symp. Digest*, 1971, pp. 22–23.

Chapter 10
FET Phase Shifters

10.1 INTRODUCTION

The use of GaAs *metal semiconductor field-effect transistor* (MESFET) as control elements in phase shifters is now well established [1–3]. Compared to the *p-i-n*, the GaAs MESFET has several advantages. They are its ultrafast switching speed (sub-nanosecond), negligible dc power consumption, and compatibility for monolithic integration. As a switching device, MESFET operates on negligible drive current and therefore requires a simple driver circuit. While the switching speed of a MESFET switch in a practical circuit is limited by the driver circuit because of the low drive current requirement, switching time as low as one nanosecond is possible [4]. A further advantage of MESFET over the *p-i-n* diode is that it can be used as an amplifying switch. The device can therefore offer insertion gain rather than insertion loss. It may, however, be noted that MESFET as an amplifier needs a drain current, and this current can be comparable to the current required to drive a typical *p-i-n* diode. This switching-*cum*-amplifying property of MESFETs is used for realizing active phase shifters.

The GaAs MESFET principles, device technology, and circuit properties are extensively covered in several books [5–8]. MESFET devices are fabricated using monolithic technology. Salient aspects of MESFET composition and monolithic fabrication techniques are reviewed in Chapter 12 as a background to *monolithic phase shifters*. In this chapter, we first introduce the basic operating principles and equivalent circuit parameters of MESFETs. We describe the properties of single-gate and dual-gate MESFETs as switches and amplifiers, and we present the design aspects of MESFET amplifier. We also review various types of passive and active FET phase shifters reported in the literature.

10.2 SINGLE-GATE MESFET CHARACTERISTICS

10.2.1 Basic Operation and Equivalent Circuit

The MESFET is a three-terminal device. The basic semiconductor structure consists of an *n*-type active layer either epitaxially grown or ion-implanted on a semi-insulating GaAs substrate. The source and drain are formed by alloyed ohmic contacts, and the gate is realized using a Schottky metal. Figure 10.1(a) shows a typical cross section of a GaAs MESFET. The Schottky contact creates a depletion layer beneath the gate even with zero gate bias. With a negative bias on the gate, the depletion layer penetrates deeper into the n layer.

Consider a MESFET biased as shown in Figure 10.1(a). Supposing the gate voltage is set to zero ($V_{gs} = 0$) and the positive bias voltage V_{ds} at the drain is increased from zero. This creates a longitudinal electric field in the channel and causes electrons to drift through the channel from the source towards the drain. Equivalently, a dc current I_{ds} flows through the conducting layer causing a voltage drop $V(x)$ along its path. As a result, a reverse potential develops between a point

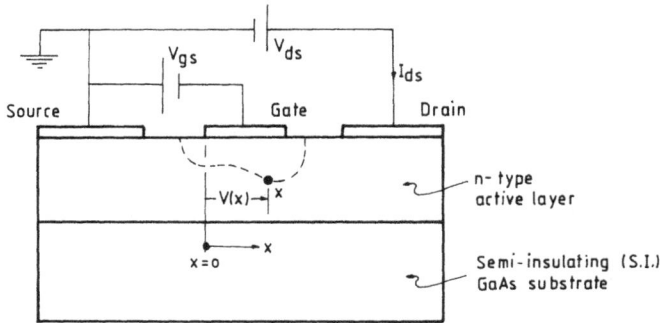

(a)

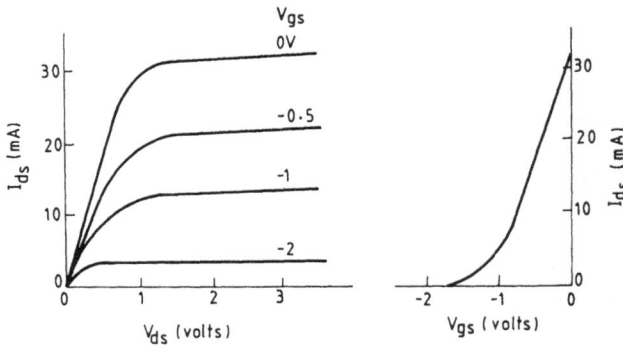

(b)

(c)

Figure 10.1 Typical current-voltage characteristics of GaAs MESFET: (a) cross section of MESFET showing depletion region; (b) drain current I_{ds} versus drain voltage V_{ds}; (c) saturation drain current I_{ds} versus gate voltage V_{gs} for fixed V_{ds}.

x in the conducting channel and the Schottky gate, and this potential becomes progressively larger as x is moved towards the drain. The charge depletion region therefore becomes thicker towards the drain, as illustrated in Figure 10.1(a). As the drain voltage V_{ds} is increased, $V(x)$ increases; the electron drift velocity increases and hence the drain current I_{ds} increases. For a sufficiently large drain voltage, when the electric field in the channel reaches a critical value, the electrons reach a maximum limiting velocity. At this drain voltage, the drain current begins to be saturated. It may be noted that this current saturation occurs due to saturation in electron drift velocity towards the drain end of the gate. Figure 10.1(b) shows typical characteristics of drain current versus drain voltage for various gate voltages. The slope of the I_{ds}–V_{ds} curve represents the output conductance of the MESFET and is given by

$$g_d = \frac{1}{r_d} = \frac{dI_{ds}}{dV_{ds}}\bigg|_{V_{gs}\text{const.}} \text{mhos} \tag{10.1}$$

If V_{ds} is fixed and the negative bias voltage V_{gs} of the gate is increased, the depletion layer becomes thicker and I_{ds} decreases. This effect is illustrated in Figure 10.1(c). When V_{gs} is sufficiently large, the channel below the gate gets completely depleted of charge carriers and I_{ds} reduces to zero. The gate reverse voltage corresponding to $I_{ds} = 0$ is called the pinch-off voltage. The slope of the I_{ds}–V_{gs} curve represents the transconductance of the device. It is defined as

$$g_m = \frac{dI_{ds}}{dV_{gs}}\bigg|_{V_{ds}\text{const.}} \text{mhos} \tag{10.2}$$

Figure 10.2 shows the high-frequency equivalent circuit representation of a GaAs MESFET, which is commonly reported in the literature [7–9]. The various intrinsic and extrinsic elements appearing in the equivalent circuit are as follows:

Intrinsic Elements

$$
\begin{aligned}
y_m &: \text{transadmittance} \\
g_m &: \text{transconductance} \\
g_d = 1/R_{ds} &: \text{output conductance} \\
R_i &: \text{input resistance} \\
C_{gs} &: \text{gate-source capacitance} \\
C_{gd} &: \text{gate-drain capacitance} \\
\tau &: \text{phase delay}
\end{aligned}
$$

Extrinsic Elements

$$
\begin{aligned}
R_s &: \text{source-gate resistance} \\
R_d &: \text{drain-gate resistance}
\end{aligned}
$$

R_g: gate metallization resistance
C_{ds}: source-drain capacitance
L_s, L_g, L_d: source, gate, drain lead inductances

Each of the above elements represents the electrical nature of a particular region of the device (see Figure 10.2(a)). For example, $(C_{gs} + C_{gd})$ represents the total gate-to-channel capacitance; R_i and R_{ds} show the effects of channel resistance; and $i_{ds} = y_m V_{gs}$ defines the voltage-controlled current source. The transadmittance y_m is given

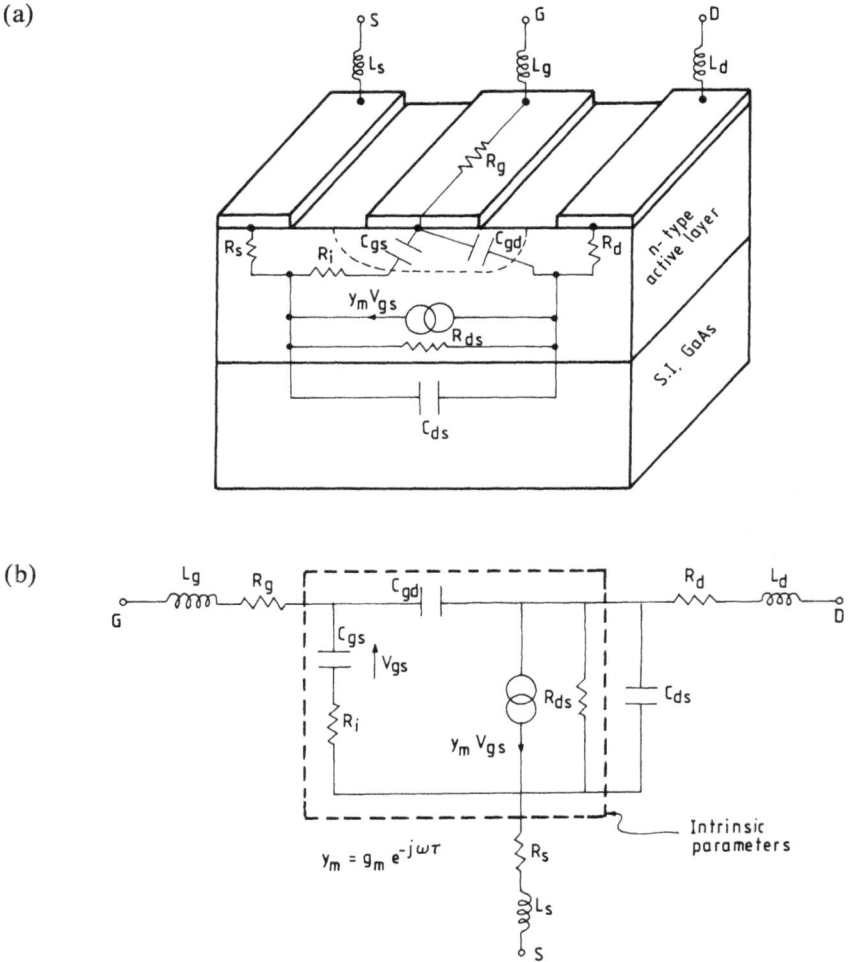

(a)

(b)

Figure 10.2 (a) GaAs MESFET structure showing the electrical parameters associated with different regions; (b) small signal equivalent circuit of MESFET.

by $g_m\,e^{-j\omega\tau}$, where g_m is the transconductance (frequency independent) and τ is the phase delay due to the carrier transit time in the channel (in the high field region). It may be noted that the intrinsic parameters depend on bias conditions. The extrinsic (parasitic) elements are the substrate capacitance C_{ds}, the source, gate, and drain resistances, represented by R_s, R_g, and R_d, and the lead inductances L_s, L_g, and L_d.

10.2.2 MESFET as a Passive Switch [4, 10–14]

The GaAs MESFET is operated as a switch by controlling the gate voltage. In the passive mode of operation ($I_{ds} \simeq 0$), the switch is in a low-impedance or on-state when $V_{gs} = 0$, and it is in a high-impedance or off-state when a negative bias voltage greater than the pinch-off voltage ($|V_{gs}| > |V_p|$) is applied. Figure 10.3 shows the two linear operating ranges of a GaAs MESFET switch. Since there is practically no drain current, the power consumption is negligible in both switching states. The drive power required to charge or discharge the input gate capacitance during the transition between the two states is negligibly small. The switch is bidirectional.

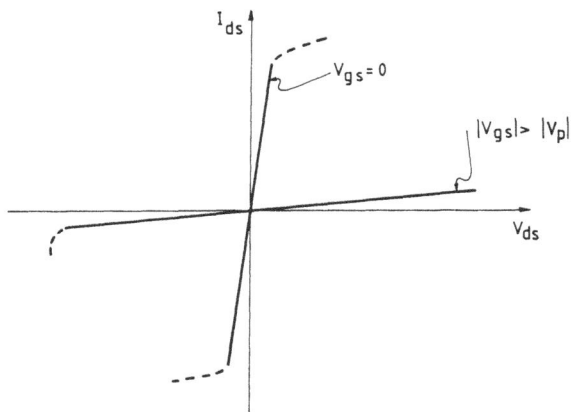

Figure 10.3 Linear operating regions of a MESFET switch.

The schematics of a GaAs MESFET illustrating the equivalent circuit parameters associated with the various regions for the two switching states are shown in Figure 10.4 [9]. The source is at ground potential and the drain is left floating (with no potential applied to it). The equivalent circuits shown do not include the lead inductances and the gate metallization resistance.

In the on-state ($V_{gs} = 0$), the depletion thickness under the gate is small. Because the drain is left floating, the gate-source and gate-drain capacitances (C_{gs} and C_{gd}) are unequal. The values of these capacitances, however, are very small and can

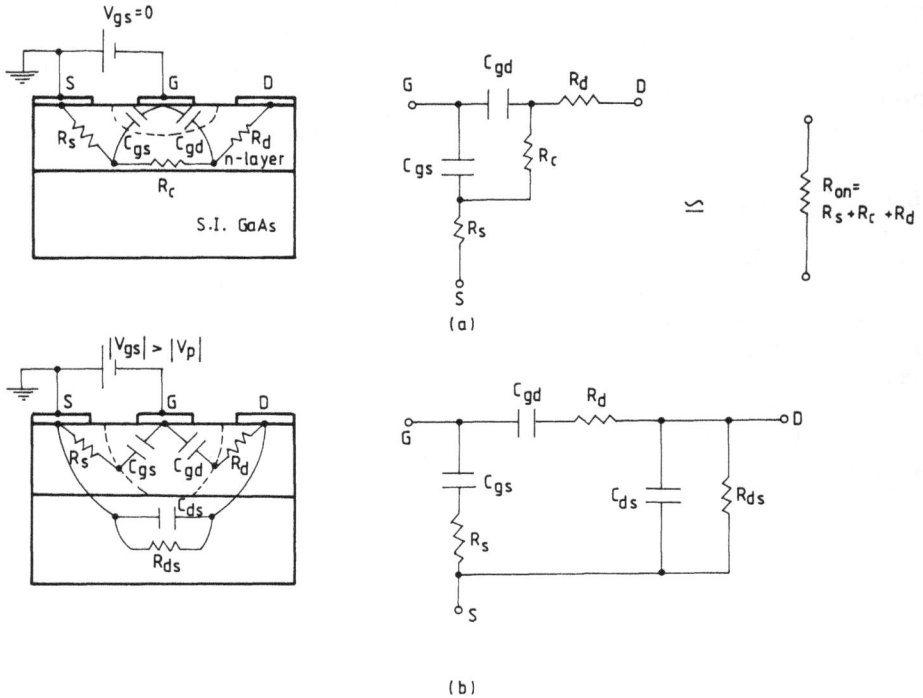

Figure 10.4 Equivalent circuit of MESFET switch in (a) on-state and (b) off-state. No potential at drain terminal.

therefore be neglected. The equivalent circuit, as a result, is reduced to a simple resistance between the source and drain (Figure 10.4(a)). It is given by

$$R_{on} = R_c + R_s + R_d \qquad (10.3)$$

where R_c is the resistance of the portion of the channel below the gate, and R_s and R_d are due to the remaining portions of the channel on either side of the gate region. The value of R_{on} is typically in the range of 2Ω to 3Ω.

When the gate is negatively biased such that $|V_{gs}| > |V_p|$, the channel is almost completely depleted of charge carriers. The FET can then be modeled in terms of an equivalent circuit, as shown in Figure 10.4(b). The capacitance C_{sd} represents the fringing capacitance between the source and drain, and R_{sd}, which is in parallel with C_{sd}, accounts for the losses associated with C_{sd}.

If the source and drain are both grounded, the depletion layer forms symmetrically about the gate. Therefore, in the on-state, we have $R_s = R_d$, so that

$$R_{on} = R_c + 2R_s \qquad (10.4)$$

Similarly, in the off-state, $R_s = R_d$ and $C_{gs} = C_{gd}$. Since the source and drain are at the ground potential, the drain is not isolated from the gate terminal. The RF impedance of the gate bias circuit affects the equivalent drain-source impedance. In practice, the gate bias circuit is configured in order to present an effective RF open at the gate terminal. Under this condition, the equivalent source-drain capacitance can be approximated by $(C_{ds} + C_{gs}/2)$. Typical values of equivalent circuit parameters for a GaAs MESFET switch with 1-mm gate periphery are $R_{on} = 2.7\Omega$, $C_{ds} = 0.14$ pF, $R_{ds} = 3\ k\Omega$, $R_s = R_d = 1.4\Omega$, and $C_{gs} = C_{gd} = 0.22$ pF [4].

Maximum Power Capability of the Switch: The maximum RF power that the MESFET can switch is related to the breakdown voltage (V_B) and the pinch-off voltage V_p. If the gate bias circuit offers an effective RF open at the gate terminal, then with gate-drain and gate-source impedances being equal, half of the drain RF voltage swing appears at the gate terminal. In the off-state, the net voltage on the gate should not fall below the pinch-off voltage during the positive half of the RF cycle, and should not exceed the breakdown voltage during the negative half of the cycle. These limiting conditions are expressed as [4]

$$-V_{gs} + \frac{V_{dmax}}{2} = -V_p \qquad (10.5)$$

$$-V_{gs} - \frac{V_{dmax}}{2} = -V_B \qquad (10.6)$$

From these relations, the maximum allowable drain voltage and the required gate bias voltage are obtained as

$$V_{dmax} = V_B - V_p \qquad (10.7)$$

$$V_{gs} = (V_B + V_p)/2 \qquad (10.8)$$

If Z_0 is the impedance level that the switch sees in its off-state, the maximum power that the switch can handle is given by

$$P_{max} = \frac{(V_B - V_p)^2}{2Z_0} \qquad (10.9)$$

10.2.3 MESFET as an Active Switch

As discussed in Section 10.2.1, the transconductance g_m of a MESFET is the gradient of drain current I_{ds} with respect to gate bias voltage V_{gs}. When I_{ds} becomes zero for $|V_{gs}| > |V_p|$, g_m also goes to zero. Since g_m can be controlled by varying the negative gate bias, the MESFET can be used as a switching amplifier (active switch). The operation of MESFET as an active switch is as follows: In the on-state, the drain terminal is forward biased with respect to source (V_{ds} positive) and the gate is at zero volts ($V_{gs} = 0$). The MESFET then functions as an amplifier. With suitable matching networks at the input and output ports, the MESFET can be operated as a matched amplifier. If the gate is now switched to a negative voltage greater than the pinch-off voltage, the FET gain reduces to zero. This is the off-state of the switch. The reduction in gain becomes the attenuation of the switch. Additional attenuation is also contributed by the input and output reflections because of mismatch.

10.3 SINGLE-GATE MESFET SWITCHING AMPLIFIER

10.3.1 Amplifier Circuit [9, 15]

In active phase shifters, MESFET is used both as an amplifier and a switch. As discussed in Section 10.2.3, the amplifying function is performed in the on-state of the switch. In order to ensure maximum on-off ratio as a switch, the amplifier should be designed for maximum gain. In the following, we present the circuit design aspects of an FET amplifier for maximum gain.

Consider a typical schematic diagram of a MESFET amplifier as shown in Figure 10.5. The FET is modeled as a two-port network. Let S_{11}, S_{12}, S_{21}, and S_{22} denote the scattering parameters of the FET device measured with its input and output terminals matched. The scattering parameters for any bias condition for the FET can be measured using an *automatic network analyzer,* such as the HP 8510A system. The input and output matching networks form part of the amplifier circuit and are used for maximizing the power gain. With arrangement shown in Figure 10.5, the overall power gain of the circuit is given by

$$G_t = \frac{P_l}{P_{as}}$$

$$= \frac{|S_{21}|^2(1 - |\Gamma_s|^2)(1 - |\Gamma_l|^2)}{|(1 - S_{11}\Gamma_s)(1 - S_{22}\Gamma_l) - S_{12}S_{21}\Gamma_l\Gamma_s|^2} \quad (10.10)$$

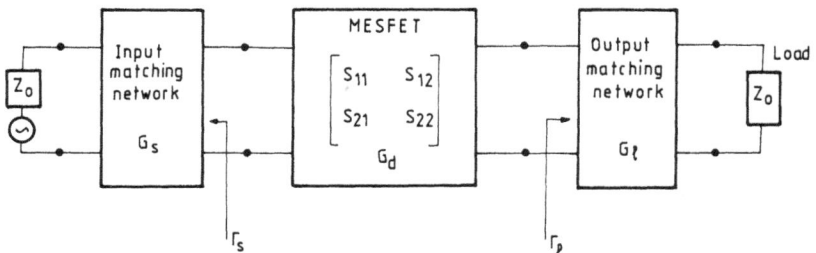

Figure 10.5 Typical schematic of a MESFET amplifier connected between source and load.

where P_l is the power delivered to the load and P_{as} is the power available from the source to the FET.

Unilateral Power Gain

For most MESFETs, $|S_{12}|$ is negligibly small. If we set $|S_{12}| = 0$, then the unilateral power gain of the amplifier is given by

$$G_u = \frac{(1 - |\Gamma_s|^2)}{|(1 - S_{11}\Gamma_s)|^2} |S_{21}|^2 \frac{(1 - |\Gamma_l|^2)}{|(1 - S_{22}\Gamma_l)|^2}$$

$$= G_s \cdot G_d \cdot G_l \qquad (10.11)$$

where

$$G_s = \frac{(1 - |\Gamma_s|^2)}{|(1 - S_{11}\Gamma_s)|^2} \qquad (10.12a)$$

$$G_d = |S_{21}|^2 \qquad (10.12b)$$

$$G_l = \frac{(1 - |\Gamma_l|^2)}{|(1 - S_{22}\Gamma_l)|^2} \qquad (10.12c)$$

G_s is the power gain due to the input matching network, G_d is the power gain of the FET device with its input and output terminals connected to matched loads, and G_l is the power gain due to the output matching network. Maximum unilateral power gain is achieved when the impedance matching networks are chosen such that $\Gamma_s = S_{11}^*$ and $\Gamma_l = S_{22}^*$. The expression for maximum unilateral power gain G_{um} is given by

$$G_{um} = \frac{1}{(1 - |S_{11}|^2)} |S_{21}|^2 \frac{1}{(1 - |S_{22}|^2)} \qquad (10.13)$$

Figure 10.6 shows typical variation in G_d and G_{um} as a function of I_{ds}/I_{dss} for a GAT6 chip at 8 GHz. I_{dss} is the saturation drain current at $V_{ds} = 5V$ with $V_{gs} = 0$. It can be seen that as I_{ds} increases, both G_d and G_{um} increase and attain maximum values when $I_{ds} = I_{dss}$.

Bilateral Case. When S_{12} is nonzero, the reflection coefficients looking into the input and output ports of the FET are different from those in the unilateral case. Let Γ_{ms} and Γ_{ml} denote the new reflection coefficients referred to the input and output ports, respectively. In order to achieve maximum gain, the input and output matching networks must be designed to provide a conjugate match to Γ_{ms} and Γ_{ml}; that is, $\Gamma_s =$

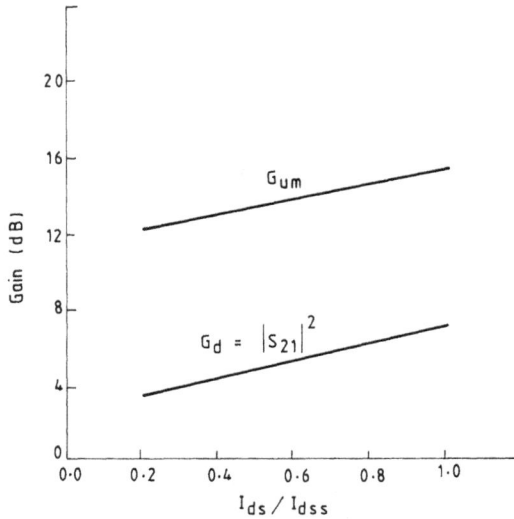

Figure 10.6 FET device gain G_d and maximum unilateral gain G_{um} of FET amplifier *versus* I_{ds}/I_{dss}. FET: GAT 6 chip, frequency: 8 GHz, $V_g = 0$, $I_{dss} = I_{ds}$ at $V_{ds} = 5$V.

Γ_{ms}^* and $\Gamma_l = \Gamma_{ml}^*$ (see Figure 10.7). The formula for computing Γ_{ms}, Γ_{ml}, and the maximum gain G_{bm} are given below [15]:

$$\Gamma_{ms}^* = \frac{C_1^*}{2|C_1|^2} [B_1 \pm (B_1^2 - 4|C_1|^2)^{1/2}], \qquad (10.14)$$

$$\Gamma_{ml}^* = \frac{C_2^*}{2|C_2|^2} [B_2 \pm (B_2^2 - 4|C_2|^2)^{1/2}],$$

$$+ \text{ sign if } B_1, B_2 < 0$$

$$- \text{ sign if } B_1, B_2 > 0 \qquad (10.15)$$

$$G_{bm} = \left|\frac{S_{21}}{S_{12}}\right| \left\{K \pm \sqrt{K^2 - 1}\right\};$$

$$+ \text{ sign if } B_1 < 0$$

$$- \text{ sign if } B_1 > 0 \qquad (10.16)$$

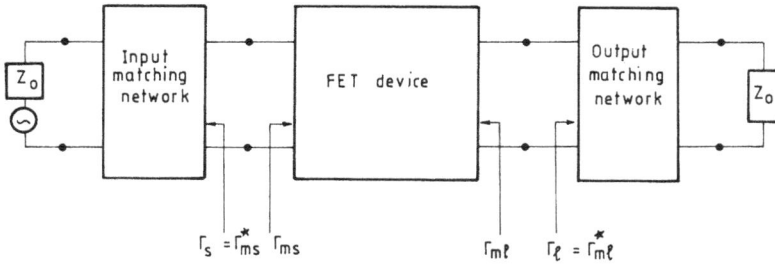

Figure 10.7 Schematic of MESFET amplifier showing matching conditions in the bilateral case $(S_{12} \neq 0)$.

where

$$C_1 = S_{11} - \Delta S_{22}^* \tag{10.17a}$$

$$B_1 = 1 + |S_{11}|^2 - |S_{22}|^2 - |\Delta|^2 \tag{10.17b}$$

$$C_2 = S_{22} - \Delta S_{11}^* \tag{10.17c}$$

$$B_2 = 1 - |S_{11}|^2 + |S_{22}|^2 - |\Delta|^2 \tag{10.17d}$$

$$\Delta = S_{11}S_{22} - S_{12}S_{21} \tag{10.17e}$$

$$K = \frac{1 - |S_{11}|^2 - |S_{22}|^2 - |\Delta|^2}{2|S_{12}S_{21}|} \tag{10.17f}$$

The parameter K is called the stability factor. In order for the amplifier to be unconditionally stable, the stability factor K must be greater than one.

Bilateral Design Example [16]

Consider the design of an amplifier using GAT6 (P109C) at 10 GHz. The S parameters of the MESFET are

$$S_{11} = 0.69 \,\underline{/144.6^\circ}$$

$$S_{21} = 1.19 \,\underline{/-31^\circ}$$

$$S_{12} = 0.05 \,\underline{/10.9^\circ}$$

$$S_{22} = 0.76 \,\underline{/-154.4^\circ}$$

We note that for these parameters, the condition $K > 1$ is satisfied. Using the formulas (10.14) through (10.17), the values of Γ^*_{ms}, Γ^*_{ml}, and maximum gain G_{bm} are calculated. They are

$$\Gamma^*_{ms} = 0.9758 \underline{/-142.93°}$$

$$\Gamma^*_{ml} = 0.9812 \underline{/155.57°}$$

$$G_{bm} = 14.25 \text{ dB}$$

Since the reflection coefficients are close to unity, a slight mismatch can cause the amplifier to oscillate. It is therefore desirable to design the amplifier for a gain less than the maximum gain. Supposing the gain is chosen to be 10 dB. In order to determine Γ^*_{ms} and Γ^*_{ml} for a 10-dB gain condition, constant gain circles are drawn on the Smith chart. The following formulas are used for determining the radius r_p and center c_p of the gain circle [15]:

$$r_p = \frac{[1 - 2K|S_{12}S_{21}|G + |S_{12}S_{21}|^2G^2]^{1/2}}{1 + D_2G} \tag{10.18}$$

$$c_p = \left\{ \frac{G}{1 + D_2G} \right\} C_2^* \tag{10.19}$$

where

$$G = \frac{G_{\text{desired}}}{G_d} \tag{10.20a}$$

$$G_d = |S_{21}|^2 \tag{10.20b}$$

$$\Delta = (S_{11}S_{22} - S_{12}S_{21}) \tag{10.20c}$$

$$D_2 = |S_{22}|^2 - |\Delta|^2 \tag{10.20d}$$

$$C_2 = S_{22} - \Delta S_{11}^* \tag{10.20e}$$

Figure 10.8 shows the gain circles for an overall gain of 10 dB, and also for 12 dB. The gain circle for a 14.25-dB gain is the point represented by Γ^*_{ml} (14.25 dB). The advantage of the gain circle is that any load reflection coefficient Γ^*_{ml} on the gain circle will result in an overall gain equal to that represented by the gain circle. Referring to Figure 10.8, Γ^*_{ml} is chosen as

$$\Gamma^*_{ml}(10 \text{ dB}) = 0.507 \underline{/155.57°} \tag{10.21}$$

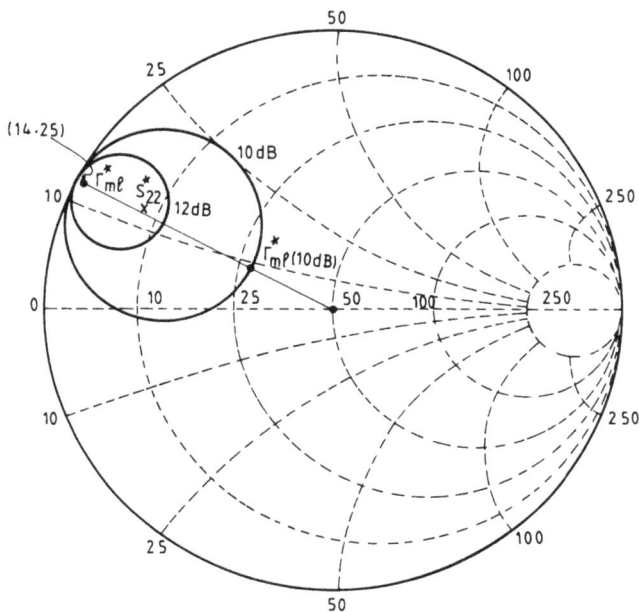

Figure 10.8 Smith chart showing constant gain circles for 10-dB and 12-dB gain of FET amplifier. Γ_{ml}^* (10 dB) = 0.507 $\underline{/155.57°}$, Γ_{ms}^* (10 dB) = 0.75 $\underline{/-143.916°}$.

The reflection coefficient Γ_{ms}^* is obtained from the formula

$$\Gamma_{ms}^* = \frac{S_{11} - \Gamma_{ml}^* \Delta}{1 - \Gamma_{ml}^* S_{22}} \tag{10.22}$$

Substituting for Γ_{ml}^*, the value of Γ_{ms}^* is obtained as

$$\Gamma_{ms}^*(10 \text{ dB}) = 0.75 \underline{/-143.916°} \tag{10.23}$$

Several schemes exist for designing the input and output matching circuits. The circuit may use transmission line sections, coupled-line sections, stubs, or a combination of these. Figure 10.9 shows an optimized amplifier incorporating straight transmission line sections. The input and output circuits are designed to provide a match to $\Gamma_{ms}^* = 0.75 \underline{/-143.916°}$ and $\Gamma_{ml}^* = 0.507 \underline{/155.57°}$. Figure 10.10(a) shows the variation in gain and Figure 10.10(b) shows the variation in the reflection coefficient ($|S_{11}|$ and $|S_{22}|$) of the amplifier as a function of frequency.

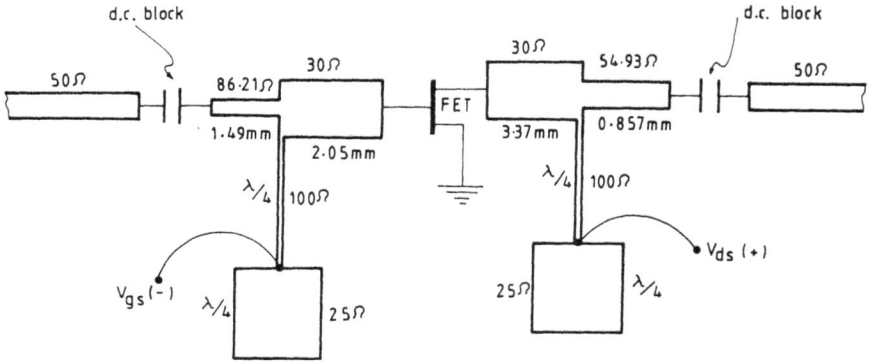

Figure 10.9 MESFET amplifier circuit.

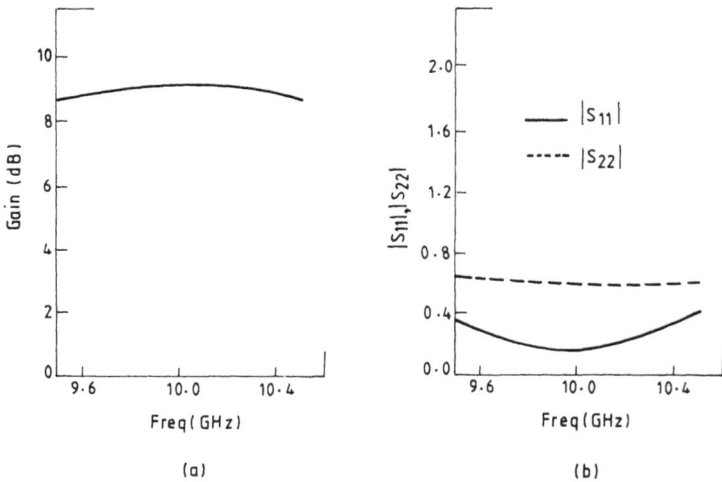

(a)

(b)

Figure 10.10 (a) Gain and (b) reflection coefficients of amplifier circuit shown in Figure 10.9.

10.3.2 Considerations as Amplifier-*Cum*-Switch

For phase-shifter applications, the amplifier must be designed to give stable forward gain at the desired frequency. As discussed in the preceding section, the magnitudes of the reflection coefficients at the input and output ports of the FET must be small in order to facilitate easy matching over the desired frequency band. As an amplifier, the switch is in the on-state. For the off-state, the FET gate is biased negatively to a voltage greater than pinch-off. In this state, the gain reduces to zero. The scattering parameters of the FET now correspond to the pinched-off state and must be measured separately. The reflections caused by a change of FET parameters give rise to mismatch loss. The on-off ratio of the switch is then the sum of the attenuation caused by a fall in gain and the mismatch loss (expressed in dB).

10.4 DUAL-GATE MESFET CHARACTERISTICS

10.4.1 Basic Operation [17–22]

The dual-gate MESFET is a four-terminal device with two control gate electrodes between source and drain. Figure 10.11(a) shows the symbol of MESFET. For understanding the basic operation and dc characteristics, a simpler representation of the device as a cascade connection of two single-gate FETs is useful [19–21]. This is shown in Figure 10.11(b). The drain terminal of FET1 is connected to the source terminal of FET2. Both the single-gate FETs can be driven to saturation using the two gate control voltages separately. In accordance with Figure 10.11, the voltages and currents can be related by the following equations:

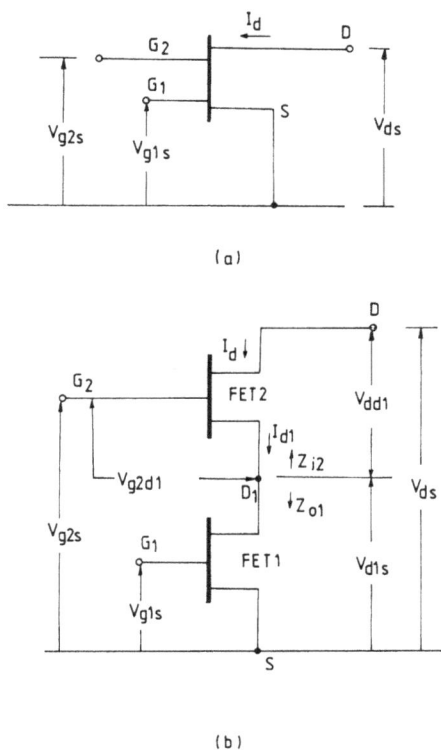

(a)

(b)

Figure 10.11 (a) Symbol of dual-gate MESFET; (b) representation as a cascade connection of two single-gate MESFET.

$$I_d = I_{d1} \qquad (10.24)$$

$$V_{ds} = V_{d1s} + V_{dd1} \qquad (10.25)$$

$$V_{g2d1} = V_{g2s} - V_{d1s} \qquad (10.26)$$

These relations can be used with the dc output characteristics of the two single-gate FETs to construct the transfer characteristic of the dual-gate MESFET. A typical transfer characteristic reported by Tsironis and Meierer [22] is reproduced in Figure 10.12. The characteristics of the two single-gate FETs are inversely superimposed because of (10.24) and (10.25). The curves show I_d versus V_{d1s} with V_{g1s} as the parameter for FET1, and I_d versus V_{dd1} with V_{g2d1} as the parameter for FET2. The curves corresponding to constant V_{g2s} are the transfer characteristics of the dual-gate FET for $V_{ds} = 5V$. Figure 10.12 thus facilitates determination of operating conditions for the dual-gate FET. For example, at a point marked P_1, corresponding to external bias $V_{ds} = 5V$, $V_{g1s} = -1V$ and $V_{g2s} = 2V$, both FET parts are saturated. Points P_2 and P_3 correspond to operation in the ohmic region for FET1 and FET2, respectively. These bias points can be used for measuring the S parameters of each FET part. Bias region in the typical range $-2V < V_{g2s} < 0.5V$, $-2.5V < V_{g1s} < -1V$ can be used for operation of a dual-gate MESFET as a gain-controlled amplifier.

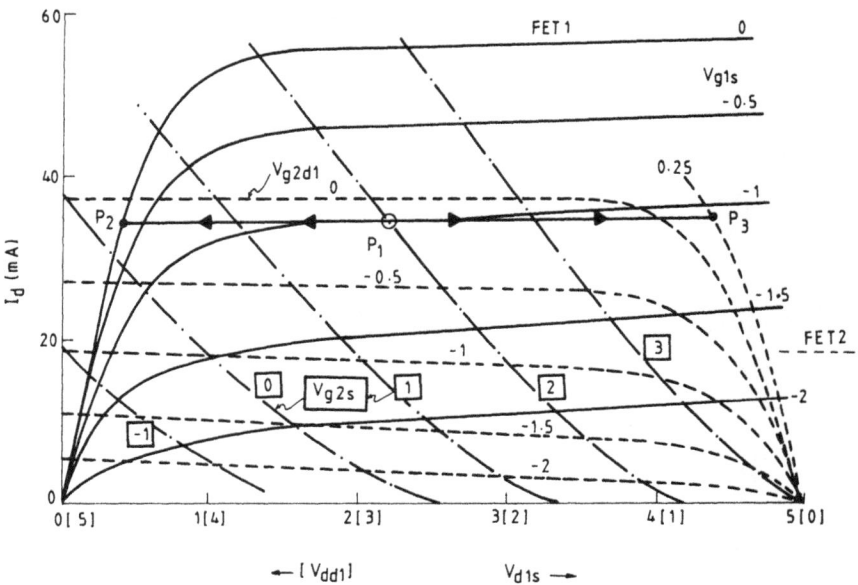

Figure 10.12 DC bidimensional transfer characteristic of GaAs dual-gate MESFET. Gate 1: 0.8 μm, gate 2: 2 μm, gate width: 200 μm, $V_{ds} = 5V$. (From Tsironis and Meierer [22], copyright © 1982 IEEE, reprinted with permission.)

10.4.2 Equivalent Circuit [22, 23]

Figure 10.13(a) shows a schematic cross section of a dual-gate MESFET identifying the small-signal equivalent circuit parameters with the various regions. The equivalent circuit is drawn by modeling the dual-gate FET as a cascade of two single-gate FETs (Figure 10.13(b)). The dotted rectangles include the intrinsic elements of the two FET parts. R_{12} denotes the resistance of the active layer between the two gates.

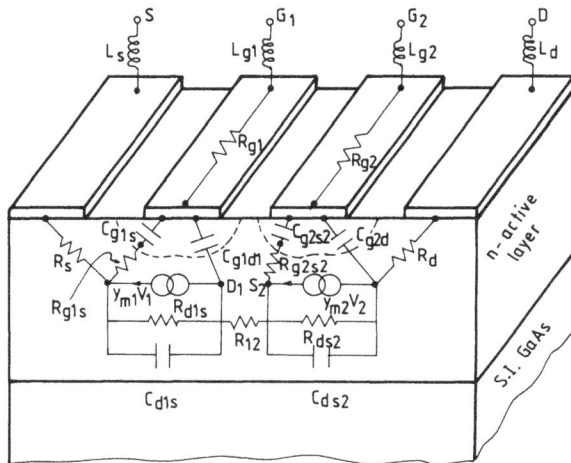

(a)

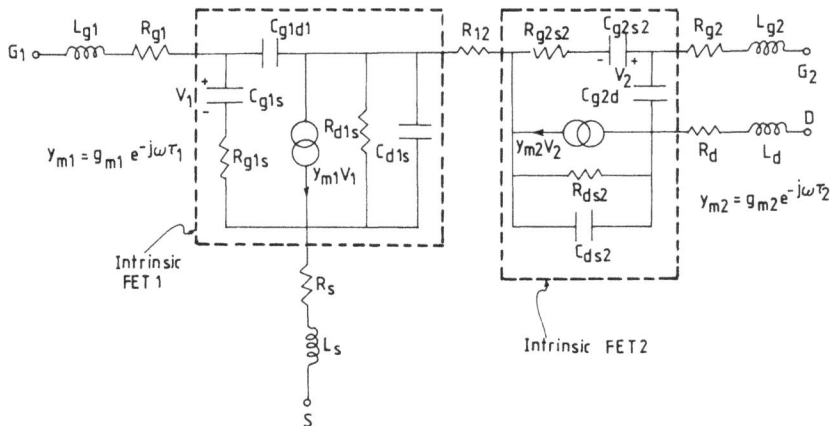

(b)

Figure 10.13 (a) Dual-gate MESFET showing various intrinsic and extrinsic elements; (b) small-signal equivalent circuit.

Other elements have the same meaning as that described in Section 10.2.1 in the case of a single-gate MESFET. In this case the channel-source capacitance C_{d1s} and the channel-drain capacitance C_{ds2} are included as part of intrinsic FETs. Tsironis and Meierer [22] have reported an accurate modeling method for characterizing each FET part by employing dc and microwave measurements under its actual bias conditions. Based on the same cascade representation and assuming a simplified equivalent circuit in terms of only the intrinsic elements (see Figure 10.14), Scott and Minasian [23] have presented a simple and efficient modeling procedure for the dual-gate MESFET. This uses dc data and three-port S-parameter measurements at a few frequencies. The method provides dual-gate MESFET parameters when both FET parts are in saturation. In order to have an idea of the magnitudes of various parameters, typical values [23] are listed, along with the equivalent circuit in Figure 10.14.

Typical Values of Elements ($i = 1, 2$).

	C_{di} (fF)	C_{gi} (pF)	C_{si} (fF)	R_{di} (Ω)	R_{si} (Ω)	g_{mi} (mmho)	τ_i (ps)
FET1	20	0.181	51	350	17	24.3	0.7
FET2	17	0.288	48	400	17	15.0	5.6

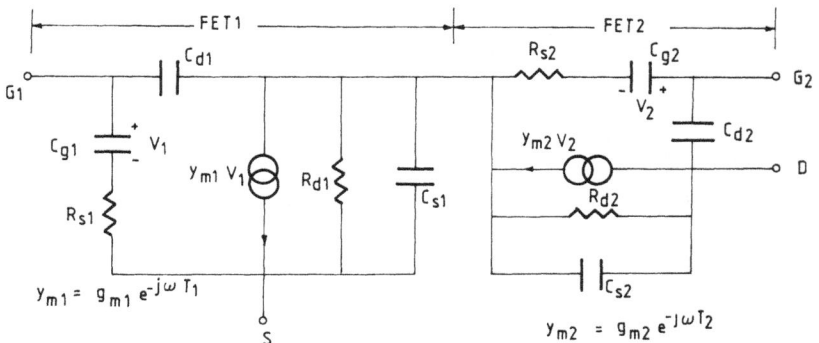

Figure 10.14 Simplified equivalent circuit model of dual-gate MESFET and typical elements values. (After Scott and Minasian [23].)

10.4.3 Operation as Amplifier-*Cum*-Switch

The dual-gate MESFET is normally operated in common-source configuration. The input signal is fed between the first gate and source and the output signal is taken between drain and source. The second gate is RF grounded. The dc bias at the second gate serves as the control voltage. With a positive bias, the device yields high gain, and with a negative bias beyond the pinch-off voltage, it offers large insertion loss. The device can therefore be operated as an active switch with a high on-off ratio.

For circuit implementation, the three-port scattering parameters of the dual-gate MESFET are measured at the two bias conditions corresponding to the on- and off-states of the switch. The three-port scattering parameters are converted to two-port scattering parameters by considering the second gate to be terminated in an impedance Z_3. The schematic is shown in Figure 10.15 [19]. The scattering parameters S_{ik} of this two-port network are related to the parameters S'_{ik} of the original three-port network by [24]

$$S_{ik} = S'_{ik} + \frac{S'_{i3}S'_{3k}}{\left(\dfrac{1}{\Gamma_3} - S'_{33}\right)} \tag{10.27}$$

where Γ_3 is the reflection coefficient of the termination Z_3 with reference to impedance Z_0.

$$\Gamma_3 = \frac{Z_3 - Z_0}{Z_3 + Z_0} \tag{10.28}$$

Knowing S_{ik}, the input and output reflection coefficients and the maximum forward gain can be calculated using the formulas given by (10.14) to (10.17). Using the notation G_f to denote the maximum available forward gain, we have from (10.16)

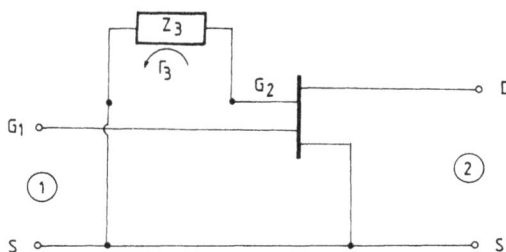

Figure 10.15 Dual-gate MESFET as a two-port device with G_2 terminated in an impedance Z_3. (After Liechti [19].)

$$G_f = \left| \frac{S_{21}}{S_{12}} \right| \left\{ K \pm \sqrt{K^2 - 1} \right\} \qquad (10.29)$$

where K is the stability factor given by (10.17f). The associated reverse gain G_r can be calculated from

$$G_r = \left| \frac{S_{12}}{S_{21}} \right| \left\{ K \pm \sqrt{K^2 - 1} \right\} \qquad (10.30)$$

Figure 10.16 shows the variation of G_f, G_r, and the input voltage standing wave ratio (VSWR) as a function of the second gate voltage V_{g2s} for the different reactive

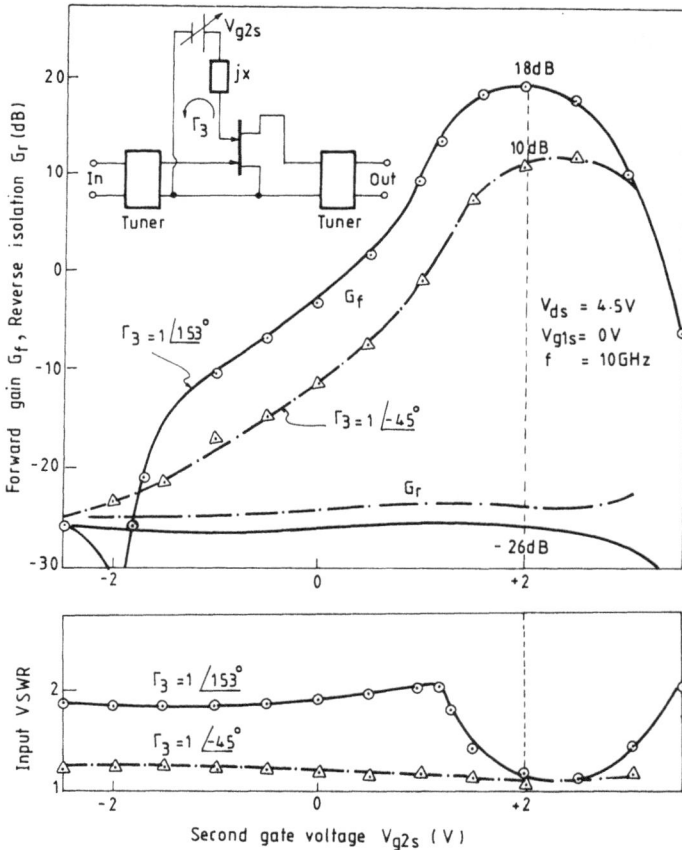

Figure 10.16 Forward gain G_f, reverse isolation G_r, and input VSWR *versus* second-gate voltage V_{g2s} for two reactive terminations at the third port. The first gate and the drain are image matched at $V_{g2s} = 2.0$V. (From Liechti [19], copyright © 1975 IEEE, reprinted with permission.)

terminations at port 3 [19]. The drain voltage V_{ds} is 4.5V, first gate voltage V_{g1s} is 0, and the frequency of operation is 10 GHz. The first gate and drain are image matched at $V_{g2s} = 2$V. It can be seen that with $\Gamma_3 = 1 \underline{/153°}$, the maximum forward gain at $V_{g2s} = 2$V is 18 dB and at $V_{g2s} = -2.5$V, the forward gain falls to -26 dB, thus giving an on-off ratio of 44 dB. It may be noted that at -2.5V, the channel under gate 2 gets completely depleted, thereby reducing the drain current to zero. The dual-gate MESFET then behaves like a passive, reciprocal device, and G_f is equal to G_r. In this bias state, the drain-to-first-gate capacitance dominates the coupling between the ports. It can be seen from the graph that the reverse isolation G_r remains fairly constant over the entire gate-bias range from 2V to -2.5V.

It is worthwhile to note that a dual-gate MESFET has considerably higher gain as an amplifier and a higher on-off ratio as a switch in comparison with a single-gate MESFET. The dual-gate MESFET is therefore more suitable for the realization of active digital phase shifters. It is also used as a variable gain amplifier [25] in the design of analog phase shifters.

10.5 PASSIVE PHASE SHIFTERS

10.5.1 Single-Gate MESFET Biasing Considerations

The operation of a single-gate MESFET as a passive switch is described in Section 10.2.2. Like the p-i-n diode, it is a bidirectional switch and can be used in either the shunt- or series-mounted configuration with respect to the transmission line. These mounting schemes are shown in Figure 10.17. The source and drain are dc grounded. In the off-state, therefore, the gate-source and gate-drain capacitances are equal. Because of this, the drain is not isolated from the gate terminal. In practical circuits, the bias network is configured so as to provide high impedance for the RF at the gate terminal. This is achieved by using a low-pass filter such that it presents an effective RF open at the gate. This biasing arrangement is shown in Figure 10.17. Alternatively, the gate bias may be fed through a resistor of high value, on the order of 5 kΩ. With the gate terminated in a high impedance, the off-state impedance of the switch may be represented as a series combination of a capacitor C_p (pinch-off capacitance between source and drain) and a resistor R_p (which is the total residual resistance at pinch-off). The RF impedance in the off-state may be increased by parallel resonating the RC combination with an inductor connected between source and drain (see Figure 10.18). The off-state-effective RF resistance (R_{off}) at resonance is approximately given by [26]

$$R_{off} = \left(\frac{1}{\omega R_p C_{off}}\right)^2 R_p \qquad (10.31)$$

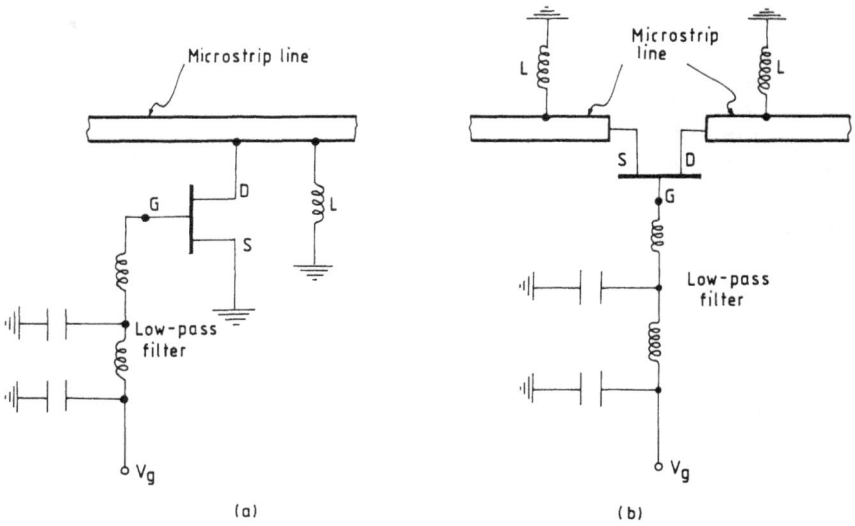

Figure 10.17 Biasing arrangement for MESFET switch in (a) shunt mode of operation and (b) series mode of operation. Source and drain are dc grounded.

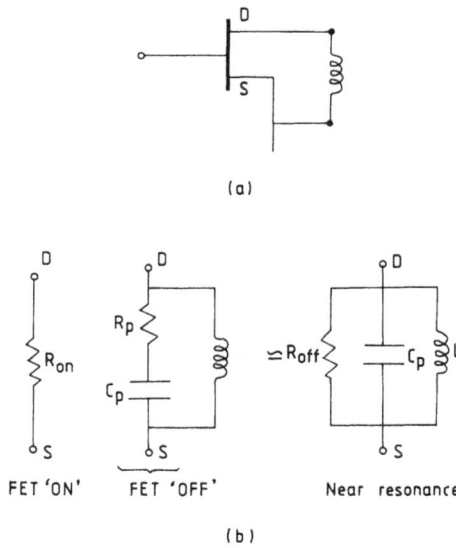

Figure 10.18 (a) MESFET with a parallel resonating inductor; (b) equivalent circuits in the on-state and off-state of the FET switch.

The figure of merit $\hat{Q}$ of the switching FET is given by

$$\hat{Q}^2 = \frac{\left| R_{\text{on}} - \left(R_p - \dfrac{j}{\omega C_{\text{off}}} \right) \right|^2}{R_{\text{on}} R_p}$$

$$= \frac{(R_{\text{on}} - R_p)^2 + \left(\dfrac{1}{\omega C_{\text{off}}} \right)^2}{R_{\text{on}} R_p} \qquad (10.32)$$

With the approximation $(R_{\text{on}} - R_p) \ll 1/\omega C_{\text{off}}$, which is normally satisfied for switching FETs, (10.32) reduces to

$$\hat{Q}^2 \simeq \frac{1}{R_{\text{on}} R_p (\omega C_{\text{off}})^2} \qquad (10.33)$$

Using the relation given by (10.31) in (10.33), $\hat{Q}^2$ can be written as [26]

$$\hat{Q}^2 \simeq R_{\text{off}}/R_{\text{on}} \qquad (10.34)$$

Since R_{off} reduces inversely as the square of frequency, the figure of merit of the MESFET switch degrades with an increase in frequency.

10.5.2 Single-Gate MESFET Phase Shifters [1–3, 26–29]

Four basic types of phase-shifter circuits are described in Chapter 8: the *switched line, hybrid coupled, loaded line,* and *low-pass high-pass.* The same circuit forms can be used to realize single-gate MESFET phase shifters, with FETs operating as bidirectional passive switches. The operating principles remain the same. Several *p-i-n* diode phase-shifter configurations based on the above basic forms are described in Chapter 9. The same circuit schemes are applicable to passive FET phase shifters. It is, however, important to note that a MESFET device offers much lower off-state impedance compared to a *p-i-n* diode. The value is typically $-j50\Omega$ at X-band. For good switching action, this capacitive reactance must be resonated with an inductive reactance, or its effect must be suitably taken into account in the design of impedance matching sections. Another important consideration is the gate bias circuit, which must ensure a high impedance for RF at the gate terminal.

MESFET phase shifters can be realized either in hybrid microwave integrated circuit (MIC) or monolithic form. In a hybrid MIC, discrete FET devices are mounted on the substrate after the circuit pattern is photolithographically etched. In a monolithic circuit, the device is formed *in situ* on a GaAs substrate as part of the monolithic process. Because of this compatibility with the monolithic technique, FET phase shifters have become attractive in monolithic version. Several passive FET phase-shifter circuits realized in monolithic chip form have been reported. For example, Sokolov *et al.* [26] have reported a Ka-band 180° phase-shifter bit in switched line configuration. The circuit uses four MESFET passive switches, each shunt mounted at a distance of $\lambda/4$ from the input or output T-junction. Ayasli *et al.* [1] have reported a 4-bit X-band phase shifter consisting of a switched line circuit for the 180° and 90° phase bits and a loaded line circuit for the 45° and 22.5° bits. Another example is the 6-bit C-band phase shifter reported by Andricos, Bahl, and Griffin [2]. This phase shifter consists of a cascade of five fixed phase bits and an analog phase bit for 0 to 11° variation. The larger bits (180° and 90°) are of the switched line type, and the smaller phase bits (45°, 22.5°, 11.25°), as well as the variable bit, are of the loaded line type. Of the four basic forms of phase-shifter circuits, the low-pass high-pass type is the most extensively used for monolithic implementation. Multibit high-pass low-pass types of FET phase shifters have been realized up to Ka-band [3, 27–29]. The details of monolithic implementation are discussed in Chapter 12. Here we present a few illustrative circuits to show the mounting and biasing schemes.

Figure 10.19 shows the schematic of a switched line phase-shifter bit with shunt-mounted FET switches. This switched line scheme is the same as that adopted by Ayasli *et al.* [1] for the 180° and 90° bits of a 4-bit monolithic phase shifter. The circuit uses three switching single-gate FETs instead of the conventional four switching elements. The insertion losses in the two switching states are equalized by choosing different impedance levels for the two switching paths. The FETs are shunt mounted with the drain connected to the microstrip and the source grounded. The drain terminal is also dc grounded through a shorted stub. Since the input and output microstrip lines do not carry any dc voltage, the need for dc blocking capacitors is eliminated. The gate-control bias is provided through a two-section low-pass filter. This type of distributed form of bias network is more commonly used in hybrid MIC phase shifters. In monolithic phase shifters, lumped inductors and capacitors are generally used for bias networks.

Figure 10.20 shows typical circuit schematics of hybrid coupled and loaded line FET phase shifters. A different biasing scheme is shown here. The gate is biased through a large resistor (typically 5 kΩ). This provides sufficiently large impedance for the RF at the gate.

Figure 10.21(a) shows the schematic of a low-pass high-pass phase shifter [28]. The circuit uses six FETs. The series-connected FETs in one T-section and the shunt-mounted FET in the other T-section are controlled by a common bias voltage. When the control voltage $V_g = 0$ and $|\overline{V}_g| > |V_p|$, the circuit reduces to a low-pass filter, as shown in Figure 10.21(b). The on-state resistances of FETs F_3, F_4, and F_5 are assumed to be small compared to the impedances in parallel with them. When the switching states are reversed ($\overline{V}_g = 0$, $|V_g| > |V_p|$), the circuit reduces to the form shown in Figure 10.21(c), provided the on-state resistances of FETs F_1, F_2, and F_6 are negligible. If the capacitors C_4 and C_5 offer sufficiently large impedances compared to the impedances due to L_3 and L_2, respectively, the circuit behaves like a high-pass filter. This phase-shifter circuit facilitates the taking into account of FET parameters as part of the filter network so that the need to parallel resonate the off-state capacitance can be eliminated.

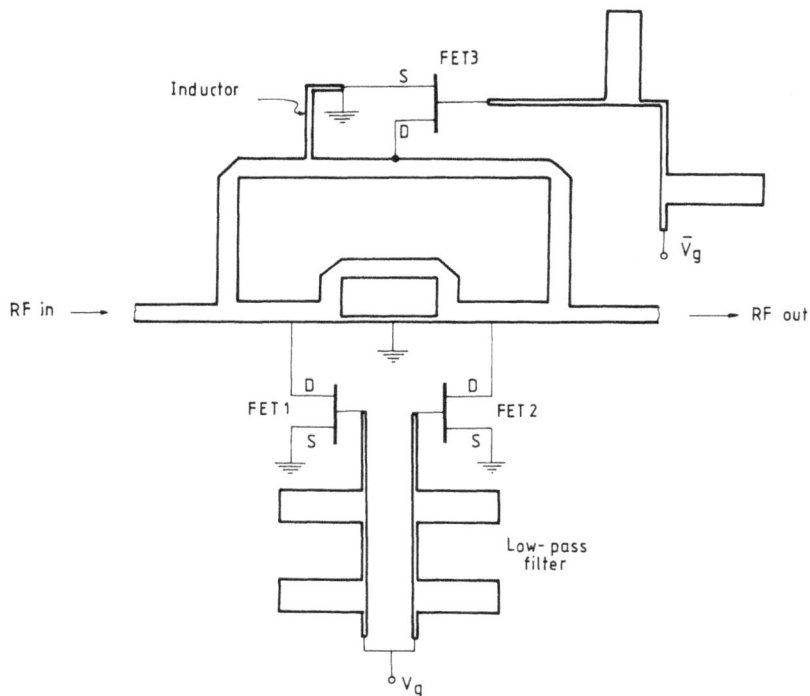

Figure 10.19 Schematic of a switched line MESFET phase shifter bit.

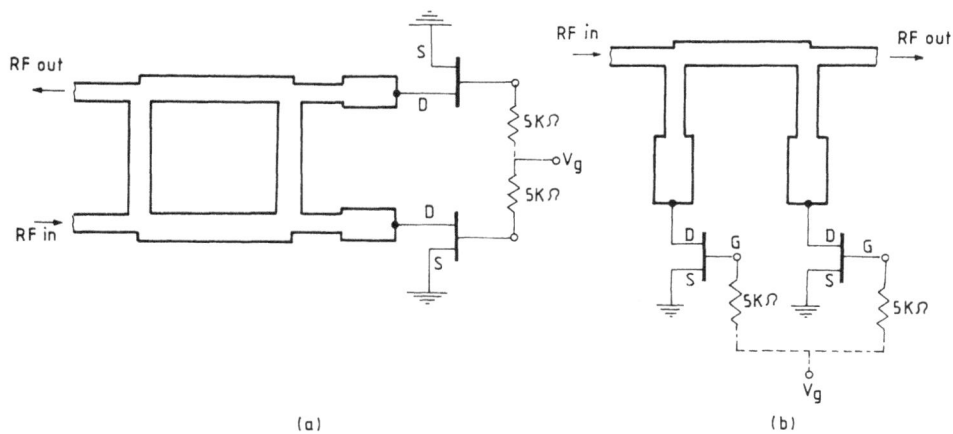

Figure 10.20 Schematics of (a) hybrid coupled reflection-type and (b) loaded-line-type MESFET phase-shifter bits.

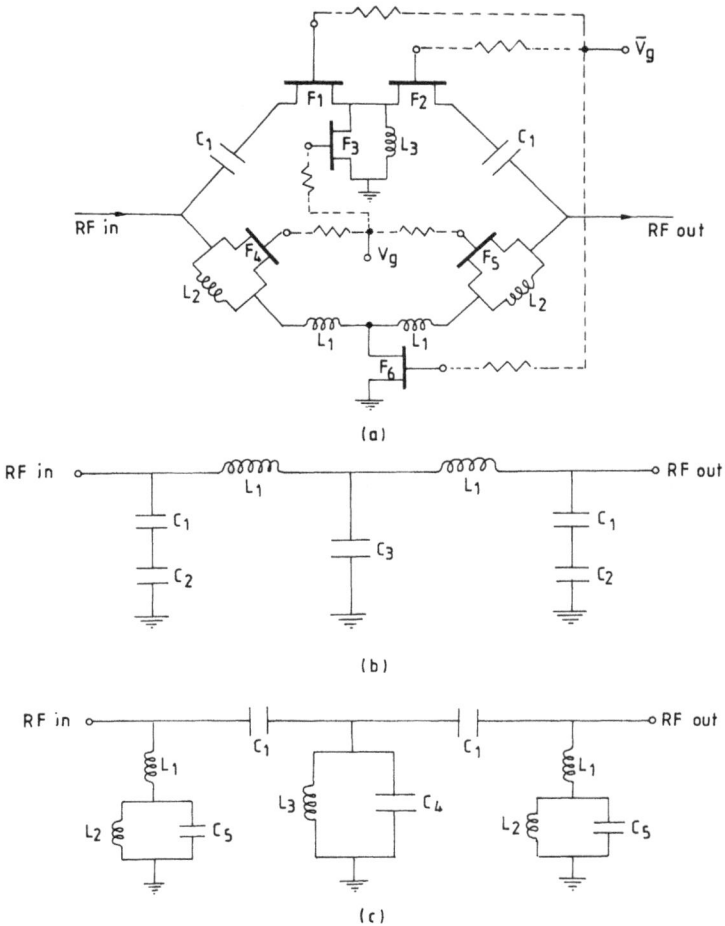

Figure 10.21 (a) Schematic of a single-bit low-pass high-pass MESFET phase shifter; (b) equivalent circuit when $V_g = 0$, $|\overline{V_x}| > |V_p|$; (c) equivalent circuit when $\overline{V_g} = 0$, $|V_x| > |V_p|$. (After Ayasli *et al.* [28].)

10.6 ACTIVE FET PHASE SHIFTERS

10.6.1 Single-Gate MESFET Digital Phase Shifter [30]

As an example of a single-gate MESFET active phase shifter, Figure 10.22 shows the circuit layout of a 180° phase bit reported by Collier and Evans [30]. The input circuit uses a slotline-to-microstrip series Tee for dividing the input into two equiamplitude antiphase signals. Since the phase split occurs in a narrow slot-microstrip

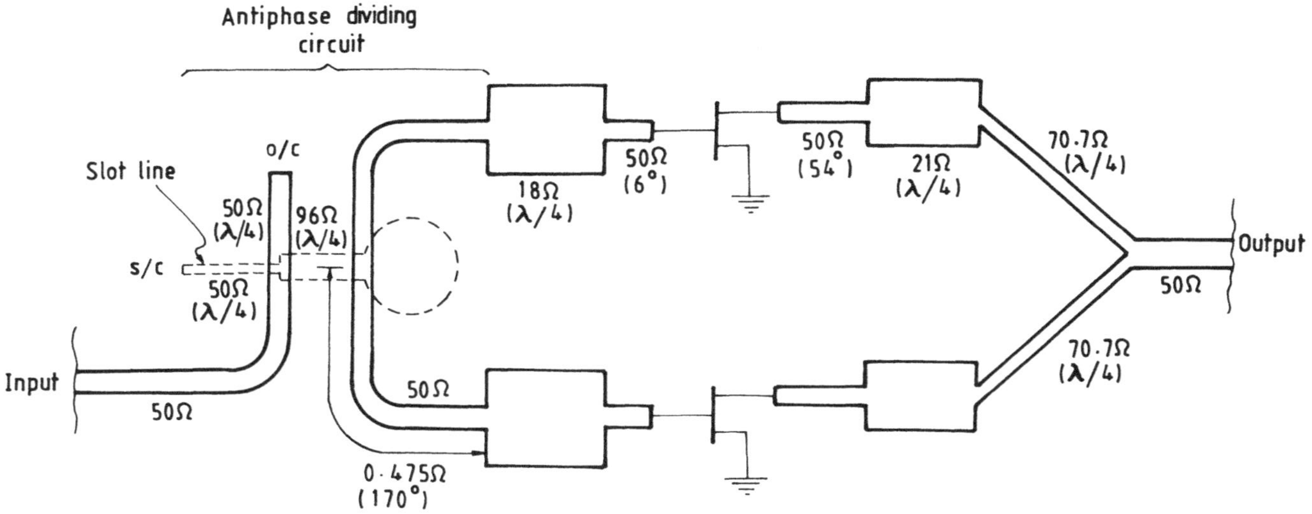

Figure 10.22 MESFET active phase shifters using a slotline-to-microstrip series Tee as an antiphase dividing circuit. (From Collier and Evans [30], copyright © 1983, reprinted with permission of IEE.)

junction, the phenomenon is inherently broadband in nature [31, 32]. Practical bandwidth, however, gets limited because of the open- and short-circuited quarter-wave stubs, which form part of the transition.

The function of microwave switching and amplification is performed by a pair of Plessey GAT4 single-gate MESFET devices, with one amplifier in each of the two signal paths. Each amplifier has a two-element matching circuit at the gate as well as at the drain terminal, with the source being common. The two out-of-phase signals form the input to the two amplifiers. When one FET is in the on-state (gate bias set to zero), the other FET is in the off-state (gate bias at pinch-off voltage of $-3V$). The input signal arriving at the FET that is off is largely reflected back. The input circuit is arranged such that this reflected signal combines with the signal routed to the FET that is on. Similarly, the output Tee network is designed such that the amplified output signal mostly reaches the output port with negligible leakage into the drain terminal of the FET that is off. This 180° phase shifter is reported to offer a gain of approximately 3 dB at 9.5 GHz [30].

10.6.2 Dual-Gate MESFET Digital Phase Shifter

For active digital phase shifting, *the dual-gate FET* (DGFET) is preferred over the single-gate FET because of its much higher on-off switching ratio. Figure 10.23 shows the schematic of an active digital phase shifter using two DGFETs. It is assumed that the two devices have the same gain and transfer phase. A Wilkinson

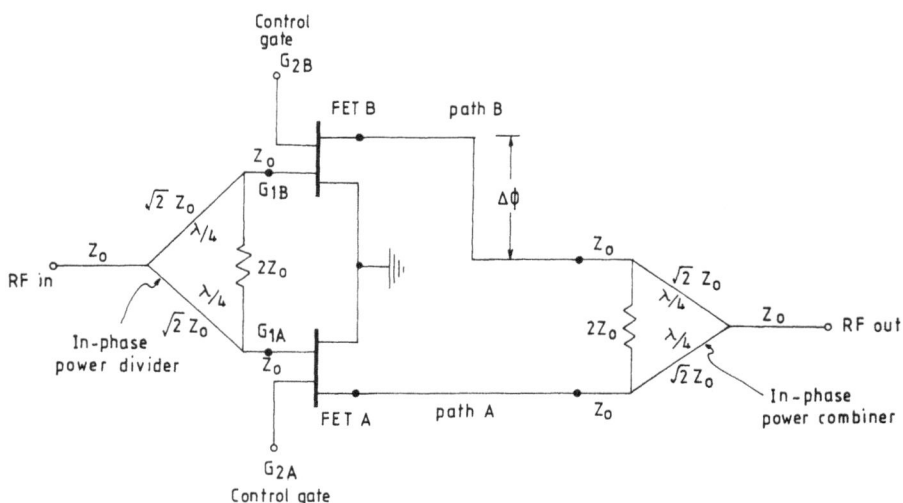

Figure 10.23 Schematic of a dual-gate FET digital phase shifter.

power divider at the input end divides the input signal equally and feeds the two in-phase outputs to the first gates (G_{1A} and G_{1B}) of the two FETs. The control voltages are applied to the second gates G_{2A} and G_{2B}. The two DGFETs are operated in a complementary manner. That is, when FET A operates as an amplifier (on-state), FET B is in the pinched-off state, and vice-versa. Thus by switching between the two amplifiers, the signal is allowed to pass through alternate transmission line paths. The relative phase delay between the two paths gives the differential phase shift ($\Delta\phi$). The Wilkinson power combiner recombines the two output paths into a single output terminal. It may be noted that there is a power loss of 3 dB each in the power divider and the power combiner. Thus, if G dB is the gain of the DGFET, the overall gain of the phase shifter is ($G - 6$) dB, assuming the mismatch and other circuit losses to be negligible.

Vorhaus, Pucel, and Tajima [33] have reported an X-band-active digital phase-shifter bit which offers a gain of 3 dB over 10% bandwidth. This circuit is realized in fully monolithic form (details are presented in Chapter 12).

10.6.3 Analog Phase-Shifting Schemes [34–37]

There are basically two approaches reported for achieving analog phase shift using DGFETs. One is based on the variation of phase of S_{21} of the DGFET circuit by controlling the bias conditions of one [34, 35] or both gates [37]. Using this scheme, a continuous linear phase variation of 120° with a 4-dB gain has been reported for a single device at X-band [34]. Using a three-device assembly, phase variation of 140° with gain up to 30 dB has been reported [35]. These phase shifters are inherently narrowband devices.

In the second approach, the DGFETs are used as variable gain amplifiers, and phase shift is achieved through complex addition of two orthogonal variable vectors. This technique offers broadband operation.

The vector summation phase shifter [37, 38] makes use of two DGFETs as variable gain amplifiers. The FETs are fed in phase-quadrature by means of a 3-dB, 90° hybrid coupler, as shown in Figure 10.24. The amplified output signals available at the two drain terminals are combined through an in-phase power combiner. Let A and B be the voltage gains of the two DGFETs corresponding to two different control voltages (V_a and V_b in Figure 10.24), and let θ_0 be the transmission phase through each amplifier. The vector sum of the two quadrature signals can be written as

$$C = |C|e^{j\phi} = e^{j\theta_0}(A + jB)$$
$$= \sqrt{A^2 + B^2} \, e^{j(\theta_0 + \tan^{-1}(B/A))} \tag{10.35}$$

The magnitude of the output signal is

$$|C| = \sqrt{A^2 + B^2} \tag{10.36}$$

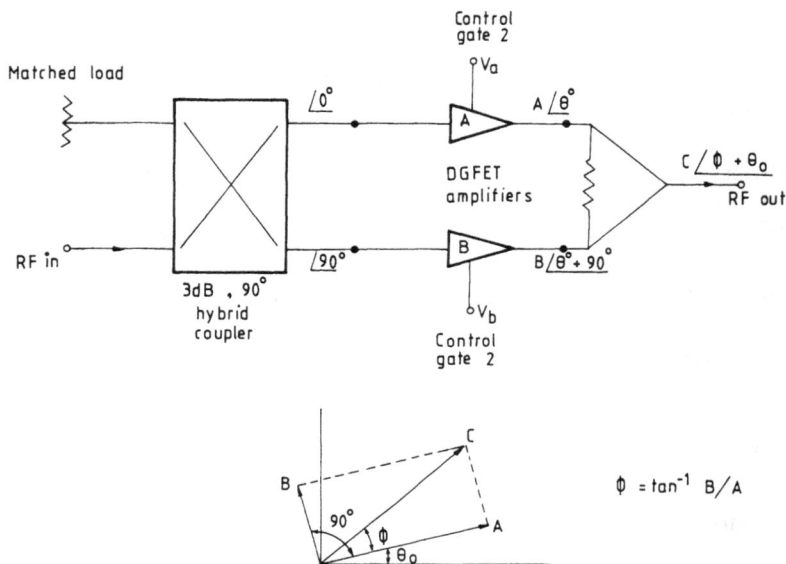

Figure 10.24 90° analog FET phase shifter using vector summation.

For all practical purposes, θ_0 may be assumed to be independent of bias voltage. The voltage-dependent differential phase shift is then given by

$$\Delta\phi = \tan^{-1}\left(\frac{B}{A}\right) \tag{10.37}$$

Thus, by independently adjusting the gain of each of the amplifiers, any phase shift between 0 and $\pi/2$ can be achieved. The amplitude of the phase shifted signal, that is, $\sqrt{A^2 + B^2}$, can be made independent of the phase shift setting by partially biasing the two amplifiers such that the output is at 0.707 level [37]. It is also possible to keep the phase shift constant and vary the overall amplitude by changing the gate voltages of both amplifiers simultaneously.

10.6.4 360° Analog Vector Phase Shifters

The 90° analog vector phase shifter described above has been used as the basis for realizing a 360° analog phase shifter [37]. The scheme is illustrated in Figure 10.25. This circuit requires four DGFET amplifiers. The input signal is first divided into four equiamplitude quadrature signals having phases 0°, 90°, 180°, and 270° by means of a 3-dB, 180° hybrid, and two 3-dB, 90° hybrids. These form the input signals to

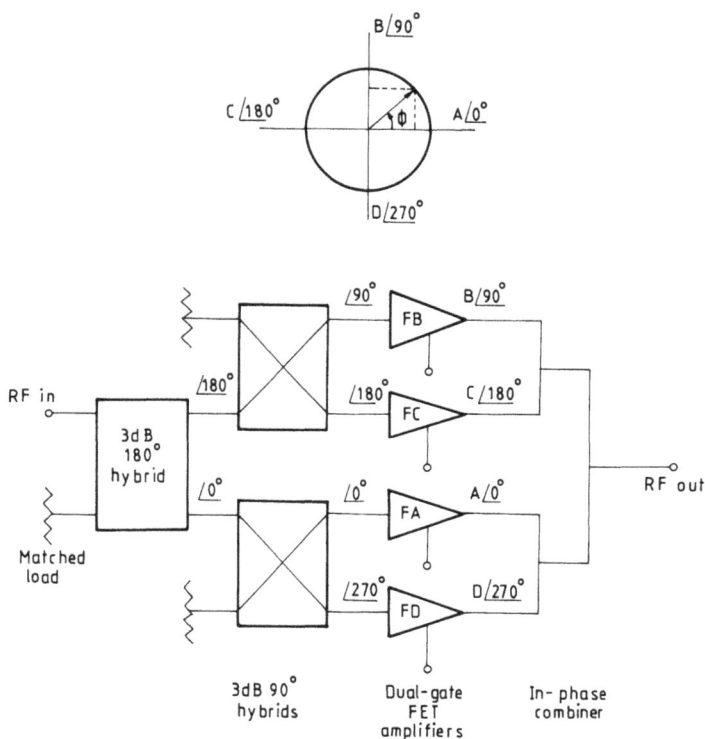

Figure 10.25 Schematic of a 360° dual-gate FET analog phase shifter. (From Kumar, Menna, and Huang [37], copyright © 1981 IEEE, reprinted with permission.)

the four DGFET amplifiers. The phase shift in the FET device is assumed to be the same for all four devices. Furthermore, the value is assumed to be invariant with respect to gate control voltage. Therefore, this constant phase factor may be neglected when the phase-shifting operation is described. If A, B, C, and D denote the voltage gains of the four FETs marked FA, FB, FC, and FD, respectively, the outputs appearing at the drain terminals are four quadrature vectors given by $A \underline{/0°}$, $B \underline{/90°}$, $C \underline{/180°}$, and $D \underline{/270°}$. These signals are combined through a four-way in-phase power combiner to obtain a phase-controlled output.

Continuously variable phase shift over the range of 0 to 360° is obtained by choosing a combination of two vectors for each of the four quadrants. This combination is shown in Figure 10.25. For example, for phase shift variation in the first quadrant (0 to 90°), vectors $A \underline{/0°}$ and $B \underline{/90°}$ are summed up. This is achieved by using FETs FA and FB as variable gain amplifiers, while the other two FETs FC

and *FD* are biased to pinch-off state. Similarly, for phase-shift variation in the second quadrant (90° to 180°), FETs *FB* and *FC* are operated as variable gain amplifiers, while *FA* and *FD* are switched to pinch-off state. Thus, by switching two FETs to off-state and controlling the bias voltages of the other two to provide variable gain, a total of 360° continuous phase shift is obtained. Kumar, Menna, and Huang [37] have reported a broadband active phase shifter of this type, which offers 360° continuous phase shift over 4 to 8 GHz.

There are other schemes that have been reported [39, 40] for implementing the vectorial principle to realize 360° analog phase shifters. Levy, Noblet, and Bender [39] have reported the use of a single FET device in conjunction with high-pass and low-pass filters to provide the four quadrature signals. This circuit can replace the three passive couplers in Figure 10.25. The circuit schematic by Levy, Noblet, and Bender [39] is shown in Figure 10.26. A single-gate FET is used in common-gate configuration. The 180° phase difference is achieved between source and drain terminals. The additional 90° phase shifts are achieved using first order high-pass and low-pass filter structures.

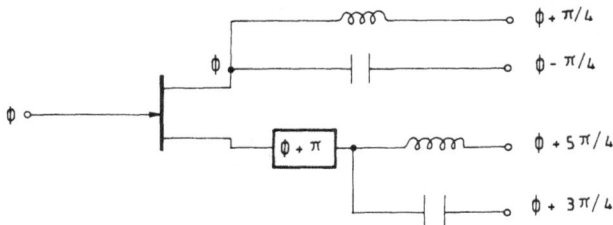

Figure 10.26 Quadrature phase shifter circuit. (From Levy, Noblet, and Bender [39], first presented at the 17th European Microwave Conference, reprinted with permission.)

Another scheme for achieving continuous 360° phase shift is to use a single pair of variable gain amplifiers in quadrature and cascade with 180° and 90° fixed phase-shifter bits [40]. The fixed bits are used to select the desired quadrant, and the variable gain amplifier pair provides the variable phase control within the chosen quadrant.

REFERENCES

1.   Y. Ayasli *et al.,* "A Monolithic Single Chip X-Band Four-Bit Phase Shifter," *IEEE Trans. on Microwave Theory and Tech.,* Vol. MTT-30, December 1982, pp. 2201–2206.

2.   C. Andricos, I.J. Bahl, and E.L. Griffin, "C-Band 6-Bit GaAs Monolithic Phase Shifter," *IEEE Trans. on Microwave Theory and Tech.,* Vol. MTT-33, December 1985, pp. 1591–1596.

3.   M.J. Schindler and M.E. Miller, "A 3-Bit K/Ka Band MMIC Phase Shifter," *IEEE Microwave and Millimeter Wave Monolithic Circuits Symp. Digest,* 1988, pp. 95–98.

4. Y. Ayasli, "Microwave Switching with GaAs FETs," *Microwave J.*, Vol. 25, November 1982, pp. 61–74.

5. R.S. Pengelly, *Microwave Field Effect Transistors-Theory, Design and Applications*, Research Studies Press, Chichester, 1982.

6. J.V. DiLorenzo and D.D. Khandelwal, *GaAs FET Principles and Technology*, Artech House, MA, 1982.

7. R. Soares, J. Graffeuil, and J. Obregon (eds.), *Applications of GaAs MESFETs*, Artech House, MA, 1983.

8. P.H. Ladbrooke, *MMIC Design-GaAs FETs and HEMTs*, Artech House, MA, 1989.

9. C.A. Liechti, "Microwave Field-Effect Transistors—1976," *IEEE Trans. on Microwave Theory and Tech.*, Vol. MTT-24, June 1976, pp. 279–300.

10. R.J. Gutmann, D. Fryklund, and D. Menzer, "Characterization and Design of GaAs MESFETs for Broadband Control Applications," *IEEE Int. Microwave Symp. Digest*, 1986, pp. 389–392.

11. R.A. Gaspari and H.H. Yee, "Microwave GaAs FET Switching," *IEEE Int. Microwave Symp. Digest*, 1978, pp. 58–60.

12. D. Fryklund and R. Walline, "Low Cost, Low Drain, High Speed, Wide Band GaAs MMIC Switches," *Microwave J.*, Vol. 28, December 1985, pp. 121–125.

13. L.C. Upadhyayulu *et al.*, "Passive GaAs FET Switch Models and Their Applications in Phase Shifters," *IEEE Int. Microwave Symp. Digest*, 1987, pp. 903–906.

14. A. Gopinath and J.B. Rankin, "GaAs FET RF Switches," *IEEE Trans. on Electron Devices*, Vol. ED-32, July 1985, pp. 1272–1278.

15. T.T. Ha, *Solid State Microwave Amplifier Design*, John Wiley and Sons, New York, 1981.

16. S.K. Koul, "Computer Aided Design of GaAs FET Amplifiers," Technical Report, School of Electronic Eng. Science, Univ. College of North Wales, U.K., 1984.

17. C. Tsironis and R. Meierer, "DC Characteristics Aid Dual-Gate FET Analysis," *Microwaves*, Vol. 20, July 1981, pp. 71–73.

18. R.A. Minasian, "Modelling DC Characteristics of Dual-Gate MESFETs," *IEEE Proc. Solid State Electronics*, Vol. 130, 1983, pp. 182–186.

19. C.A. Liechti, "Performance of Dual-Gate GaAs MESFETs as Gain Controlled Low-Noise Amplifiers and High-Speed Modulators," *IEEE Trans. on Microwave Theory and Tech.*, Vol. MTT-23, June 1975, pp. 461–469.

20. S. Asai, F. Murai, and H. Kodera, "GaAs Dual-Gate Schottky Barrier FETs for Microwave Frequencies," *IEEE Trans. on Electron Devices*, Vol. ED-22, October 1975, pp. 897–904.

21. T. Furutsuka, M. Ogawa, and N. Kawamura, "GaAs Dual-Gate MESFETs," *IEEE Trans. on Electron Devices*, Vol. ED-25, June 1978, pp. 580–586.

22. C. Tsironis and R. Meierer, "Microwave Wide-Band Model of GaAs Dual-Gate MESFETs," *IEEE Trans. on Microwave Theory and Tech.*, Vol. MTT-30, March 1982, pp. 243–251.

23. J.R. Scott and R.A. Minasian, "A Simplified Microwave Model of the GaAs Dual-Gate MESFET," *IEEE Trans. on Microwave Theory and Tech.*, Vol. MTT-32, March 1984, pp. 243–248.

24. G. Bodway, "Circuit Design and Characterization of Transistors by Means of Three-Port Scattering Parameters," *Microwave J.*, Vol. 11, May 1968, pp. 55–63.

25. M. Kumar and H.C. Huang, "Dual-Gate MESFET Variable-Gain Constant Output Power Amplifier," *IEEE Trans. on Microwave Theory and Tech.*, Vol. MTT-29, March 1981, pp. 185–189.

26. V. Sokolov *et al.*, "A Ka-band GaAs Monolithic Phase Shifter," *IEEE Trans. on Microwave Theory and Tech.*, Vol. MTT-31, December 1983, pp. 1077–1083.

27. R.S. Pengelly, "GaAs Monolithic Microwave Circuits for Phased Array Applications," *IEE Proc.*, Pt. F., Vol. 127, 1980, pp. 300–311.

28. Y. Ayasli *et al.*, "Wide-Band Monolithic Phase Shifter," *IEEE Trans. on Microwave Theory and Tech.*, Vol. MTT-32, December 1984, pp. 1710–1714.

29. M.J. Schindler *et al.*, "Monolithic 6 to 18 GHz 3-Bit Phase Shifter," *IEEE GaAs IC Symp. Digest,* 1985, pp. 129–132.

30. R.J. Collier and D.H. Evans, "An F.E.T. 180 Degree Phase Shifter for X-Band," *IEE* Coll. on Microwave Integrated Circuits (London), Digest No. 1983/71, October 1983, pp. 4/1–4/4.

31. F.C. DeRonde, "A New Class of Microstrip Directional Couplers," *IEEE MTT-S Microwave Symp. Digest,* 1970, pp. 184–189.

32. G.J. Loughlin, "A New Impedance-Matched Wide-Band Balun and Magic Tee," *IEEE Trans. on Microwave Theory and Tech.,* Vol. MTT-24, March 1976, pp. 135–141.

33. J.L. Vorhaus, R.A. Pucel, and Y. Tajima, "Monolithic Dual-Gate GaAs FET Digital Phase Shifter," *IEEE Trans. on Microwave Theory and Tech.,* Vol. MTT-30, July 1982, pp. 982–992.

34. C. Tsironis and P. Harrop, "Dual-Gate GaAs MESFET Phase Shifter with Gain at 12 GHz," *Electronics Letters,* Vol. 16, No. 14, July 1980, pp. 553–554.

35. C. Tsironis, P. Harrop, and M. Bostelmann, "Active Phase Shifters at X-Band Using GaAs MESFETs," *IEEE Int. Solid State Circuits Conf. Digest,* 1981, pp. 140–141.

36. R. Pengelly, C. Suckling, and J. Turner, "Performance of Dual-Gate GaAs MESFETs as Phase Shifters," *IEEE Int. Solid State Circuits Conf. Digest,* 1981, pp. 142–143.

37. M. Kumar, R.J. Menna, and H.C. Huang, "Broad-Band Active Phase Shifter Using Dual-Gate MESFET," *IEEE Trans. on Microwave Theory and Tech.,* Vol. MTT-29, October 1981, pp. 1098–1102.

38. Y. Gazit and H.C. Johnson, "A Continuously Variable Ku-Band Phase/Amplitude Control Module," *IEEE MTT-S Int. Microwave Symp. Digest,* 1981, pp. 436–438.

39. D. Levy, A Noblet, and Y. Bender, "A 2-18 GHz Continuously Variable 0-360° Phase Shifter," *17th Eur. Microwave Conf. Digest,* 1987, pp. 125–130.

40. J.R. Selin, "Continuously Variable L-Band Monolithic GaAs Phase Shifter," *Microwave J.,* Vol. 30, September 1987, pp. 211–218.

Chapter 11
Varactor Diode Phase Shifters

11.1 INTRODUCTION

Varactor diode phase shifters are basically analog devices in which the varactors function as variable reactance elements. This variable reactance is achieved through voltage-tuned capacitance of the diode under reverse-bias condition.

Varactor diode phase shifters commonly make use of reflection-type circuits [1–9], of which the hybrid coupled type is the most popular [5–9]. Since the phase shift is achieved by continuously varying the capacitance of the varactor, the switched line, the high-pass low-pass, and the switched network types of circuits described in Chapter 8 are not suitable. The loaded line type of transmission circuit can be used by employing shunt-mounted varactors [10, 11]. This configuration, however, has not proved to be as practical as the hybrid coupled phase shifter. In the following, we present the design aspects of some typical varactor diode phase shifters. The characteristics of some of the reported hybrid microwave integrated circuit (MIC) phase shifters are reviewed. The monolithic versions of varactor diode phase shifters are separately covered in Chapter 12.

11.2 VARACTOR DIODE EQUIVALENT CIRCUIT [1, 2]

A microwave varactor diode element consists of a p^+-n-n^+ sandwich structure with a typical doping profile as illustrated in Figure 11.1. The depletion layer associated with the p^+-n junction is similar to the intrinsic region of the p-i-n diode, except that the depletion layer thickness is much smaller and can be easily expanded into the n region by applying a reverse voltage. The junction layer capacitance $C_j(V)$ of the varactor is a function of the space charge layer width w, which in turn depends on the applied reverse-bias voltage. This capacitance may be expressed as

$$C_j(V) = \frac{\varepsilon A}{w(V)} \qquad (11.1)$$

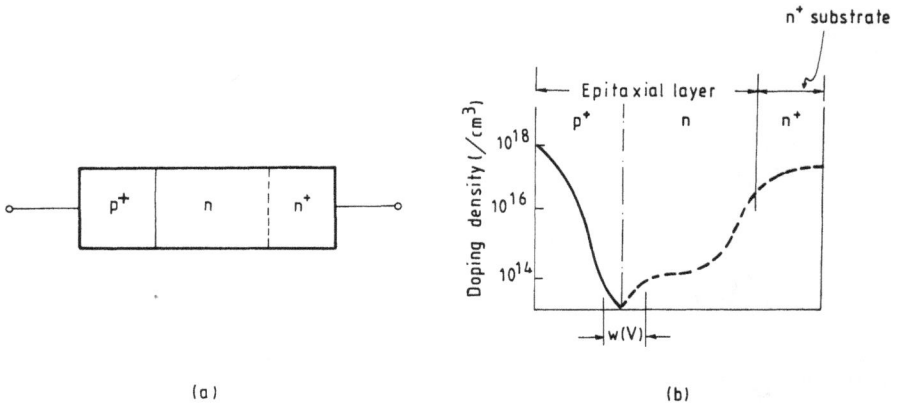

Figure 11.1 (a) Varactor diode cross section; (b) typical doping density profile.

where A is the cross-sectional area, and ε and $w(V)$ are the dielectric constant and voltage-dependent width of the depletion region, respectively. As the reverse-bias voltage is increased from 0 volts, the depletion layer width w increases and, consequently, the capacitance decreases. The remainder of the n region constitutes a series resistance R_s whose value also decreases with an increase in the reverse-bias voltage. In a varactor, the doping density at the contact edges is made very high in order to reduce this resistance to as small a value as possible. Figure 11.2 shows the equivalent circuit of a packaged varactor diode, which includes the lead inductance L_s and the package capacitance C_p. Since the main function of the varactor is

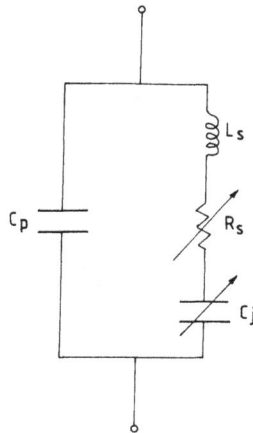

Figure 11.2 Varactor diode equivalent circuit.

to act as a variable capacitance, the capacitance range with bias must be large enough to yield the desired phase variation, and the reactance $1/\omega C_j$ at the highest frequency of operation must be high compared to R_s. In order to keep R_s small, the varactor is operated in the reverse-biased region.

11.3 REFLECTION-TYPE PHASE SHIFTERS

Reflective-type analog phase shifters have been reported that make use of the reflective properties of a varactor terminating a transmission line in conjunction with either a circulator [3, 4] or a 3-dB hybrid coupler [2, 5–9]. The basic circuit configurations are essentially the same as those described in Chapters 8 and 9 for *p-i-n* diode phase shifters. While the *p-i-n* diode is switched between low- and high-impedance states, the varactor is voltage-tuned to provide a continuous reactance change. The reflective circuit must, however, be designed suitably to achieve the desired phase shift response, while maintaining as low an insertion loss as possible with voltage tuning.

11.3.1 Variable Reactance as Reflective Termination [5]

Consider a 3-dB, 90° hybrid symmetrically terminated in variable reactances jX, as shown in Figure 11.3. If Z_0 is the characteristic impedance of the input and output arms, the reflection coefficient Γ of the reactive termination is given by

$$\Gamma = e^{j\phi} = \frac{(j\bar{X} - 1)}{(j\bar{X} + 1)}, \quad \bar{X} = \frac{X}{Z_0} \tag{11.2}$$

so that

$$\phi = \pi - 2\tan^{-1}(\bar{X}) \tag{11.3}$$

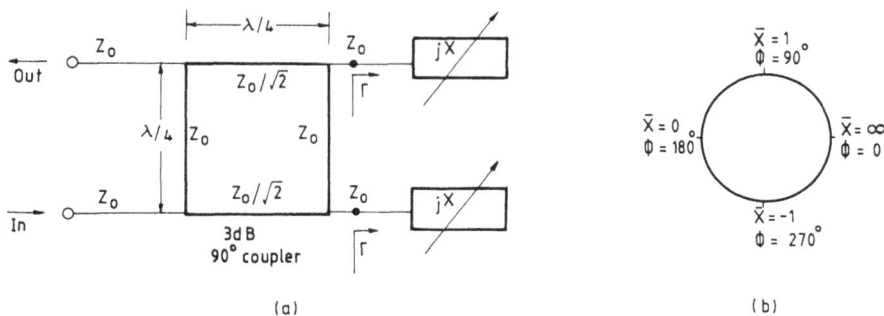

Figure 11.3 (a) Schematic of a coupler terminated in variable reactance; (b) phase of reflected wave as a function of normalized reactance; $\bar{X} = X/Z_0$, $\Gamma = \exp(j\phi)$.

As illustrated in Figure 11.3, for a phase variation over the range $0°$ to $360°$, $\bar{X}$ must vary from $+\infty$ through 0 to $-\infty$. The rate of change of phase shift with reactance is maximum when $\bar{X} = 0$. Hence, in order to achieve maximum phase change, the capacitive reactance of the diode must be series resonated with an inductive reactance. This can be achieved by means of a stub.

11.3.2 Termination Using a Varactor in Series with a Shorted Stub [2, 6]

Figure 11.4 shows a reflective termination consisting of a series combination of a varactor and an inductor. Let R_d and C_d denote the resistance and variable capacitance, respectively, of the diode. Although the resistance is also voltage dependent, since its value is small in comparison with the capacitive reactance, it may be assumed to be a constant. The impedance of the reflective termination is given by

$$Z_{in} = R_d + jX \tag{11.4}$$

where

$$X = \left(\omega L - \frac{1}{\omega C_d} \right) \tag{11.5a}$$

$$\omega L = Z_s \tan \beta l_s \tag{11.5b}$$

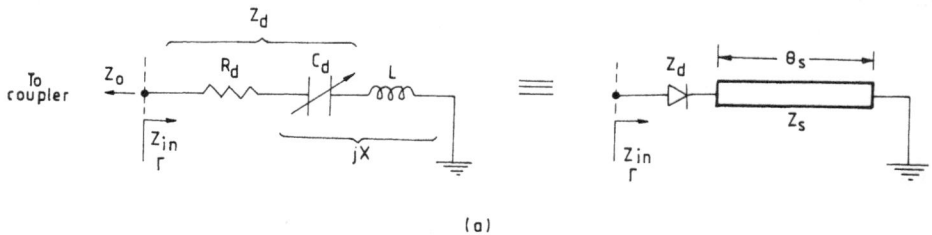

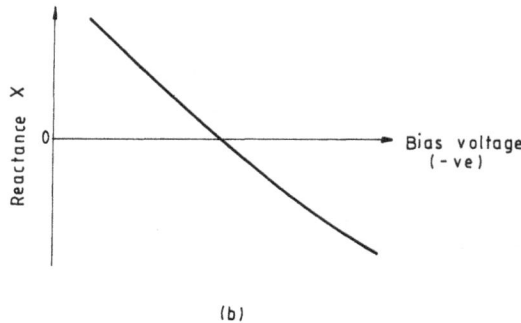

(a)

(b)

Figure 11.4 (a) Series-tuned varactor circuit as reflective termination; (b) variation of reactance X with bias voltage.

The inductive reactance given by (11.5b) is realized as the input reactance of a short-circuited stub of characteristic impedance Z_s and length l_s. The reflection coefficient Γ due to the termination is given by

$$\Gamma = |\Gamma| e^{j\phi} = \frac{(R_d - Z_0) + jX}{(R_d + Z_0) + jX} \tag{11.6}$$

Maximum phase change can be achieved by resonating the capacitive reactance of the diode at the center of its capacitance range with the inductive reactance of the shorted stub. As the bias voltage of the varactor is varied from 0 volts to a large negative value close to its breakdown voltage, the capacitance C_d decreases from a maximum value C_{max} to a low value C_{min}, and the net reactance X of the circuit changes from an inductive reactance through zero to capacitive reactance. It may be noted that maximum insertion loss also occurs at series resonance corresponding to $X = 0$. The expression for insertion loss is given by

$$\alpha(\text{dB}) = 20 \log_{10}\left(\frac{1 + R_d/Z_0}{1 - R_d/Z_0}\right)$$

$$\simeq 17.4 \, R_d/Z_0 \tag{11.7}$$

For a small excursion $\pm\Delta X/2$ about $X = 0$, the phase shift can be expressed (neglecting R_d) as [2]

$$|\Delta\phi| = 4 \tan^{-1}(\Delta X/2Z_0)$$

$$\simeq 2\Delta X/Z_0 \text{ rad} \tag{11.8a}$$

or

$$|\Delta\phi| \simeq 115\Delta X/Z_0 \text{ deg} \tag{11.8b}$$

The figure of merit for the phase shifter is expressed as

$$F = \frac{|\Delta\phi| \, (\text{deg})}{\alpha(\text{dB})} \tag{11.9}$$

Substituting for α and $|\Delta\phi|$ from (11.7) and (11.8b), respectively, in (11.9) yields

$$F \simeq 6.6 \, \Delta X/R_d \text{ deg/dB} \tag{11.10}$$

For a capacitance variation from C_{max} to C_{min}, the change in reactance is

$$\Delta X = \frac{1}{\omega C_{min}} - \frac{1}{\omega C_{max}} \tag{11.11}$$

Dividing by R_d, we can write

$$\frac{\Delta X}{R_d} = \frac{f_c}{f} \left(1 - \frac{1}{K} \right) \tag{11.12}$$

where

$$f_c = \frac{1}{2\pi R_d C_{min}} \tag{11.13a}$$

$$K = \frac{C_{max}}{C_{min}} \tag{11.13b}$$

With (11.12) in (11.10), the expression for figure of merit becomes [2]

$$F \simeq 6.6 \frac{f_c}{f} \left(1 - \frac{1}{K} \right) \text{ deg/dB} \tag{11.14}$$

It has been pointed out by White [2] that practical phase shifters designed for phase shifts up to 180° achieve about 50% to 75% of the figure of merit predicted by (11.14), though the expression is derived for small phase shift approximations. For a capacitance ratio $K \geq 5$, the expression for F can be approximated to 6.6 f_c/f deg/dB. As an example, if we choose a varactor having $f_c \simeq 100$ GHz, then at a frequency of 10 GHz, the value of F is approximately 66 deg/dB.

11.3.3 Reflective Termination Using Two Series-Tuned Varactors [4, 8]

As compared with a single varactor-tuned circuit, a termination consisting of two series-tuned varactor diodes in parallel connection is reported to offer a much higher figure of merit [4, 8]. The concept of using two series-tuned circuits was introduced by Henoch and Tamm [4] to design a 360° varactor diode phase modulator. The circuit makes use of series resonance of one diode, with an inductance at low bias voltage, and the series resonance of the second diode with another inductance at high bias voltage. The termination is connected to a circulator through a quarter-wave transformer. The modulator is reported to give 360° phase shift at a center frequency of 2 GHz, with an attenuation ripple of 1.3 dB over 10% bandwidth, and a decrease of phase shift by 7° at the band edges [4]. The phase shifter is, however, reported to have large frequency dependence for phase shifts less than 360°.

By introducing a slight change in one of the series resonant circuits, Ulriksson [8] has demonstrated the design of a continuously variable 180° phase shifter having optimum frequency response for all phase shifts up to 180°. Figure 11.5(a) shows the reflection circuit using lumped elements, as well as its distributed element version. In the following section, we summarize the design approach from Ulriksson [8].

Referring to Figure 11.5(a), we note that Z_0 is the impedance of the 3-dB hybrid coupler (or circulator), and the reflective circuit includes a section of transmission line (of characteristic impedance Z_t and length θ_t) as a transformer. The varactor diode is represented as a variable reactance X_d. The distributed element version of the circuit uses a shorted stub in place of the lumped tuning inductor L_1, and an open-circuited stub in place of the series combination of L_2 and C. The total input reactance X of the reflective circuit and the phase ϕ of the reflection coefficient Γ are given by

$$X = Z_t \frac{(X_l + Z_t \tan \theta_t)}{(Z_t - X_l \tan \theta_t)}$$

$$= \pi - 2 \tan^{-1} \left[\frac{Z_t(X_l + Z_t \tan \theta_t)}{Z_0(Z_t - X_l \tan \theta_t)} \right] \qquad (11.15)$$

where

$$X_l = \frac{X_1 X_2}{X_1 + X_2} \qquad (11.16a)$$

$$X_1 = X_d + X_{s1}; \quad X_{s1} = Z_1 \tan \theta_1 \qquad (11.16b)$$

$$X_2 = X_d + X_{s2}; \quad X_{s2} = -Z_2 \cot \theta_2 \qquad (11.16c)$$

If $\phi^{V_{max}}$ denotes the value of ϕ corresponding to the maximum reverse-bias voltage V_{max} and ϕ^V denotes its value at any other voltage between V_{min} and V_{max}, the differential phase shift is given by

$$\Delta\phi = \phi^V - \phi^{V_{max}} \qquad (11.17)$$

For optimum frequency response, we require $d(\Delta\phi)/df = 0$ at the center frequency.

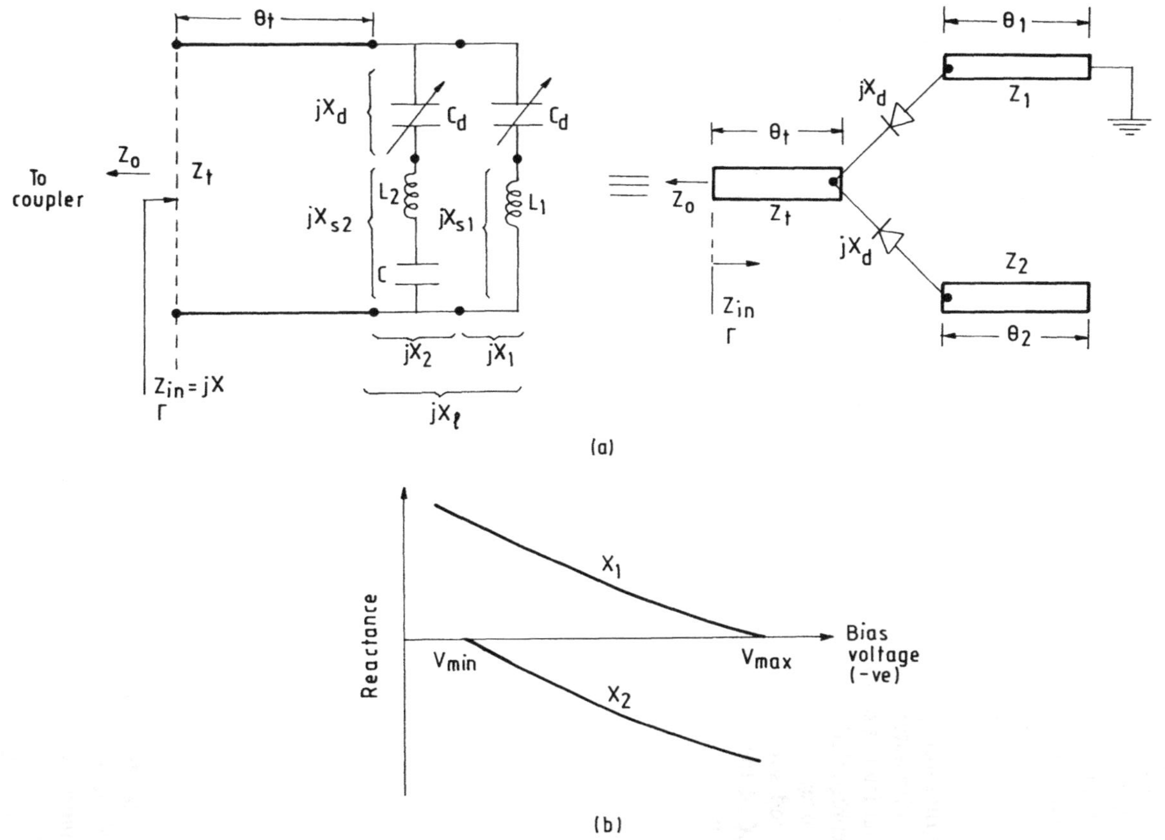

Figure 11.5 (a) Reflective termination employing two series-tuned varactors in parallel; (b) variation of diode circuit reactances X_1 and X_2 with bias voltage. (After Ulriksson [8].)

This gives

$$\frac{\partial \phi^V}{\partial f} = \frac{\partial \phi^{V_{max}}}{\partial f} \tag{11.18}$$

where

$$\frac{\partial \phi}{\partial f} = -\frac{2Z_0 Z_t \sec^2 \theta_t \left[Z_t \dfrac{\partial X_l}{\partial f} + (Z_t^2 + X_l^2)\dfrac{\theta_t}{f} \right]}{[Z_0^2(Z_t - X_l \tan \theta_t)^2 + Z_t^2(X_l + Z_t \tan \theta_t)^2]} \tag{11.19}$$

and

$$\frac{\partial X_l}{\partial f} = \frac{\left[X_2^2 \dfrac{\partial X_1}{\partial f} + X_1^2 \dfrac{\partial X_2}{\partial f} \right]}{(X_1 + X_2)^2} \tag{11.20}$$

The value of X_{s1} can be chosen so that the reactance X_1 is zero at maximum voltage. From (11.16b) we have

$$X_d^{V_{max}} + Z_1 \tan \theta_1 = 0 \tag{11.21}$$

By choosing the characteristic impedance Z_1 of the inductive stub as large as possible, the value of θ_1 can be calculated from (11.21). For determining Z_2 and θ_2, we specify the maximum phase shift and use the condition that this maximum phase shift should have an optimum frequency response. For a specified set of values of Z_t and θ_t, we can make use of the following relations:

$$\Delta\phi(max) = \phi^{V_{min}} - \phi^{V_{max}} \tag{11.22}$$

$$\frac{\partial \phi^{V_{min}}}{\partial f} = \frac{\partial \phi^{V_{max}}}{\partial f} \tag{11.23}$$

Substituting (11.15), (11.16), (11.19), and (11.20) in (11.22) and (11.23) and setting $X_1^{V_{max}} = 0$ yields the following relations:

$$X_l^{V_{min}} = \frac{Z_t[\xi^{V_{min}} - Z_t \tan \theta_t]}{[Z_t + \xi^{V_{min}} \tan \theta_t]} \tag{11.24}$$

$$\xi^{V_{min}} = Z_0 \tan\left[\frac{\Delta\phi(max)}{2} + \tan^{-1}\left(\frac{Z_t}{Z_0} \tan \theta_t\right) \right] \tag{11.25}$$

$$X_2^{V\min} = \left(\frac{1}{X_l^{V\min}} - \frac{1}{X_1^{V\min}}\right)^{-1} \tag{11.26}$$

$$\frac{\partial X_2^{V\min}}{\partial f} = (X_2^{V\min})^2 \left[\frac{1}{Z_t(X_l^{V\min})^2} \left\{\frac{\zeta\left(\dfrac{X_1^{V\max}}{\partial f} + Z_t\right)\dfrac{\theta_t}{f}}{Z_t(Z_0^2 + Z_t^2\tan^2\theta_t)}\right.\right.$$

$$\left.\left. - ((X_l^{V\min})^2 + Z_t^2)\frac{\theta_t}{f}\right\} - \frac{1}{(X_1^{V\min})^2}\frac{\partial X_1^{V\min}}{\partial f}\right] \tag{11.27}$$

where

$$\zeta = Z_0^2(Z_t - X_l^{V\min}\tan\theta_t)^2 + Z_t^2(X_l^{V\min} + Z_t\tan\theta_t)^2 \tag{11.28}$$

Ulriksson [8] has suggested using the parameters Z_t and θ_t to optimize the frequency behavior at phase shifts less than $\Delta\phi(\max)$. This is achieved by a nonlinear minimization of a variable P defined as

$$P(Z_t, \theta_t, \Delta\phi(\max)) = \sum_{n=1}^{k}\left[f\frac{\partial\phi^{V_n}}{\partial f} - f\frac{\partial\phi^{V\max}}{\partial f}\right]^2 \tag{11.29}$$

where the voltages V_ns are chosen such that the corresponding phase shifts form approximately equal steps from 0 to $\Delta\phi(\max)$. An L-band phase shifter realized using the above design approach is reported to offer less than 5° variation over 1.5 to 1.7 GHz for all phase shifts less than 180° [8].

11.4 TRANSMISSION-TYPE PHASE SHIFTERS

11.4.1 Varactor-Loaded Transmission Line Circuits

Dawirs and Swarner [10] have reported a phase shifter that makes use of voltage-controlled varactors shunted at intervals across a transmission line. The basic circuit incorporates two identical capacitances C_1 at the ends of a transmission line of length $2l$ and characteristic admittance Y_1, with another variable capacitance C_2 located at the middle of the line, as shown in Figure 11.6. The end capacitances are used to control the phase shift, whereas the middle capacitor C_2 is adjusted to provide a good input match.

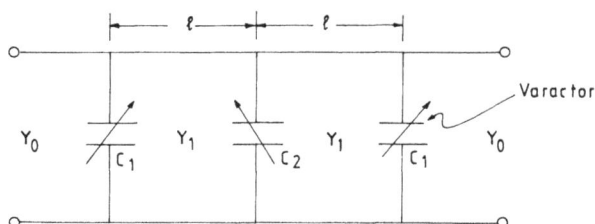

Figure 11.6 Transmission line with shunt-loaded varactors. (After Dawirs and Swarner [10].)

Liechte and Epprecht [11] have described a varactor-controlled phase shifter in coaxial configuration for wideband operation. The phase shifter is a three-port device. It makes use of a power divider at the input end to feed two coaxial lines, each series loaded at periodic intervals with varactors. The coaxial line section in series with a varactor is equivalent to an inductance in series with a voltage variable capacitor. The reactance is positive or negative depending on whether the diode resonance is below or above the operating frequency. In one coaxial line, the reactance is made inductive, and in the other it is made capacitive such that the phase-frequency responses of the two lines have equal gradients, resulting in broadband performance.

11.4.2 Varactor-Controlled FET Phase Shifter

A different type of transmission phase shifter in the form of a varactor-controlled MESFET has been reported by Coupez and Perichon [12]. The phase shifter consists of a MESFET with a series-tuned varactor circuit in the feedback loop between the drain and gate, and also at either the gate input or drain output terminal. The basic configurations are shown in Figure 11.7. Varying the reverse voltage of the varactors causes a change in the phase of the transmission coefficient T_{21} of the phase shifter. For circuit A, the expressions for T_{21} and for the input and output reflection coefficients Γ_1 and Γ_2 are given by [12]

$$T_{21} \simeq \frac{2}{3} e^{-j2\tan^{-1}x} \tag{11.30}$$

$$\Gamma_1 \simeq -\frac{1}{3}\frac{(1 - j3x)}{(1 + jx)} \tag{11.31}$$

$$\Gamma_2 \simeq -\frac{1}{3}\frac{(1 - jx)}{(1 + jx)} \tag{11.32}$$

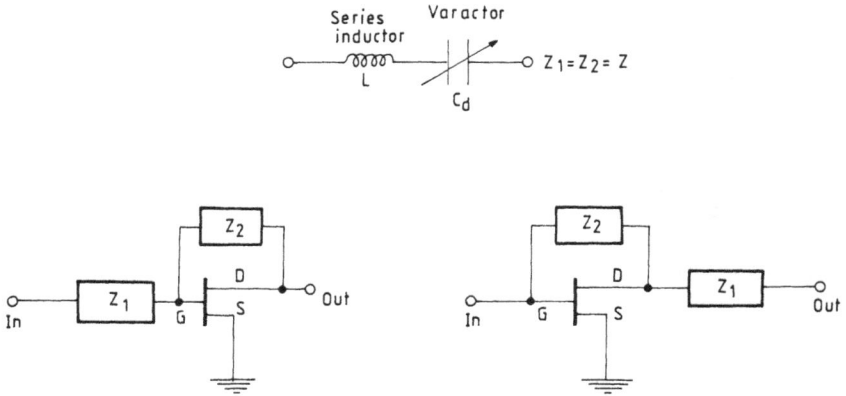

Figure 11.7 Varactor-controlled MESFET as a phase shifter. (From Coupez and Perichon [12], first presented at the 18th European Microwave Conference, reprinted with permission.)

where

$$x = \frac{1}{Z_0} \left(\omega L - \frac{1}{\omega C_d} \right) \tag{11.33}$$

The above expressions assume the output conductance g_d of the FET to be negligible and the product of the conductance g_m and the characteristic impedance Z_0 to be approximately unity. The inductance L of the resonant circuit can be realized by a short length of high-impedance microstrip line, and C_d denotes the junction capacitance of the varactor. It may be noted from (11.30) that the phase shift varies in the same manner as that of a reflective-type hybrid phase shifter having the termination shown in Figure 11.5(a). Another interesting feature of the circuit is that its insertion loss is insensitive to change in frequency and phase shift. An experimental phase shifter of this type realized at 900 MHz is reported to offer a linear phase-shift variation over 0 to 180° (within ±1°) and a flat insertion loss (≈5 dB) response as a function of varactor bias over approximately 10% bandwidth [12].

11.5 COMPARISON WITH *P-I-N* DIODE PHASE SHIFTERS

The basic circuit forms used in analog varactor diode phase shifters are common in the digital *p-i-n* diode phase shifters. Since the reflective-type hybrid coupled form is most commonly used in varactor diode phase shifters, we compare these with the reflective-type *p-i-n* diode phase shifters.

The continuous phase-shift range in an analog phase shifter is basically limited by the variable capacitance range as well as by the insertion loss variation. On the other hand, in a digital *p-i-n* diode phase shifter, the size of the phase bit is governed by the line length and is not limited by the diode parameters. The peak power rating of a *p-i-n* diode phase shifter is larger than that of a varactor diode phase shifter by more than a factor of 100. This is because the breakdown voltage of *p-i-n* diodes (−200V to −1000V) is typically an order of magnitude larger than that of varactors. Secondly, in varactors having nonlinear variation of capacitance with voltage, the RF voltage swing that can be applied at any bias point without affecting the average capacitance observed by the microwave signal is limited, whereas in a *p-i-n* diode, the full reverse-bias range can be used for the RF voltage swing. In terms of average power-handling capability also, the *p-i-n* diode phase shifter is superior because of its smaller insertion loss and the ability of the *p-i-n* diode to dissipate a greater amount of power. Typically, the average power-handling capability at L-band is on the order of 100W for a *p-i-n* diode phase shifter and about 25 mW for a varactor diode phase shifter.

Both varactor and *p-i-n* diode phase shifters exhibit rapid response time. The switching time of *p-i-n* diode phase shifters is typically in the range 0.05 to 0.5 μs, whereas the response time of varactor diode phase shifters is on the order of a few nanoseconds. The drive power requirement of both phase shifters is low, typically on the order of 100 mW per phase bit in a *p-i-n* diode phase shifter, and less than 1 mW for a varactor diode phase shifter.

REFERENCES

1. H.A. Watson (ed.), *Microwave Semicondutor Devices and Their Circuit Applications*, McGraw-Hill, New York, 1969.
2. J.F. White, *Microwave Semiconductor Engineering*, Van-Nostrand, NJ, 1982.
3. R.V. Garver, "360° Varactor Linear Phase Modulator," *IEEE Trans. on Microwave Theory and Tech.*, Vol. MTT-17, March 1969, pp. 137–147.
4. B.T. Henoch and P. Tamm, "A 360° Reflection Type Diode Phase Modulator," *IEEE Trans. on Microwave Theory and Tech.*, Vol. MTT-19, January 1971, pp. 103–105.
5. S. Hopfer, "Analog Phase Shifter for 8–18 GHz," *Microwave J.*, Vol. 22, March 1979, pp. 48–50.
6. R.H. Hardin, E.J. Downey, and J. Munushian, "Electronically-Variable Phase Shifters Utilizing Variable Capacitance Diodes," *Proc. IRE* (Lett.), Vol. 48, May 1960, pp. 944–945.
7. G. Apsley, L. Coltun, and M. Rabinowitz, "Quickly Device a Fast Diode Phase Shifter," *Microwaves*, Vol. 18, May 1979, pp. 67–68.
8. B. Ulriksson, "Continuous Varactor-Diode Phase Shifter with Optimum Frequency Response," *IEEE Trans. on Microwave Theory and Tech.*, Vol. MTT-27, July 1979, pp. 650–654.
9. R.V. Garver, "Broadband Binary 180° Diode Phase Modulators," *IEEE Trans. on Microwave Theory and Tech.*, Vol. MTT-13, January 1965, pp. 32–38.

10. H.N. Dawirs and W.G. Swarner, "A Very Fast Voltage Controlled Microwave Phase Shifter," *Microwave J.*, Vol. 5, June 1962, pp. 99–107.

11. C.A. Liechte and G.W. Epprecht, "Controlled Wideband Differential Phase Shifters Using Varactor Diodes," *IEEE Trans. on Microwave Theory and Tech.*, Vol. MTT-15, October 1967, pp. 586–589.

12. J.Ph. Coupez and R.A. Perichon, "A Novel Microwave Transmission Phase Shifter," *18th Eur. Microwave Conf. Digest*, 1988, pp. 1023–1027.

Chapter 12
Monolithic Phase Shifters

12.1 INTRODUCTION

The p-i-n diode, GaAs FET, and varactor diode phase shifters described in earlier chapters (9, 10, and 11) are all based on the hybrid microwave integrated circuit (MIC). It may be recalled that in hybrid MIC, all passive components are deposited on the surface of a low-loss dielectric substrate, and discrete semiconductor devices are either bonded or soldered onto the passive circuit. In the *monolithic microwave integrated circuit* (MMIC) technique, the entire circuit consisting of passive circuit elements, active devices, and interconnections are formed *in situ* on or within a semi-insulating, semiconducting substrate. The main advantages of MMIC over the hybrid MIC are its small size and weight, improved reliability and reproducibility through elimination of wire bonds, and its ability to incorporate multifunctional performance on a single chip. Elimination of wire bonds and embedding of active devices within the semiconducting substrate minimize the undesired parasitics and, therefore, improve the bandwidth performance. Furthermore, the multifunctional capability on a single chip offers scope for realization of integrated receiver front end or transmit-receive modules, including the phase shifter. While the MMIC approach offers all the advantages of integration, it loses the flexibility of circuit tuning and trouble-shooting available in hybrid MICs. This problem can, however, be circumvented by the use of *computer-aided design* (CAD) techniques so that the need for circuit adjustments is minimized.

This chapter is devoted to the description of monolithic phase shifters. Most monolithic microwave MMIC and *monolithic millimeter-wave integrated circuit* (MMWIC) phase shifters are based on GaAs technology, with GaAs *metal semiconductor field-effect transistors* (MESFET) as control elements for digital phase shifting, and varactors for analog phase shifting. Other control elements such as Schottky barrier diodes and p-i-n diodes are also used, but to a limited extent. Another potential active element, the technology of which is currently emerging, is the GaAs-based *high electron mobility transistor* (HEMT). The subjects of GaAs material

---

technology and the fabrication of GaAs FETs and HEMTs have been extensively reported in the literature [1–20]. As a background to the subsequent sections on monolithic phase shifters, we first provide a brief summary of MMIC materials and process technology. The main parts of monolithic phase-shifter circuits in general, including the types of active devices used, are described. Various types of monolithic phase-shifter configurations reported in the literature are reviewed.

12.2 MATERIALS AND PROCESS TECHNOLOGY

12.2.1 Materials [1–8]

MMICs are now predominantly based on GaAs technology. It may be mentioned that the silicon (Si) technology, which is much older and extremely well developed, has been extensively applied to low-frequency circuits. Its extension to microwave frequencies, particularly beyond S-band, has not proved practical because of the inadequate semi-insulating property of Si. The approach based on *silicon-on-sapphire* (SOS) is reported to offer definite improvement over the silicon MMICs [8]. In this approach, the high-resistivity sapphire substrate replaces the semi-insulating Si, thus eliminating the losses associated with the low-resistivity Si substrate at microwave frequencies. However, with rapid advances in GaAs process and device technologies over the last decade, GaAs has emerged as the most versatile and superior material for MMICs as well as MMWICs.

In order to understand the advantages of GaAs for MMICs and MMWICs, we compare the properties in the semiconducting and semi-insulating states with those of Si. Table 12.1 lists their main properties and also the properties of sapphire, which is known as a low-loss dielectric. The resistivity of semi-insulating GaAs (10^7 to 10^9 Ω-cm) is four orders of magnitude higher than that of semi-insulating Si. A semi-insulating GaAs can therefore be used as a low-loss dielectric substrate for realizing passive circuitry and also for providing electrical isolation between active devices in the circuit. The electron mobility in GaAs ($\simeq 4300$ cm^2/V-s) is over six times higher than in Si, thus making it suitable for realizing active semiconductor devices operable at higher frequencies. GaAs varactors have a higher cut-off frequency than Si varactors. The minority carrier life time for GaAs is nearly five orders of magnitude less than for Si, leading to fast-switching GaAs *p-i-n* diodes.

Although the potential capabilities of GaAs material for MMICs were recognized in the early 1960s, the material technology was in its infancy. During the 1970s, the development of GaAs FET as a microwave solid-state device received considerable focus. Since GaAs FETs are planar devices and are formed on semi-insulating substrates, the process technology developed for these directly benefited the fabrication of MMICs. This brought in a revival of interest in GaAs MMICs and stimulated their development in the 1980s.

Table 12.1
Properties of Substrates Used in MMICs

Material	ε_r (approx.)	$Tan\delta \times 10^4$ at 10 GHz	Surface Roughness (μm)	Thermal Conductivity ($W/cm/°C$)	Resistivity (Ω-cm)	Elec. Mobility (cm^2/V-s)	Minority Carrier Life Time (sec)	Energy Gap (eV)	Density (gm/cc)	Saturation Electron Velocity (cm/s)
Si	11.7–12	40–150	<0.025	0.9	—	700*	2.5×10^{-3}	1.12	2.33	9×10^6
Semi-insulating Si	11.7–12	10–100	<0.025	0.9	10^3–10^5	—	—	—	2.33	—
GaAs	12–13	10–60	<0.025	0.3	—	4300*	10^{-8}	1.42	5.32	1.3×10^7
Semi-insulating GaAs	12–13	16	~0.025	0.3	10^7–10^9	—	—	—	5.32	—
Sapphire	11.6 (C-axis)	0.4–0.7	~0.005	0.4	>10^{14}	—	—	—	3.98	—

*At $10^{17}/cm^3$ doping

12.2.2 GaAs Technology [1–7, 9–10]

The first process in GaAs MMIC fabrication is the formation of an active n-type layer on a GaAs single-crystal substrate. Important requirements for this base substrate are that it must have high resistivity ($\geq 10^7 \, \Omega$/square), low background doping, and good thermal stability, and must be free from crystalline defects.

For active layer formation, two important techniques are available—namely, epitaxy and ion implantation. Significant among the epitaxial growth techniques are the *molecular beam epitaxy* (MBE) and the *metal organic chemical vapor deposition* (MOCVD).

MBE—Molecular beam epitaxy is essentially an advanced form of vacuum evaporation carried out under an ultrahigh vacuum (typically 10^{-9} torr) environment. Material compounds or elements are evaporated from heated Knudsen cells to form molecular beams. The cells are arranged to direct the molecular beam flux towards the single-crystal GaAs substrate. By accurately controlling the temperatures of the cells, the substrate, and the beams by means of suitable shutters, epitaxial film of the desired chemical composition is grown on the substrate. With the facility for producing multiple molecular beams in the same chamber, multilayer semiconductor films can be accurately grown. The technique provides well-controlled thickness of each layer, extremely abrupt interfaces, good composition, and smooth surface.

MOCVD—Growth of epitaxial films using the MOCVD technique involves the reaction of two or more chemically reactive gases in a quartz reaction cell. The reaction cell houses the substrate on a graphite susceptor and has an RF heating coil surrounding it. For growing GaAs film, the metal organic can be trimethylgallium (TMG), and arsenic is derived from arsine. The metal organic is stored in the form of a liquid at low temperature (typically $-10°C$ for TMG) and is delivered to the reaction cell in a controlled manner by a hydrogen carrier gas, where it is mixed with arsine (AsH_3). By means of the heating coil, gases are heated to about 600° to 800°C above the GaAs substrate, thereby inducing chemical reaction. The basic reaction can be expressed in the form

$$Ga(CH_3)_3 + AsH_3 \rightarrow GaAs + 3CH_4 \tag{12.1}$$

If arsine is well in excess of the metal organic (typically five to fifty times more), then carbon is incorporated into the gallium sublattice, giving rise to n-type doping. The rate of film growth depends on the concentration of metal organic, and the generation of abrupt interface is governed by the speed with which the composition of the gas in the cell can be changed.

Ion Implantation—The ion implantation technique involves the formation of an active layer by bombarding the surface of the substrate to be implanted with an accelerated beam of ions. The ion beam is produced in a source which is held at a high dc potential and is directed towards the substrate maintained at ground potential. The majority of ions penetrates the substrate and is slowed down by random inter-action with the nuclei and electrons of the material until they reset at some depth. The doping profile as a function of depth can be controlled by controlling the energy of ions. The material is then annealed to recover its crystallinity and enable implanted ions to find correct lattice sites where they can act as dopants. The annealing tem-perature for GaAs is in the range of 800° to 900°C. One major problem with ion implantation into GaAs is that the material decomposes above a temperature of about 650°C. In order to solve this problem of decomposition, it is a common practice to use an encapsulant consisting of thin layers of dielectrics (≈ 0.1 μm) such as SiO_2, Si_3N_4, and Al_2O_3. Another commonly used technique of preventing decomposition is known as *capless annealing*. In this method, annealing of GaAs substrates is car-ried out in an atmosphere of arsenic overpressure.

The choice of ion species to be implanted depends to a large extent on the operating voltage and, hence, the implant energy. Ions suitable for formation of the *n*-type layer are Si, S, and Se, the preferred ion being Si. For example, MESFETs can be formed by using low doses of Si ions for the FET channel and high doses for the source and drain contacts. Active ion implantation can also be used for form-ing resistors and doped regions of a Schottky barrier diode. Proton implantation can be used to achieve electrical isolation between closely spaced GaAs integrated circuit devices.

As compared with epitaxial techniques, ion implantation offers better control of doping profiles over large area wafers and improved reproducibility.

12.3 CIRCUIT ELEMENTS FOR MONOLITHIC PHASE SHIFTERS

In this section, we describe the general features and circuit design considerations of various parts constituting MMIC phase shifters. The passive circuitry part consists of planar transmission lines and lumped elements, including grounding connections and cross-overs. Among the active devices, the GaAs MESFET is the most exten-sively used, particularly as a switching element in digital phase shifters. MESFET is also used as a variable capacitance for analog phase shifting, but to a very limited extent. Other control elements that are far less popular compared to MESFET include the planar Schottky barrier diode and *p-i-n* diode as switches, and the planar varactor diode and HEMT as variable capacitance analog devices.

12.3.1 Planar Transmission Lines [21–25]

Figure 12.1 shows the geometries of four planar transmission lines—namely, mi-crostrip, slotline, coplanar waveguide, and coplanar strips, which are structurally

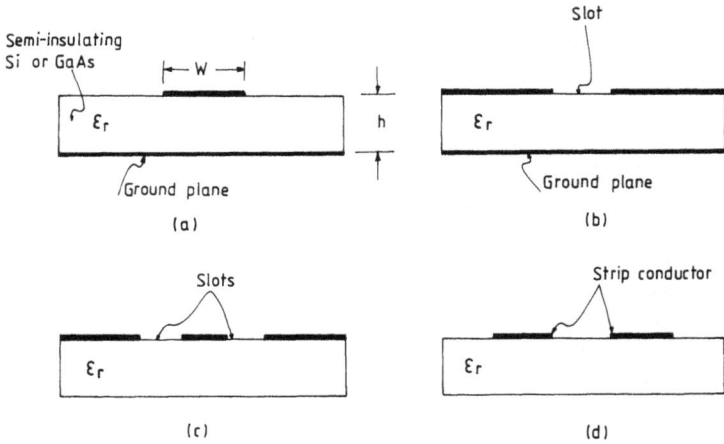

Figure 12.1 Planar transmission lines for MMICs: (a) microstrip, (b) slotline, (c) coplanar waveguide, (d) coplanar strips.

compatible for monolithic integration. Of these, the slotline is a non-transverse electromagnetic (TEM) line and the remaining three are quasi-TEM lines. Because of its higher dispersion and attenuation loss, the slotline is the least preferred. Practical MMICs adopt the less dispersive quasi-TEM lines, the microstrip being the most popular followed by the coplanar waveguide. With the introduction of "via" grounding techniques, the microstrip circuit design on GaAs has become even more flexible. The ground plane of the microstrip eases the problem of heat dissipation when high-power active devices are mounted in the circuit.

The analysis and propagation characteristics of the above transmission lines, including the multilayer versions, are well documented in the literature [21–25]. The impedance range achievable in a microstrip is approximately 10Ω to 100Ω. The coplanar line offers somewhat higher impedances; its typical range is 25Ω to 120Ω. Its loss is also higher than that in a microstrip. Figure 12.2 illustrates the variation in conductor loss of a microstrip on a GaAs substrate as a function of substrate thickness and characteristic impedance at 10 GHz [21]. The loss in dB/cm varies almost inversely as the thickness of the substrate. In order to keep the propagation losses low and retain reasonable linewidth at higher frequencies, a thicker substrate is desirable. On the other hand, from the point of view of thermal impedance and prevention of higher order mode propagation, the substrate must be kept sufficiently thin. The choice of substrate thickness is therefore a compromise between these conflicting requirements. A thickness of about 5 mils is common for low- to medium-power GaAs MMICs operating up to about 20 GHz.

In a typical MMIC, there is usually an additional dielectric separation layer, whether the transmission line is fabricated from first or second layer metallization.

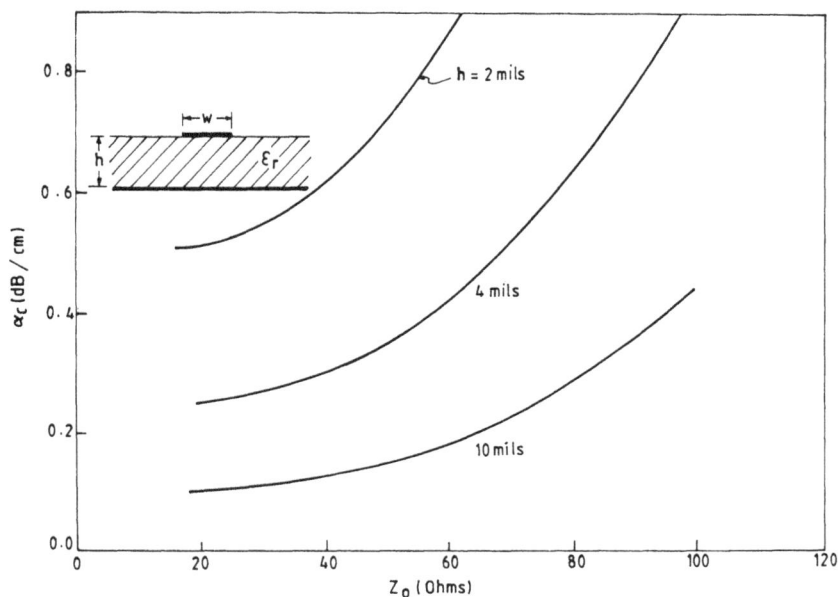

Figure 12.2 Conductor loss α_c of microstrip on GaAs substrate as a function of characteristic impedance Z_0. $f = 10$ GHz, $\varepsilon_r = 12.9$. (From Purcel [21], copyright © 1981 IEEE, reprinted with permission.)

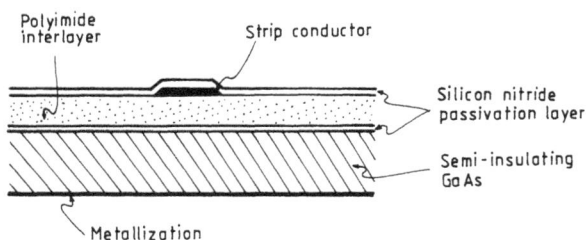

Figure 12.3 Typical cross section of a microstrip in a GaAs MMIC.

Figure 12.3 shows a typical cross section of a microstrip in MMIC showing a polyimide interlayer with silicon nitride passivation.

12.3.2 Lumped and Distributed Passive Elements [26–39]

The choice between lumped and distributed passive elements is determined by the requirements on minimum circuit size and accurate circuit modeling. Inductors, capacitors, and resistors in lumped form add considerable flexibility to MMIC design.

Because the dimensions of lumped elements must be maintained less than about 0.1 wavelength, their utility is confined to frequencies below 20 GHz. At higher frequencies, it becomes important to model these circuit elements in terms of their distributed representation to achieve the required accuracy in design. Alternatively, short sections of transmission lines, such as the open- and short-circuited stubs, can be used for realizing passive circuit elements.

Inductors—Planar inductors can be realized in many forms: a thin straight ribbon (high-impedance line), single loop, meanderline, circular spiral, and square spiral. These configurations are shown in Figure 12.4. Ribbon, single loop, and meanderline are easy to realize because they require the use of only one metallization layer. The spiral inductor requires the use of either an underpass or an airbridge to connect the center of the spiral to the external circuit.

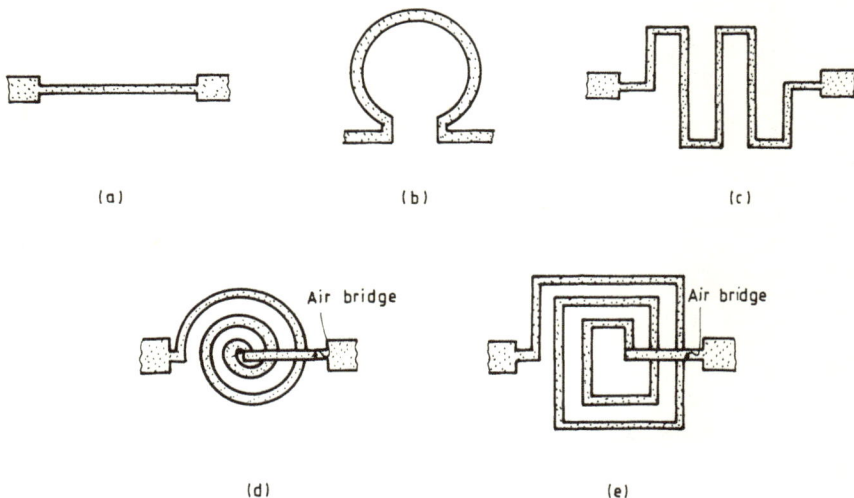

Figure 12.4 Planar inductor configurations: (a) straight ribbon, (b) single loop, (c) meanderline, (d) circular spiral, (e) square spiral.

Ribbon and single loop are normally used when inductance values up to about 2 nH are required. Higher values up to about 50 nH are achievable in meanderline and spiral inductors. These structures make use of the mutual coupling between adjacent conductors to achieve higher inductance in a small area. In order to ensure lumped behavior, the total line length must be confined to a fraction of a wavelength.

An approximate design of ribbon, single loop, and spiral inductors can be carried out using the quasi-static closed form inductance formulas reported by Grover [26] and Terman [27]. For meanderlines on GaAs substrates, Greenhouse [28] has reported a method for taking into account the coupling between adjacent strip conductors in the inductance calculations. In general, the intersegment coupling may be neglected if the spacing between the adjacent strips is greater than about five times the strip width when the strip width is much less than the substrate thickness. In the case of spiral inductors, also, more accurate models have been reported [28–31], which include the intersegment fringing capacitance and shunt capacitance with respect to ground. The effect of ground plane is to reduce the primary inductance of the inductor. This effect becomes important when the size of the spiral exceeds the substrate thickness. The resistance of the spiral is a function of the strip width and the total length of the strip conductor. With an increase in the number of turns in a spiral, the total inductance and the resistance increase. The resonant frequency of the spiral decreases, thus limiting the upper frequency of operation.

Practical MMICs generally use inductances in the range 0.5 to 10 nH. The Q factors of such inductors are on the order of 50 at X-band. For higher Q factors, distributed elements in the form of microstrip stubs are recommended.

Capacitors—Two types of planar capacitor configurations are commonly employed in MMICs. They are the interdigitated capacitor (Figure 12.5(a)) and the overlay or *metal-insulator-metal* (MIM) capacitor (Figure 12.5(b)).

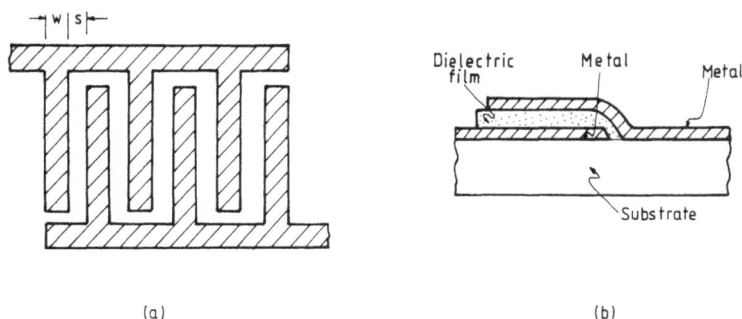

Figure 12.5 Planar capacitors for MMICs: (a) interdigitated and (b) overlay (or MIM).

In the interdigital structure, capacitance is formed by the fringing fields between two sets of interlaced metal fingers. It is used for low values of capacitances up to about 1 pF or so. In order for the capacitor to remain lumped at microwave frequencies, its overall length should be small, with the gap width between adjacent fingers typically 5 μm or less. Its Q factor is dependent on the ratio w/s, where w is the strip width and s is the gap. By maximizing w, Q factors in excess of 400 can be achieved. Care should be taken to ensure that there is no transverse resonance in the structure when several fingers are used.

The analysis of interdigital capacitors has been reported by several investigators [32–35]. Of these, the methods adopted by Hobdell [33] and Esfandiari *et al.* [34] are more general and take accurate account of the parasitic inductance and resistance.

The MIM configuration shown in Figure 12.5(b) is used for realizing larger values of capacitance (4 to 200 pF) normally required for bias and interstage coupling. The capacitor makes use of two overlapping metal layers with a dielectric film in between. The structure requires the use of a two-level metallization scheme. Dielectric films that have been found suitable for adoption in capacitors for MMICs include polyimide (ε_r = 3 to 4.5), silicon nitride (Si_3N_4, ε_r = 6 to 7), and tantalum oxide (Ta_2O_5, ε_r = 20 to 25) [21, 36–38]. The larger the dielectric constant, the higher the achievable capacitance density. For example, the capacitance density of a MIM capacitor with 2000Å-thick Si_3N_4 film is on the order of 300 pF/mm^2, whereas that with a Ta_2O_5 film of the same thickness is reported to be 1100 pF/mm^2 [21]. In contrast with the interdigital capacitors, MIM capacitors offer much higher capacitance values, typically up to about 25 pF with Si_3N_4 dielectric and up to about 200 pF with Ta_2O_5 dielectric [37–39].

Resistors—Planar resistors for MMICs have been realized over a wide range of resistivities covering approximately 50Ω to 500Ω/square. An important technique suitable for GaAs MMICs is ion implantation of the high-resistivity region. For example, an implanted n-type active layer has a resistivity of about 500Ω/square for a carrier concentration of 10^{17}/cm^3 [6]. The temperature coefficient of resistance of implanted resistors is high ($\approx$3200 ppm/°C). Another technique of implementing resistors is by the deposition of thin metal films, such as tantalum nitride (Ta_2N) and nichrome. Tantalum nitride adheres well to polyimide. It offers a resistivity of approximately 90Ω/square and has a low temperature coefficient of resistance (≈-100 ppm/°C) [6].

Low Inductance Grounds—Two types of low-inductance grounds are commonly employed in MMICs: the wrap-around and the *via* hole ground [21, 39]. The wrap-around requires active devices to be placed close to the circuit edge to minimize the grounding inductance. While this edge-ground approach offers process simplicity, the via grounding technique offers the flexibility of placing ground at any point in the circuit. In the via hole method, narrow cone-shaped holes (approximately 50 to 100 μm in diameter) are produced through the substrate, either by wet chemical etching or by reactive ion etching until the top metallization pattern is reached. These holes are then metallized in the same metallization process as that used for forming the ground plane. With its effectiveness in providing low inductance ground returns (typically 40 to 60 pH/mm of substrate thickness [21]), the via ground technique has added considerable flexibility to microstrip circuit design on GaAs substrates.

Interconnects—The requirement on low-loss, low-inductance, low-parasitic capacitance interconnects in MMICs is met by means of gold-plated *air-bridges*. They are commonly used for interconnection of FET sources, and connection of top plates of capacitors and inner ends of spiral inductors to other parts of the circuit. A typical air-bridge cross-over for a spiral inductor is shown in Figure 12.4(d). It consists of a deposited gold strap about 5 μm thick, which crosses over the conducting pattern of the circuit, leaving a narrow air gap of approximately 4 μm in between. A layer of photoresist is used to support the air-bridge while plating, and a second photoresist layer is used on top to confine the gold deposition to the desired area.

12.3.3 GaAs MESFET [3, 4, 7, 40]

The most commonly employed control element in MMIC phase shifters is the GaAs MESFET based on Schottky barrier gate. Because of its compatibility with MMIC technology, MESFET has emerged as the most viable alternative for MMIC phase shifters to the conventional *p-i-n* diodes. MESFETs are known to operate at high switching speeds on the order of subnanoseconds with negligible dc bias power. Other active devices that have been found suitable for phase-shifter applications include the planar Schottky barrier varactor diodes, planar *p-i-n* diodes, and HEMTs. Figure 12.6 shows the basic geometry of an epitaxially grown GaAs MESFET structure with a Schottky barrier gate. The structure consists of a *n*-type active layer on a semi-insulating GaAs substrate with a n^--type buffer layer in between. The buffer layer and the active layer are normally grown in a continuous epitaxial growth run. The buffer layer has high resistivity ($\approx 10^7$ Ω cm) and small carrier concentration. Besides acting as an extension of the base substrate, the buffer layer improves the sharpness of the carrier density profile and helps to maintain high electron mobility up to the interface with the base substrate. The buffer layer also helps in improving the FET parameters, particularly noise figure and gain.

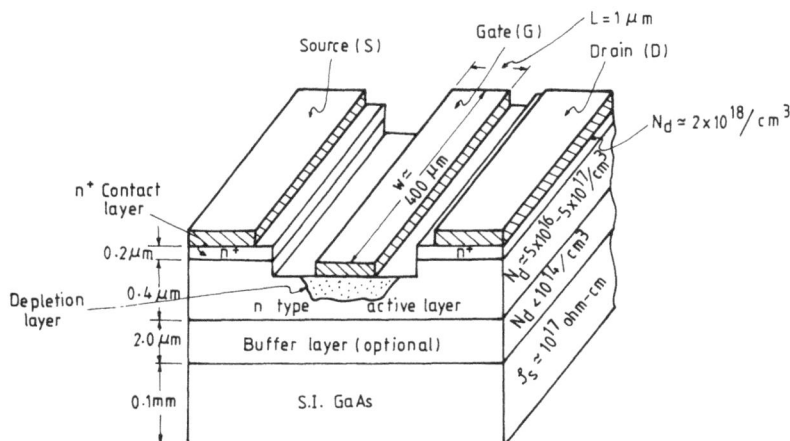

Figure 12.6 Basic schematic of GaAs MESFET showing typical parameters.

The two ohmic contacts for source and drain are realized by depositing a multilayer film of Au-Ge-Ni by simple evaporation, sputtering, or electron-beam evaporation, and then annealing the wafer to about 500°C. Germanium diffuses (to a depth of about 500Å to 1000Å) and forms a heavily doped n^+ layer under the metal layer to form a low-resistivity ohmic contact. The contact resistivity of this n^+ layer is in the typical range of 10^{-5} to 10^{-6} Ω-cm^2. The preferred metallization for the gate is Ti-Pt-Au or Ti-Pd-Au. Titanium, which forms the lowest metal layer, has very good Schottky barrier properties, and gold, which forms the top layer, has very low resistivity. The intermediate refractory barrier material serves to prevent diffusion of Au.

It may be noted that the buffer layer is not always required, as, for example, in ion-implanted devices. In ion implantation, the required doping level of the active layer is achieved by directly implanting donors onto the surface of the bulk GaAs material.

There are two commonly used techniques of MESFET fabrication: the self-aligned and the recessed-channel. Figure 12.7 shows the cross-sectional views of some typical FET structures employing these techniques. The self-aligned MESFET (Figure 12.7(a)) is a true planar structure, wherein the n^+ contacts are aligned with respect to the gate. The technology permits bringing the ohmic (usually the n^+) contacts as near to the gate as possible without producing any undesirable electrical problems. This facilitates improvement in device performance by minimizing the parasitics. In the recessed channel technique (Figure 12.7(b),(c)), a thicker n-type epitaxial layer is grown, and the channel under the gate is defined by etching a small portion of the active layer. This circumvents the problem of having to maintain tight tolerances on the n-type epitaxial layer thickness. The mushroom-shaped gate shown in Figure 12.7(c) helps in reducing the gate resistance by increasing the cross-sectional area of the gate metal while retaining the effective gate length in contact with the GaAs surface. Another technique of reducing the gate resistance is to increase the number of parallel gates within the same gate width. Reduction in gate resistance is important when considering the improvement of the high-frequency performance.

MESFET as a Switch—In a MESFET, the gate voltage controls the depletion region of the Schottky barrier. The current flow between the source and drain is therefore controlled by modulating the width of the depletion layer. The current handling capability of a FET is determined by the gate width w (see Figure 12.6). Multiple gate fingers are commonly used for reducing the gate resistance when the width is required to be kept large for higher current capability. The transit time and maximum operating frequency are determined by the ratio of gate length to saturated velocity of electrons. For high-frequency operation, the transit time of electrons moving through the region under the gate must be kept small. This is done by reducing the gate length L. As an example, FET devices with a 1-μm gate length are usable up to about 12 GHz, and with 0.5 μm they can be used up to about 20 GHz. For higher frequencies up to 40 GHz, shorter gate length of about 0.25 μm is needed.

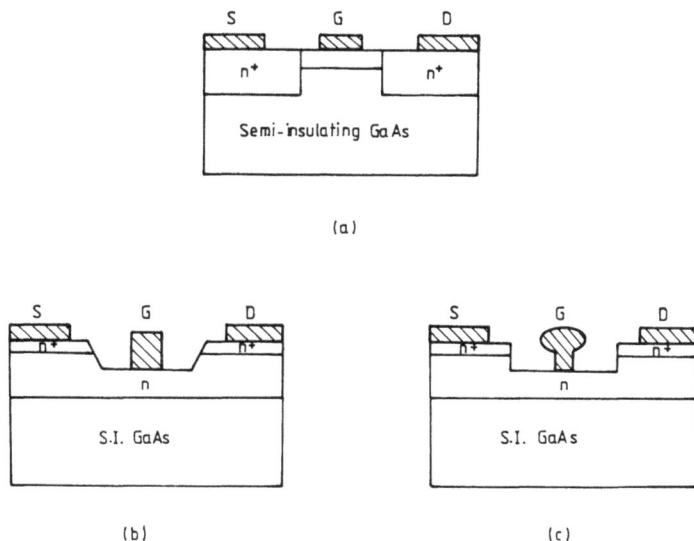

(a)

(b)

(c)

Figure 12.7 Typical cross sections of (a) self-aligned MESFET, (b) recess-channel MESFET, (c) recess-channel MESFET with mushroom gate.

MESFET can be operated as a switch by switching the gate bias between zero voltage and a negative voltage. With a negative bias, the channel width reduces, thus reducing the source-to-drain current. The voltage at which this current practically reduces to zero is called the pinch-off voltage (V_p). It may be noted that in either of the two switching states, the gate current is small, thus requiring negligible dc bias power.

Figure 12.8 shows the resistive and capacitive regions of MESFET in the two switching states. The low-impedance state (on-state) corresponds to zero gate bias ($V_g = 0$), and the high-impedance state (off-state) corresponds to negative gate bias ($V_g = -V$, $|V| > |V_p|$). In the on-state, the channel is open, except for a zero-field depletion layer thickness. Hence, for current levels less than the saturated channel current, FET can be modeled as a small resistance between source and drain. Referring to Figure 12.8(a), this resistance can be represented as ($R_c + 2R_0$) where R_c is the resistance of the open channel and R_0 is the resistance between the channel edge and the source and drain. In the off-state, the channel is completely depleted of free charge carriers, giving rise to a capacitance C_g between the gate and source and drain. Other parameters that may be included in the equivalent model are the fringing capacitance C_{sd} between the source and drain, R_{sd}, which represents the RF losses associated with C_{sd}, and R_g, which represents the changing resistance in series with C_g. When the gate terminating impedance is very high, the effective capacitance

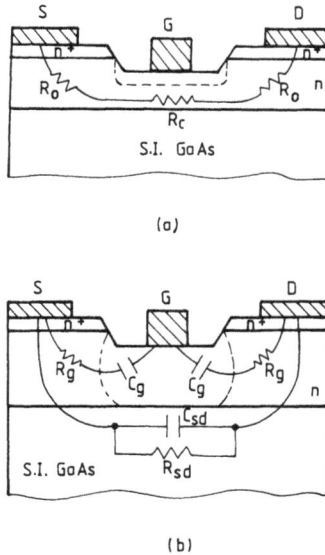

Figure 12.8 Schematic of a switching MESFET showing resistance and capacitance regions under (a) on-state (no gate bias) and (b) off-state ($|V_g| > |V_p|$).

between the source and drain can be approximated by $(C_{sd} + C_g/2)$, and the effective drain resistance can be considered as a parallel combination of R_d and $2/\omega^2 \, C_j^2 \, R_j$ [40].

12.3.4 Other Active Devices

Schottky Barrier Varactor Diodes [41–45]

Analog phase shifters commonly employ voltage-variable capacitance devices as control elements. Such devices must offer low resistance and large variation in capacitance. One method of realizing a variable capacitance device is to connect the source and drain of a GaAs MESFET and utilize the junction capacitance of the gate terminal. Dawson *et al.* [41] have reported a surface-oriented device of this type. Figure 12.9 shows the diode geometry and its fabrication sequence. The process involves selective implantation of active and n^+ layers on the semi-insulating GaAs substrate. The active layer implantation is made sufficiently deep so that diode breakdown occurs prior to punch-through. The conduction layer can thus surround the cathode side of the depletion region, enabling low series resistance to be maintained. The doping

1. **N⁺ Implant**
   Photoresist — 2 Microns
   PSG (7% P₂) — 2800 A°
   N⁺ Recess — 700 A°

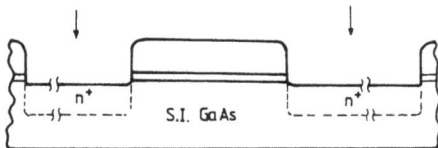

2. **N Implant**
   Photoresist — 2 Microns
   PSG (7% P₂) — 2800 A°

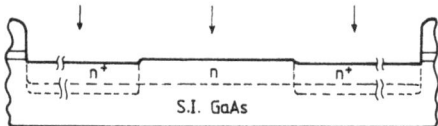

3. **Ohmic Contacts**
   AuGe/Ni/Au — 1500/400/500 A°

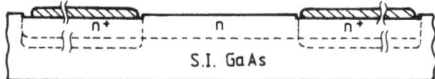

4. **Anode Contact**
   Ti/Pt/Au — 500/400/6500 A°

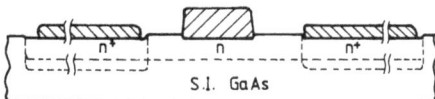

5. **Circuit Overlay**
   Cr/Pd/Au — 500/1000/10000 A°

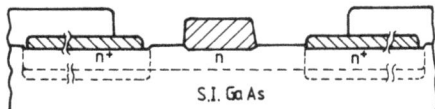

Figure 12.9 Fabrication sequence of FET-based Schottky varactor diode. (From Dawson *et al.* [41], copyright © 1984 IEEE, reprinted with permission.)

level is kept sufficiently high to achieve low series resistance, but low enough to leave 80% depletion of the active layer without breakdown to permit maximum variation in capacitance. It may be noted that the n^+ contact layer is deeper than the active layer. This helps in maintaining low resistance between the ohmic contact and the Schottky layer at all bias levels.

Another means of achieving variable capacitance is through planar varactor diodes. Several investigators have reported planar varactor diodes with special doping profiles that yield a high degree of phase linearity with respect to bias voltage [42–45]. Figure 12.10 shows the fabrication sequence of a planar varactor diode using n-on-n^+ epitaxial layers and an etched-mesa process [45]. The thick n layer provides for large capacitance ratio and the n^+ layer helps to lower the series resistance.

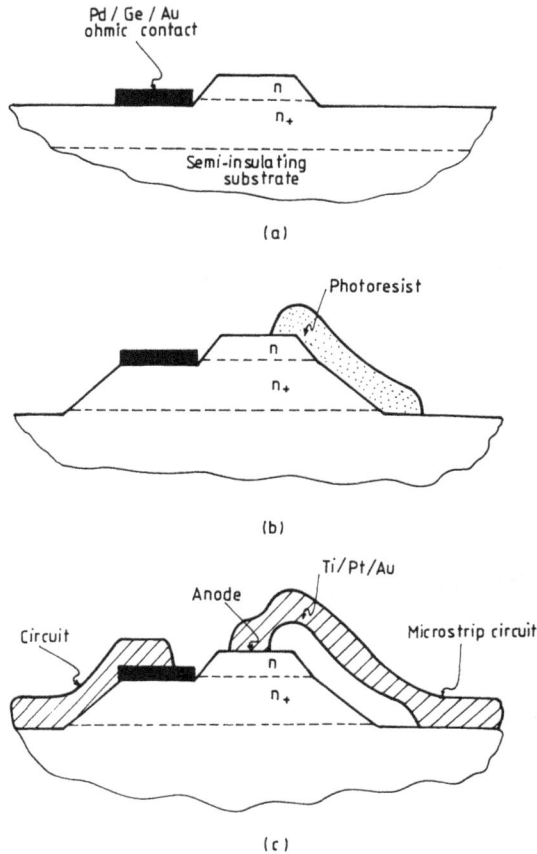

Figure 12.10 The process steps of planar varactor for monolithic analog phase shifter: (a) ohmic contact; (b) mesa etch and air-bridge definition; (c) electron-beam evaporated Ti/Pt/Au for anode, air-bridge, and microstrip circuit. (From Chen *et al.* [45], copyright © 1987 IEEE, reprinted with permission.)

Planar P-I-N Diodes [46, 47]

Planar GaAs *p-i-n* diodes have been developed for use as switching elements [46, 47]. In order to bring both contacts to the top surface, a deep n^+ layer is used below the I-layer. Both these layers are epitaxially grown to a thickness of about 2 μm each. The top p^+ contact is made by Zn diffusion. Because of the need to have thick layers, ion implantation is not practical and therefore integration with FET-based MMICs is difficult. The technology requires epitaxial growth and a mesa process for isolation. As compared with FETs, the planar *p-i-n* diodes have lower capacitance and are therefore useful for millimeter-wave monolithic digital phase shifters.

HEMT Devices [6, 48–52]

HEMTs are known to have several advantages over the MESFETs, such as higher maximum frequency of operation and lower noise [6, 48–51]. Double heterojunction HEMTs are reported that offer larger capacitance ratios (typically on the order of 7) over the MESFETs, thus making them suitable for monolithic analog phase shifters [51, 52].

Figure 12.11 shows the basic structure of a HEMT. It consists of an *n*-doped AlGaAs layer over an undoped GaAs layer, with a thin film of undoped AlGaAs as a spacer. The electrons contributed by the *n* doping move about freely through the entire crystal until they fall into the lowest energy states located on the GaAs side of the heterojunction interface. These electrons, on accumulation, are free to move in a two-dimensional plane of the interface. This electron accumulation layer has the properties of a *two-dimensional electron gas* (2 DEG). Because of the spatial separation of conduction electrons from their parent donor impurities, the rate of electron scattering by the ionized impurities is considerably reduced. This results in better coherence of electron velocity in the direction of applied electric field, leading to enhanced electron mobility as well as velocity. The use of a thin layer of undoped AlGaAs adjacent to the heterointerface helps to increase the spatial separation of electrons in the 2 DEG from the impurity scatterers, thereby enhancing the electron mobility further. The electron mobility of a 2 DEG in a HEMT is approximately 9000 cm^2/V-s, which is roughly twice the mobility of an active layer in GaAs MESFET. HEMT devices, therefore, have the potential for offering higher switching speed, higher operating frequency range, and lower noise than FETs [48–51].

The fabrication process involves first the growth of film layers on semi-insulating GaAs using molecular beam epitaxy in order to achieve atomically smooth heterojunction interfaces. The top surface is then capped by an n^+ GaAs layer to prevent

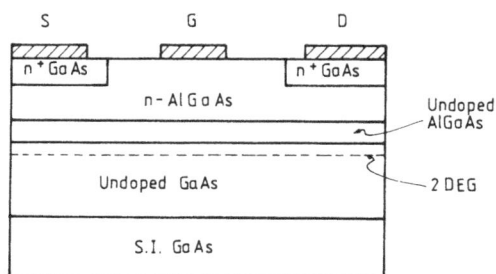

Figure 12.11 Basic structure of a HEMT.

oxidation of AlGaAs and to permit low-resistance ohmic contacts to the structure. The device fabrication after this growth is similar to that of MESFET. For monolithic integration, proton or boron implantation is used for isolation. The ohmic metal is usually a Au-Ge-Ni-Au alloy, and the Schottky gate metal is Ti-Pt-Au.

HEMTs can be realized as either enhancement (normally-off) or depletion (normally-on) devices depending on the GaAs layer thickness and doping density. In the enhancement device, the AlGaAs layer is made thinner so that the Schottky barrier extends into the GaAs, preventing the formation of 2 DEG under zero bias.

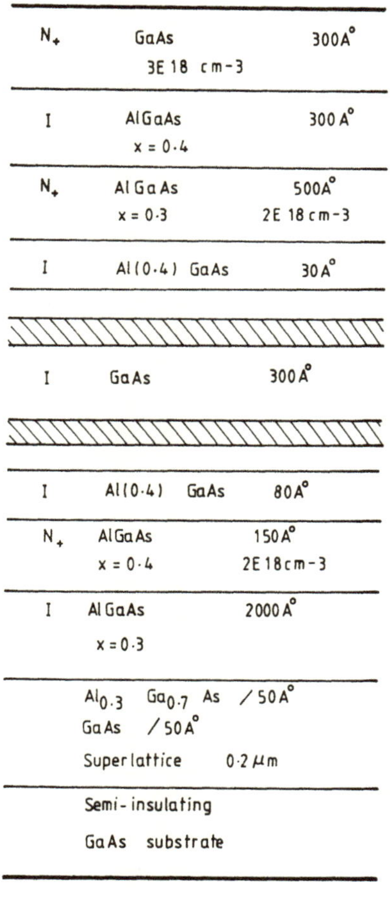

N_+	GaAs 3E 18 cm−3	300Å
I	AlGaAs x = 0·4	300 Å
N_+	Al Ga As x = 0·3	500Å 2E 18 cm−3
I	Al (0·4) GaAs	30Å
	///////////////////	
I	GaAs	300Å
	///////////////////	
I	Al(0·4) GaAs	80Å
N_+	AlGaAs x = 0·4	150Å 2E 18cm−3
I	Al GaAs x = 0·3	2000Å
	$Al_{0.3}$ $Ga_{0.7}$ As / 50Å GaAs / 50Å Superlattice 0·2 μm	
	Semi-insulating GaAs substrate	

Figure 12.12 Cross section of an optimized double-heterojunction diode. (From Cazaux *et al.* [52], first presented at the 18th European Microwave Conference, reprinted with permission.)

The donors in AlGaAs act to control the interface potential of the GaAs at the heterojunction. A positive bias is required to cause electron accumulation in the 2 DEG at the interface. In the depletion device, the AlGaAs thickness is such that the Schottky barrier depletion region just overlaps the depletion region caused by electron transfer to the 2 DEG. Thus a 2 DEG exists at zero bias and a negative gate bias is required to completely deplete the 2 DEG. The device can thus be operated as a voltage-controlled capacitor by varying the gate voltage. The capacitance range and the capacitance-*versus*-voltage profile can be controlled by optimizing the aluminum fraction (x) of the $Al_xGa_{1-x}As$ layer and varying the spacer thickness [52]. Additional flexibility is achieved by introducing an undoped region under the Schottky gate electrode [51, 52].

Large capacitance ratio on the order of 7 has been reported for HEMTs employing double-heterojunction (DH) designs [52]. The DH HEMT makes use of a normal and an inverted heterojunction (with undoped GaAs and n-AlGaAs layers interchanged) on both sides of a common undoped layer such as GaAs or InGaAs. Figure 12.12 shows the cross section along with optimized parameters of a DH device reported by Cazaux *et al.* [52]. For a bias variation over the range 0.5V to −2V, the maximum capacitance ratio reported for this device is 7:1, whereas that for a MESFET having the same pinch-off voltage (−2.1V) is around 3:1.

12.3.5 Typical MMIC Process

Monolithic circuits are fabricated on GaAs single-crystal wafers. Figure 12.13 shows a typical process flow diagram for GaAs MMICs [12]. The major process steps are described below [11–13].

Active Layer Formation and Device Isolation

The first step in MMIC process is formation of the n-type active layer. Either the epitaxial growth technique or ion implantation can be used. In order to define active channels for FETs or diodes and to provide isolation between closely spaced active devices, one of the following techniques is adopted: (1) mesa etching, which removes the n layer from areas other than those required for forming active devices; (2) selective ion implantation of the active area using a photoresist mask; and (3) proton bombardment of the area surrounding the active region to destroy the conductivity.

Ohmic Contact Formation

Ohmic contacts are formed by sequential evaporation of Au, Ge, and Ni, followed by lift-off and alloying at around 450°C.

1. **Mesa Etch**

2. **Ohmic Contact Metallization**

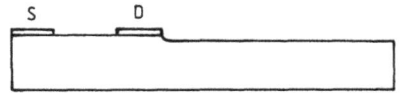

3. **Schottky Gate Metallization**

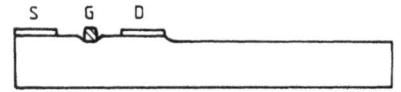

4. **First Level Metal**
   (Inductors, Capacitors, 8-D Overlay)

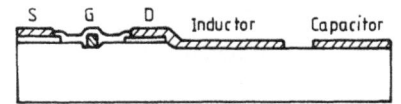

5. **Capacitor Top Plates**

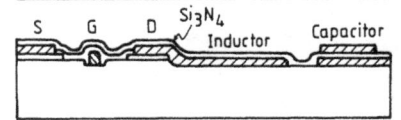

6. **Air Bridge Interconnects**

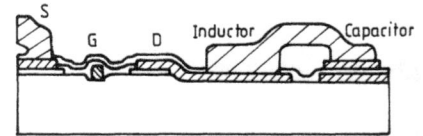

7. **Via Grounding, Backside Plating**

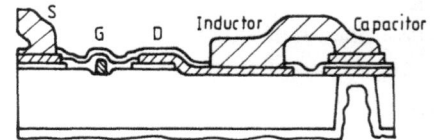

Figure 12.13 GaAs MMIC process flow diagram. (From Wisseman *et al.* [12], copyright © 1983 IEEE, reprinted with permission.)

Schottky Gate Formation

After the ohmic contact metallization, gates are defined by contact photolithography. For submicron gates, electron beam lithography is employed. The gate area is recessed using a wet etch process to achieve the desired device current. This is followed by evaporation of Ti-Pt-Au and patterned by lift-off.

First-Level Metallization

Following gate metal definition, sequential layers of Ti and Au are deposited and patterned by lift-off to form the first-level metal pattern. This forms inductors, lower plates of MIM capacitors, and provides overlays for ohmic contacts. The pattern can also be formed by ion milling through a photoresist mask to remove the unwanted areas.

Dielectric Deposition and Passivation

Next, a layer of high-quality Si_3N_4 film is deposited (using plasma-enhanced chemical vapor deposition (CVD)). This forms the dielectric for MIM capacitors, a protective layer for discrete devices, and a cross-over insulator in a two-level metallization scheme.

Second-Level Metallization and Capacitor Top Plates

A second metal layer of Ti-Au is evaporated and patterned by lift-off to form top metal plates of MIM capacitors.

Air-Bridge Interconnects

Gold electroplating to a thickness of about 15 μm is used for defining air-bridge interconnects and microwave transmission lines. A thick layer of photoresist is used to support the bridge, while electroplating and another layer confine the plating to the desired area.

Via-Hole Formation and Backside Metallization

After defining the circuit on the top surface of the substrate, the substrate is thinned (to about 100 μm), via holes are etched through the substrate (using ion etching), and the entire back surface including holes is metallized to form the RF ground plane.

12.4 MONOLITHIC PASSIVE PHASE SHIFTERS

Monolithic phase shifters adopt essentially the same circuit forms as those used for *p-i-n* diode phase shifters in hybrid MIC. The circuits are of four basic types: switched line, hybrid coupled, loaded line, and low-pass high-pass. The analysis and design of these circuits are described in Chapters 8 and 9. In this section, we describe some of the monolithic phase shifters reported in the literature and highlight the features pertaining to monolithic integration.

12.4.1 Digital Phase Shifters

GaAs FETs are the most readily available switch elements for monolithic digital phase shifters. Monolithic multibit phase shifters using FETs have been reported up to V-band [53–63]. All four circuit forms referred to above have been exploited for realizing MMIC and MMWIC phase shifters. There are certain common features that apply to all phase-shifter circuits employing FETs. We first list these features and then present a few illustrative circuit examples.

Although FET is a three-terminal device, only the source and drain form the RF terminals; the gate is used for bias control. The RF transmission lines in the circuit are isolated from the dc bias path, thus eliminating the need to use dc blocking capacitors between individual phase bits. The two switching states correspond to zero gate bias (low-impedance or on-state) and a negative bias (high-impedance or off-state). In the on-state, FET can be modeled as a small series resistance. In the off-state, the bias voltage lies between the pinch-off voltage and the breakdown voltage, and the FET is modeled as a capacitor in series with a small resistance (see Chapter 10 for the equivalent circuit). It may be noted that unlike the *p-i-n* diode, the capacitive reactance of the FET in the off-state is not very high. It is typically on the order of $-j50\Omega$ at X-band. Therefore, in the off-state, the capacitance of FET must be either resonated to present a high impedance, or its effect must be included in the design of impedance matching sections. An important feature of FET is that it requires negligible dc bias power, thus making the bias control circuitry simpler. The switch is bidirectional. The source-gate and the drain-gate capacitances are equal. The source and drain terminals are at ground potential. Since the drain terminal is not isolated from the gate, the gate circuit must be configured to present an effective RF open to the FET at the gate. This is achieved by either connecting the control signal through a resistor of high value (approximately 10 kΩ) or a low-pass filter.

Switched Line and Loaded Line Circuits With FET Switches

Monolithic digital FET phase shifters using switched line and loaded line techniques have been reported by several investigators [53–55]. As an illustration of a typical monolithic digital circuit, we describe below an X-band 4-bit FET phase shifter as reported by Ayasli *et al.* [53]. The circuit is shown in Figure 12.14. It makes use of switched line configuration for the 180° and 90° bits and the loaded line configuration for the 45° and 22.5° bits. Each switched line bit makes use of three FET switches instead of the conventional four switching diodes. Equal insertion loss between the two states is achieved by designing the two switched paths at different

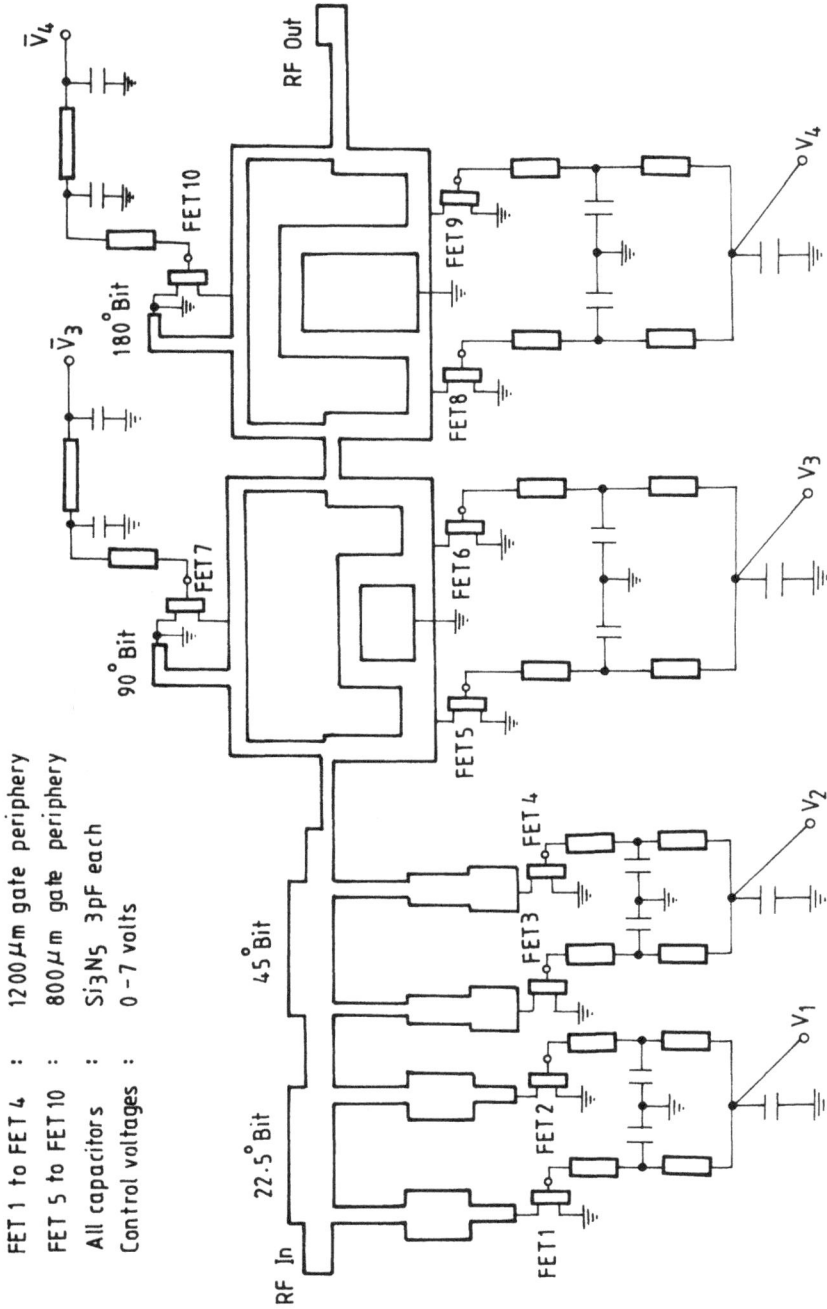

Figure 12.14 Schematic circuit diagram of a 4-bit phase shifter. (From Ayasli [40], reprinted with permission of *Microwave Journal*, copyright © 1982 Horizon House-Microwave Inc.)

impedance levels. The loaded line bits are designed to provide constant phase shift over the desired frequency band. Each loading stub consists of a suitably designed three-section transformer and a matching network that is terminated by a 1200 μm gate periphery FET switch. All four phase bits are cascaded in a single GaAs chip 6.4 mm $\times$ 7.9 mm $\times$ 0.1 mm in size (Figure 12.15). The chip incorporates three epitaxially grown layers on a semi-insulating GaAs; an undoped buffer layer ($N_d <$ 5 $\times$ 10^{13}/cm^3, t = 2.0 μm), an active layer ($N_d \simeq$ 9 $\times$ 10^{16}/cm^3, t = 0.4 μm), and a top n^+ layer ($N_d >$ 2 $\times$ 10^{18}/cm^3, t = 0.2 μm). Device isolation is achieved through a combination of a shallow mesa etch and a damaging 0^+ implant. Ohmic contacts are formed by Ni-Au-Ge metallization, and gates are formed by Ti-Pt-Au. Gates are recessed and have a length of 1 μm. There are 16 capacitors, each of value 3 pF. They use a 5000Å-thick SiN$_4$ (ε_r = 6.8) layer formed by plasma-assisted CVD process. The circuit uses via holes for ground connections and gold-plated air-bridges for connection to MIM capacitors. The phase shifter is reported to offer an insertion loss of 5.1 $\pm$ 0.6 dB at 9.5 GHz for all 16 states and a return loss of better than 10 dB over a 2.5-GHz bandwidth [53].

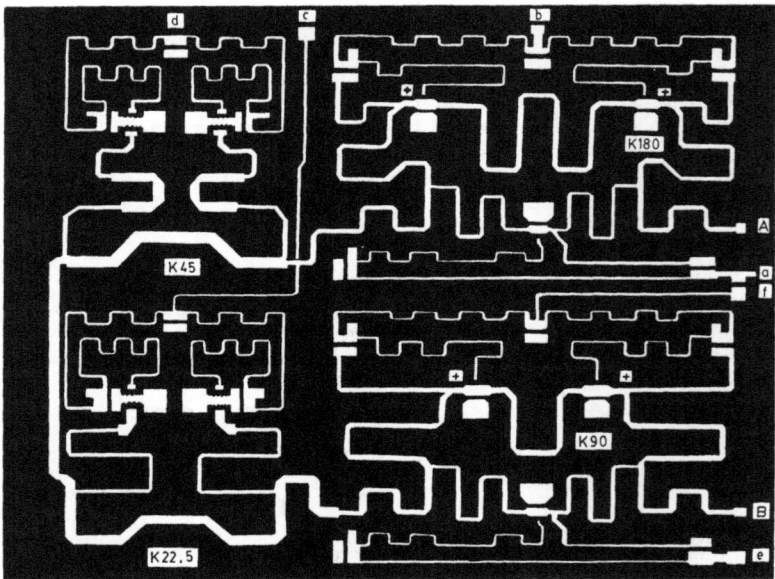

Figure 12.15 Four-bit phase-shifter chip. A = RF input/output, B = RF output/input; a = 180° bit reference arm control signal; b = 180° bit phase delay arm control signal; c = 22.5° bit control signal; d = 45° bit control signal; e = 90° bit reference arm control signal; f = 90° bit phase delay arm control signal; chip size: 6.4 $\times$ 7.9 $\times$ 0.1 mm^3; total gate periphery (1 μm gate length) = 9.6 mm; Number of air-bridges = 77; Number of via holes = 26. (From Ayasli *et al.* [53], copyright © 1982 IEEE, reprinted with permission.)

Hybrid Coupled and Loaded Line Circuits With FET Switches

Andricos, Bahl, and Griffin [56] have reported a C-band 6-bit FET phase shifter consisting of a cascade of five digital bits and a variable (0–11°) control bit for tuning. The 11° analog bit and the 11.25°, 22.5°, and 45° discrete bits are of loaded line type, and the 90° and 180° bits are of hybrid coupled type. Figure 12.16 shows the schematic of the loaded line phase bit along with the electrical and physical parameters of the three phase bits. The circuit uses two high-impedance stubs terminated in FET switches. The stubs are spaced by about a quarter-wavelength on the main line. The variable phase bit consists of an 11.25° loaded line section in which FETs are used in a variable impedance mode. For the other three discrete bits, FETs are operated as switches. Three different FET peripheries (1200, 1800, and 2400 μm) are chosen for the three phase bit circuits, each corresponding to optimum

Electrical and Physical Parameters for Loaded-Line Phase Bits

Bit Size (Degree)	Z_{OT} (Ω)	θ_T (Deg.)	W_T (μm)	l_T (μm)	Z_{OA} (Ω)	θ_A (Deg.)	W_A (μm)	l_A (μm)	FET size (μm)
11.25	50.5	88.5	88	4674	100	112.5	8	6030	2400
22.5	45	85	114	4450	90	113.5	14	6325	1200
45.0	48	78.3	100	4115	56.5	116.5	66	6223	1800

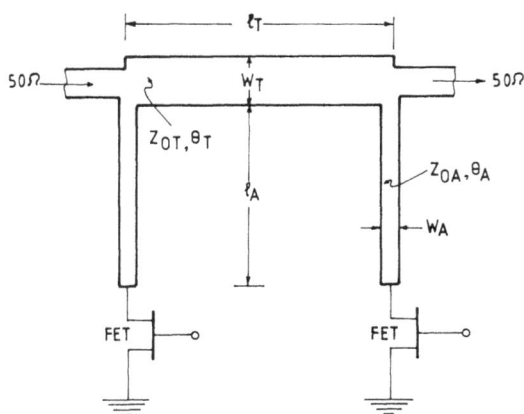

Figure 12.16 Schematic for 11.25°, 22.5°, and 45° loaded line bits. (From Andricos *et al.* [56], copyright © 1985 IEEE, reprinted with permission.)

performance in terms of best voltage standing wave ratio (VSWR) and minimum insertion loss.

Figure 12.17 shows the schematic of the 90° and 180° phase bits. The hybrid is a Lange coupler having dimensions l = 4966 μm and $w = s$ = 12 μm. The two coupled arms of the coupler are symmetrically terminated in FETs. Both 90° and 180° bits make use of 2400-μm FET peripheries to yield the best VSWR and minimum insertion loss. The design parameters of these bits are listed, along with the schematics, in Figure 12.17. Figure 12.18 shows the photograph of the 6-bit chip [56]. The chip size is 9.43 mm × 4.2 mm × 0.125 mm. The circuit uses a total of 12 FETs (1-μm gate length), 12 mesa resistors (5 kΩ each), 50 air-bridges, and 2 Lange couplers. Mesa resistors are used for making gate bias connection. The phase shifter is reported to give a total insertion loss of 8.7 dB with a maximum variation of ±1.2 dB over a frequency range of 5 to 6 GHz for all phase-shift values [56].

Pao $et\ al.$ [57] have reported a V-band 4-bit phase shifter by cascading four branchline coupled FET phase-shifter bits. The FET devices use a selective ion implant of an active n layer ($N_d \simeq 1.6 \times 10^{17}/\text{cm}^3$, t = 0.25 μm) and a heavily doped n^+ layer ($N_d \simeq 10^{18}/\text{cm}^3$) under the ohmic contacts for providing low source resistance. The device isolation is achieved through proton bombardment, and high-yield Schottky gates (0.6-μm gate length) are produced by optical contact lithography. The 4-bit phase shifter is reported to offer an insertion loss of approximately 7 ± 2.5 dB over a frequency range of 59 to 63 GHz.

Electrical and Physical Parameters for Hybrid Coupled Phase Bits

90° Bit	180° Bit
Z_{OA} = 44 Ω	Z_{OA} = 33.2 Ω
θ_A = 14.6°	θ_A = 20°
W_A = 120 μm	W_A = 204 μm
l_A = 762 μm	l_A = 1016 μm
FET Size = 2400 μm	Z_{OB} = 77.5 Ω
	θ_B = 19.5°
	W_B = 26 μm
	l_B = 1067 μm
	Z_{OC} = 70 Ω
	θ_C = 30.4°
	W_C = 36 μm
	l_C = 1650 μm
	FET Size = 2400 μm

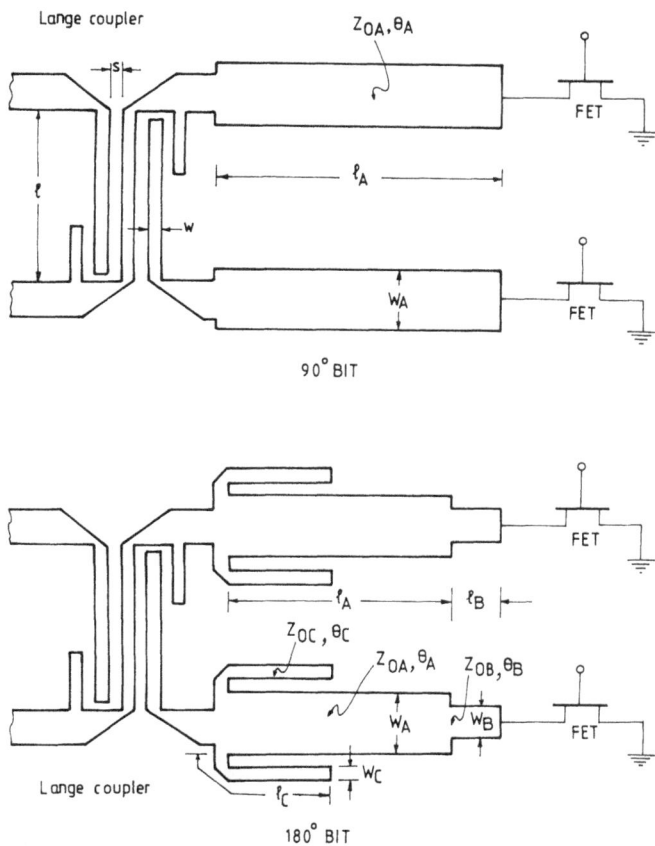

Figure 12.17 Schematic for 90° and 180° hybrid coupled phase bits. (From Andricos *et al.* [56], copyright © 1985 IEEE, reprinted with permission.)

Low-Pass High-Pass Circuits With FET Switches

The low-pass high-pass technique employs switching between a low-pass filter and a high-pass filter using FET switches [55, 58–60]. The filter networks use lumped inductors and capacitors, which can be easily realized in monolithic form. Broadband phase shifters can be realized in this configuration provided the switching parasitics are negligible. Schindler and Miller [58] have reported a 3-bit K/Ka-band (18–40 GHz) phase shifter in which the off-state capacitance of the switching FET is incorporated as an integral part of the lumped-element circuit. Figure 12.19 shows the schematic diagram of the 45° and 90° phase bits. For the 180° bit, a bridge topology [60] has been reported. Referring to the circuit shown in Figure 12.19, when the

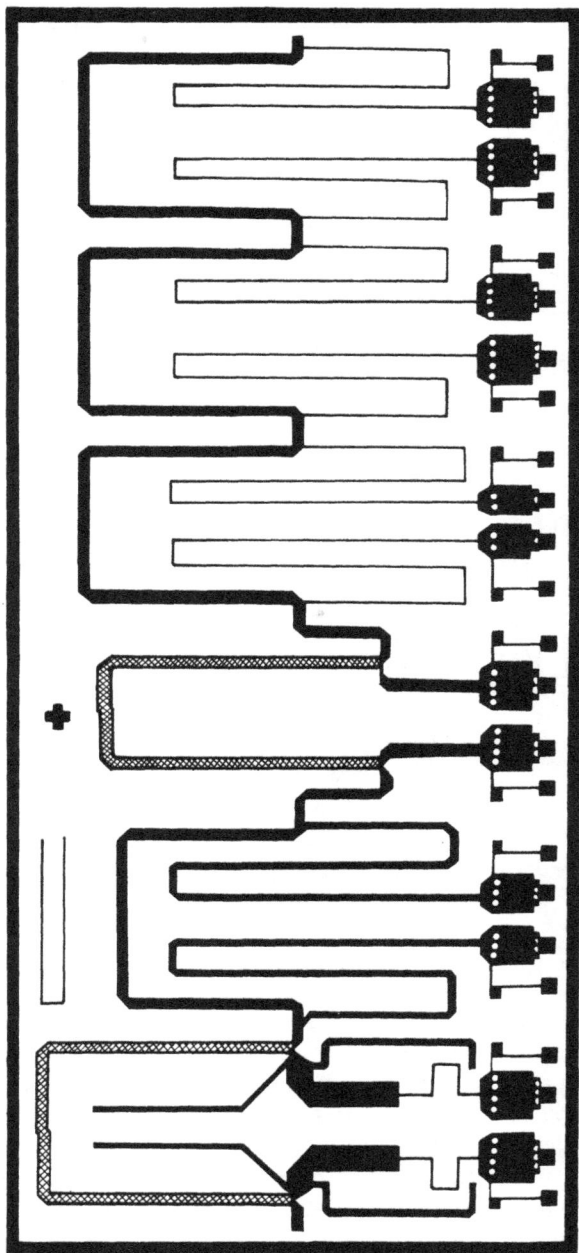

Figure 12.18 Photograph of the 6-bit FET phase shifter. (From Andricos *et al.* [56], copyright © 1985 IEEE, reprinted with permission.)

FETs marked F_1, F_2, and F_4 are on, and F_3 and F_5 are off, the circuit reduces to a three-element T low-pass filter. In this state F_5 is used as a capacitive element of the filter. The capacitance of F_3 is a parasitic that can be made small. The T high-pass filter is realized by reversing the biases. In this state, F_1 and F_2 are used as capacitors that, in parallel with C_1 and C_2, respectively, form the series elements. The capacitance of F_4 is parasitic and can be made small by minimizing the gate width of F_4. L_3 then becomes the shunt element of the high-pass filter. A 3-bit phase shifter realized on a chip 1.3×2.1 mm^2 in size is reported to offer an average loss from 9 to 10 dB over the frequency range 18 to 40 GHz.

Hybrid Coupled Switched Reflection-Type Circuit With Schottky Barrier Diodes

Hybrid coupled switched reflection-type circuit based on switched delay lines has been reported [61, 62]. Figure 12.20 shows the schematic of a 3-bit V-band phase shifter of this type employing Schottky barrier diodes as switching elements [61]. The circuit makes use of a single branchline coupler with its coupled arms terminated in symmetric reflective networks. The reflective network consists of a high-impedance line (80Ω) periodically loaded with eight shunt-mounted Schottky barrier diodes. The principle of operation of the circuit is the same as that of the conventional hybrid coupled phase shifter. Under reverse bias, the Schottky diode presents a capacitive load to the main line, and under forward bias it presents a short circuit, thus reflecting the power back to the coupler. The reflected power from the two symmetric ter-

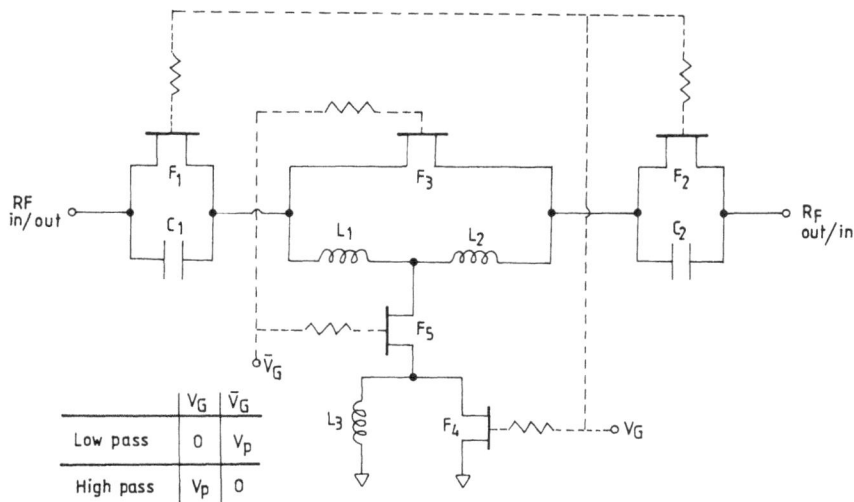

Figure 12.19 Schematic of low-pass high-pass type phase shifter for 45° and 90° bits. (From Schindler and Miller [58], copyright © 1988 IEEE, reprinted with permission.)

minations adds up and appears at the output port of the coupler. The phase of the reflected signal (or phase delay) depends on the location of the shorted diodes in the coupled arms. The circuit thus functions as a switched delay line. The diode capacitance and the parameters of the transmission line sections between diodes are chosen so that each section offers an electrical length of 22.5° at the design frequency. With eight identical equispaced diodes, phase shift can be incremented in steps of 45° up to 360°. The phase shifter is reported to offer an insertion loss of 10.8 ± 1.8 dB, rms phase error of 2.7°, and a VSWR better than 2.1 at 62.5 GHz.

12.4.2 Analog Phase Shifters

Monolithic analog phase shifters are commonly realized by integrating hybrid couplers with voltage-controlled varactors, mostly FET-based Schottky varactors or planar varactors [41, 44, 45, 57, 64–66]. Potential utility of heterojunction HEMTs as voltage-controlled capacitors in phase shifters has also been reported [52]. The operating principles and design of varactor-controlled hybrid coupled phase shifters in MIC are described in Chapter 11. The same circuit techniques also apply to monolithic circuits. In monolithic implementation, varactors with different doping profiles can be directly integrated to yield the desired capacitance-voltage characteristic (equivalently, phase-frequency characteristic).

Dawson et al. [41] have reported a monolithic X-band phase shifter using a 3-dB Lange coupler integrated with FET-based Schottky varactors. The circuit employs two pairs of varactors with the two diodes of each pair connected back to back in each coupled arm. The varactor is a surface-oriented diode using a flat-profile, ion-implanted, thick n layer with a deeply buried n^+ layer (structure shown in Figure 12.9). For a gate bias varying from 0 to -10V, the phase shifter is reported to offer a 105° phase shift with a 2.5 ± 0.5 dB insertion loss [41]. The phase shift is not a linear function of bias as expected from a square law capacitance-voltage relationship of the FET.

A Ku-band monolithic phase shifter integrating a 90° branchline coupler with planar varactor diodes has been reported by Chen et al. [45]. Figure 12.21 shows the circuit diagram as well as a scanning electron microscope (SEM) micrograph of the finished chip. Varactors are series mounted at the two coupled ports of the coupler. The 50Ω microstrip line section between the coupler and the varactor helps to minimize the phase shift variation over the desired frequency band. The tuning inductor is used to maximize the total phase shift as well as to minimize its variation over the bandwidth. With varactor diodes employing MBE-grown n on n^+ epitaxial layers (structure shown in Figure 12.10), the 1-bit phase shifter is reported to offer a maximum phase shift of 109 ± 3° with 1.8 ± 0.3 dB insertion loss in the frequency range 16 to 18 GHz. With a cascade of three phase bits, a total phase shift of 360° and an insertion loss of 4.2 ± 0.9 dB are reported for the same frequency range [45].

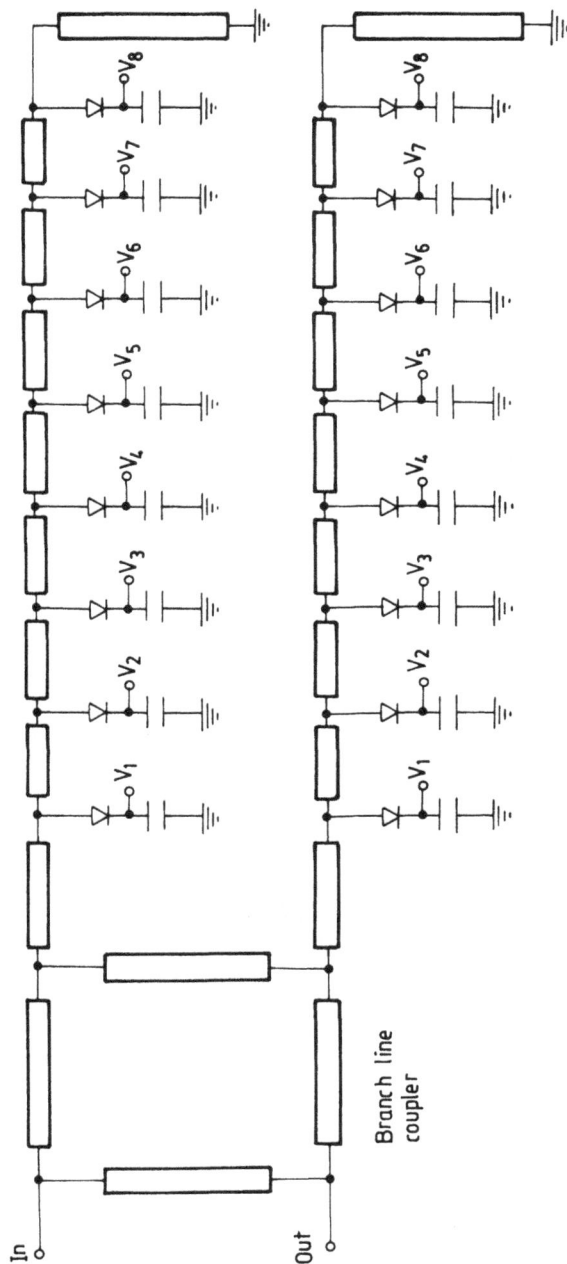

Figure 12.20 Hybrid coupled switched reflection-type V-band phase shifter. (From Jacomb-Hood, Seielstad, and Merrill [61], copyright © 1987 IEEE, reprinted with permission.)

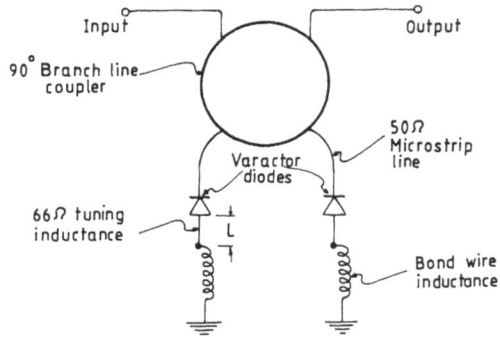

(a)

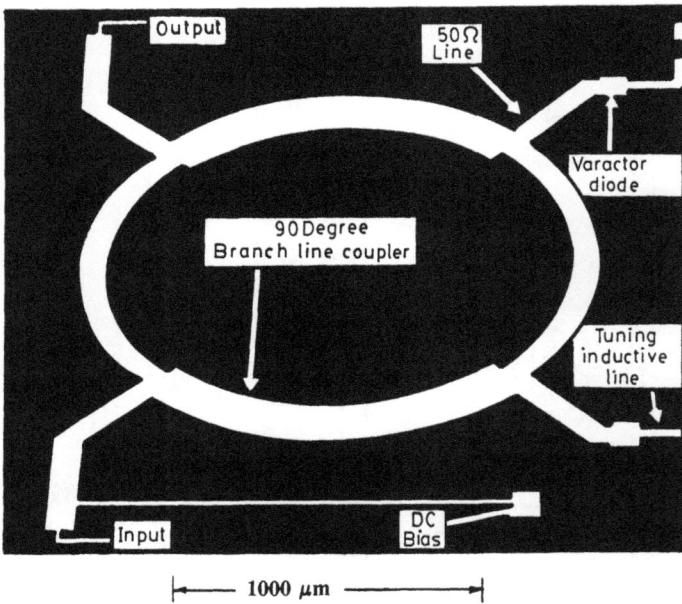

(b)

Figure 12.21 (a) Circuit diagram of an analog hybrid coupled phase shifter; (b) SEM micrograph of the finished chip. (From Chen *et al.* [45], copyright © 1987 IEEE, reprinted with permission.)

Figure 12.22 shows a broadband (6–18 GHz) monolithic phase shifter reported by Krafcsik *et al*. [64]. The circuit employs a Lange coupler and dual-varactor reflection circuits. The varactor is a surface-oriented diode structure with a hyperabrupt doping profile obtained by selective ion implantation. The capacitance ratio reported is 3:1. Under zero bias, the reflective section marked R_1 resonates at 5.4 GHz and the section marked R_2 resonates at 2.5 GHz. With reverse bias applied to varactors simultaneously, phase shift occurs due to R_1 over the range 6 to 9 GHz. However, R_1 is inductive and insensitive to varactor tuning from 9 to 18 GHz. Over the same frequency range, R_2 is capacitive and tunes the inductive circuit R_1, resulting in phase shift. The phase shifter, realized in a monolithic chip (size 0.148 $\times$ 0.068 $\times$ 0.015 in^3), is reported to offer a 160° phase shift and an insertion loss of 2.7 $\pm$ 1.3 dB over 6 to 18 GHz.

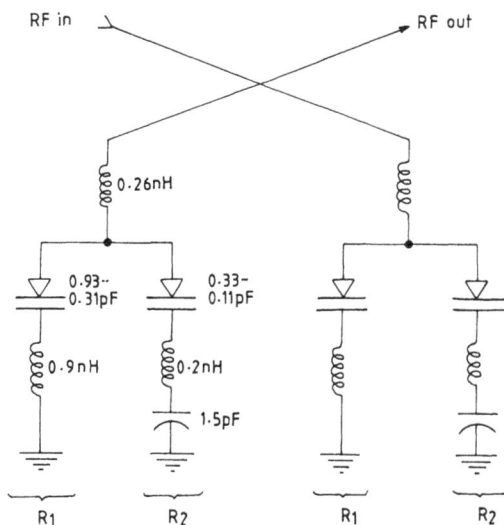

Figure 12.22 Broadband dual-varactor analog phase shifter. (From Krafcsik *et al*. [64], copyright © 1988 IEEE, reprinted with permission.)

Monolithic analog phase shifters using varactors have been realized at V-band by Pao *et al*. [57]. A single-bit branchline coupled varactor phase shifter is reported to give a phase shift of 54° with less than 1.3 dB of insertion loss at 60 GHz.

Cazaux *et al*. [52] have demonstrated the potential of using HEMT technology in realizing monolithic analog phase shifters. The advantages of HEMTs over MESFETs are summarized in Section 12.3.4. Figure 12.23 shows the equivalent circuit of a monolithic T-phase shifter employing voltage-variable DH HEMTs having large capacitance ratios (7:1) [52]. The diodes are shown as variable capacitors. The structure of a DH HEMT and its optimized parameters are shown in Figure 12.12. The diodes employ deposition of interdigitated ohmic and Schottky fingers on top of the n^+ GaAs cap and undoped AlGaAs layers, respectively. The diode offers a capacitance ratio of 7:1 for a voltage variation from about +0.8V to −2V. In the circuit shown in Figure 12.23, the shunt diode capacitance is chosen to be twice that of the series diode for optimum matching. The inductor in the shunt arm is used for matching at the center frequency. The T-phase shifter is reported to offer a phase shift of 55° with an insertion loss of 2 dB and a VSWR of less than 1.8 at 6 GHz [52].

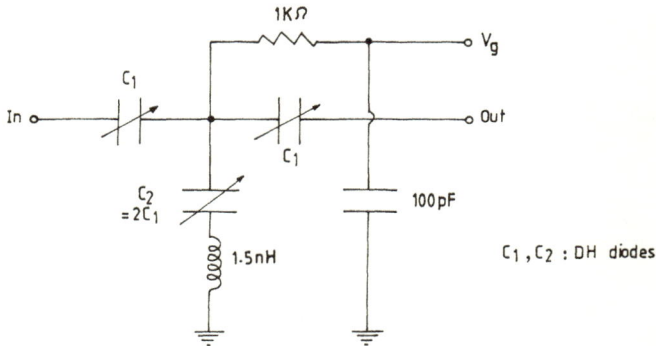

Figure 12.23 Schematic of analog T-phase shifter using double-heterojunction (DH) diodes. (From Cazaux *et al.* [52], first presented at the 18th European Microwave Conference, reprinted with permission.)

12.5 MONOLITHIC ACTIVE PHASE SHIFTERS

Active FET phase-shifter circuits that offer both phase and gain control are described in Chapter 10. This section reviews the monolithic active FET phase shifters. Recent development in optically controlled GaAs MMIC phase shifters is also reviewed.

12.5.1 Dual-Gate FET Phase Shifters

Monolithic digital active phase shifters have been realized using dual-gate GaAs FETs (DGFET) [67, 68]. Figure 12.24 shows the layout of a single-bit X-band active phase shifter and Table 12.2 presents its electrical and physical parameters as reported by Vorhaus, Pucel, and Tajima [67]. The circuit uses two DGFETs in a

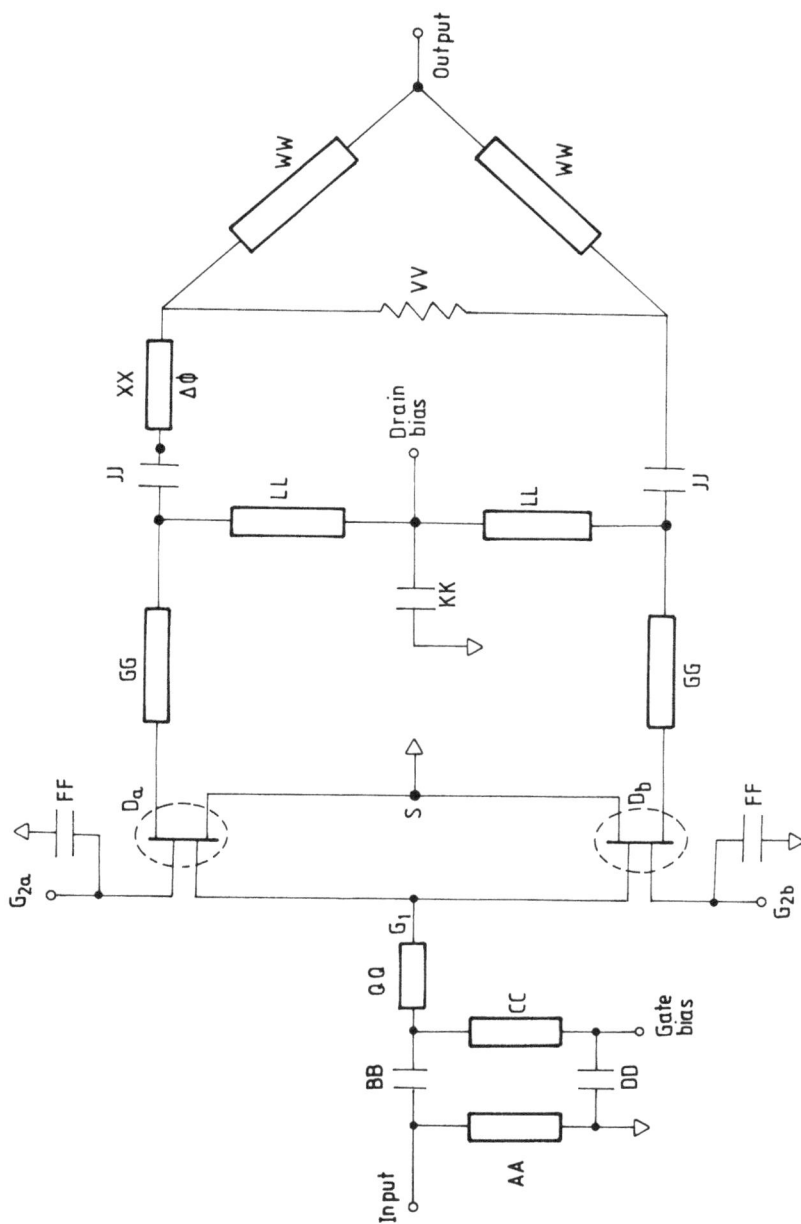

Figure 12.24 Active phase-shifter circuit layout. (From Vorhaus, Pucel, and Tajima [67], copyright © 1982 IEEE, reprinted with permission.)

Table 12.2
Active Phase-Shifter Matching Circuit Element Values
(From Vorhaus, Pucel, and Tajima [67], Copyright © 1982 IEEE, reprinted with permission.)

| | Transmission Lines | | | |
| | Electrical Characteristics | | Physical Dimensions | |
Element	Impedance (Ω)	Phase Length (deg)	Width (mm)	Length (mm)
AA	78.3	63.0	0.0220	2.006
CC	73.6	19.0	0.0270	0.603
GG	100.0	50.2	0.0086	1.616
LL	100.0	13.7	0.0086	0.440
QQ	59.0	9.0	0.0510	0.281
WW	70.7	90.0	0.0306	2.848
	50.0	22.5	0.0762	0.695
XX	50.0	90.0	0.0762	2.780
	50.0	180.0	0.0762	5.562

| Lumped Elements | |
Element	Value
BB	0.31 pF
DD	15 pF
FF	7 pF
JJ	0.50 pF
KK	20 pF
VV	100 Ω

single-pole double-throw (SPDT) configuration. The first gates (marked G_1) of the two FETs are connected together to form a common input terminal. The matching networks at the common input and at the two drain terminals transform the FET impedances to a 50Ω level. The FETs are operated in their on- or off-mode by switching the gate voltage from zero to pinch-off voltage. By selectively controlling the voltages at the second gates (G_{2a} and G_{2b}), one of the two FETs is turned on while the other is in the off-state, thus allowing the signal to pass through either the lower reference path or the upper path, which has an additional phase delay of $\Delta\phi$. The two output paths are combined into a common 50Ω output by means of a Wilkinson three-port combiner. Thus, switching between the two DGFETs results in a differential phase shift of $\Delta\phi$. The circuit includes bias input and dc bypass circuitry. The large capacitors FF (7 pF), DD (15 pF), and KK (20 pF) are used for RF bypass.

The smaller capacitors, *BB* (0.31 pF) and *JJ* (0.5 pF), form part of the matching networks. They also serve as dc blocks. The entire circuit (single bit) is realized in a chip $2.5 \times 3.0 \times 0.1$ mm^3 in size and incorporates thin-film Si_3N_4 MIM capacitors for RF tuning, dc blocking and bypassing, thin-film Ti resistors, air-bridge RF/dc cross-overs, via-hole front side grounding, and integral beam leads. Each phase-shifter bit is reported to offer a gain of about 3 dB over 10% bandwidth centered at 9.5 GHz.

12.5.2 Optically-Controlled Phase Shifters

Several investigators have reported the effect of direct optical illumination on the dc and microwave characteristics of GaAs MESFETs [69–72] and HEMTs [72]. Figure 12.25 shows the schematic of a MESFET illuminated by means of an optical fiber. The optical source is usually a semiconductor laser coupled to the fiber. Alternatively, the device can be illuminated by an LED. When light of photon energy equal to or greater than the semiconductor bandgap energy is absorbed, free charge carriers are produced within the active layer of the semiconductor. This causes photovoltaic effects in the Schottky barrier depletion region and photoconductive effects in the parasitic resistances in series with the active channel and in the substrate. These effects result in a change of basic device parameters such as the gate-to-source capacitance, transconductance, and the drain-to-source resistance [69]. Simons and Bhasin [72] have shown that the channel conductance and the gate-to-source capacitance increase, and the switching time decreases, with optical illumination. Furthermore,

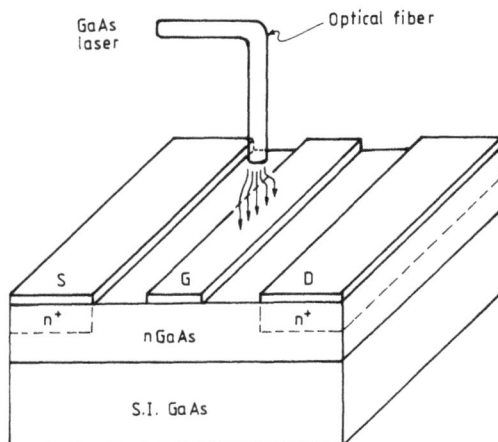

Figure 12.25 Optical illumination of MESFET.

$Al_{0.3}Ga_{0.7}As/GaAs$ HEMT has higher sensitivity to optical illumination than GaAs MESFET (with photon energy greater than the semiconductor bandgap).

For digital phase-shifter applications, optically-controlled GaAs MESFET can be used as a switch. With a reverse-bias voltage equal to the pinch-off voltage ($V_g = -V_p$), the MESFET is in the off-state. By illuminating the MESFET with sufficient light intensity, a voltage equal to $+V_p$ is generated at the Schottky gate that overcomes the pinch-off voltage and switches the MESFET to on-state.

Herczfeld et al. [73] have reported a different scheme for optical control of GaAs MMIC phase shifters. Instead of direct optical control, an optical detector is used to generate the control voltage for the phase shifter. The scheme of phase control is illustrated in Figure 12.26 [73]. The figure shows the schematic of an L-band transmit-receive module, of which the phase shifter is a part. The phase shifter is continuously variable and is implemented as a vector modulator consisting of three fixed bits to select a given 45° sector of operation, two quadrature amplifiers to produce continuous phase shift, and a final gain stage to control the overall gain of the phase shifter. The operation of the phase shifter is as follows: The optical sensing element, which is a multifinger MESFET, is first biased near pinch-off. When illuminated by an LED, it generates a voltage that is applied to the two operational amplifiers. One operational amplifier is in the inverting mode and the other is in the noninverting mode. Varying the light intensity gives rise to a change in drain-to-source voltage of the MESFET which, after amplification, provides the control voltage to the phase shifter. Phase shift is obtained by changing the bias voltages of the dual-gate FET amplifiers of the vector modulator. The outputs are combined in the power combiner to give the desired phase to the optically controlled variable gain amplifier.

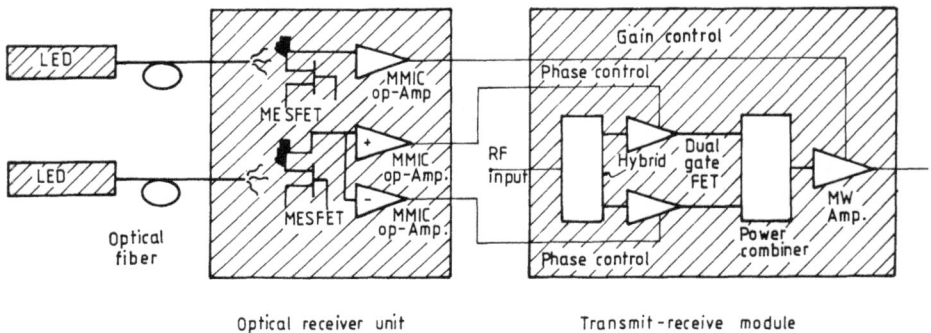

Figure 12.26 Optical control of a GaAs MMIC transmit-receive module. (From Herczfeld et al. [73], first presented at the 18th European Microwave Conference, reprinted with permission.)

Table 12.3

Performance Parameters of Monolithic Digital FET Phase Shifters

	Maloney et al. [74]	Andricos et al. [56]	Ayasli et al. [53]	Schindler et al. [75]	Schindler and Miller [58]	Pao et al. [57]
Frequency	L-band 1–2 GHz	C-band 5–6 GHz	X-band 8.5–11.5 GHz	C-Ku bands 6–18 GHz	K-Ka bands 18–40 GHz	V-band 59–63 GHz
Number of phase bits	5	6	4	3	3	4
Type of phase bit	High-pass Low-pass 180°; Loaded line 45°, 22.5°, 11.25°	Hybrid coupled 180°, 90°; Loaded line 45°, 22.5°, 11.25°; Variable 0–11° (see Figs. 12.16–12.18)	Switched line 180°, 90°; Loaded line 45°, 22.5° (see Figs. 12.14 and 12.15)	High-pass Low-pass 180°, 90°, 45°	High-pass Low-pass 180°, 90°, 45° (see Fig. 12.19)	Hybrid coupled 180°, 90°, 45°, 22.5°
Insertion loss	7.5 ± 1 dB	8.7 ± 1.2 dB	5.1 ± 0.6 dB	10.4 ± 1.1 dB	9.5 ± 2.5 dB	7 ± 2.5 dB
Return loss	10 dB min	15 dB min	10 dB min	13 dB min	10 dB min	—
Rms phase error at center freq.	±5°	±2°	±11.4°	±8.5°	±10°	—
Number of FETs	—	12	10	17	16	—
Total gate periphery of FETs	11.2 mm	25.2 mm	9.6 mm	4.9 mm	—	—
Size	4.95 × 3.45 mm²	9.43 × 4.2 mm²	6.4 × 7.9 mm²	2.21 × 1.25 mm²	2.1 × 1.3 mm²	2.3 × 1.5 mm²

Table 12.4
Performance Parameters of Monolithic Analog Varactor Phase Shifters

	Dawson et al. [41]	*Krafcsik et al. [64]*	*Chen et al. [45]*	*Pao et al. [57]*
Frequency	X-band 8–12.4 GHz	C-Ku bands 6–18 GHz	Ku-band 16–18 GHz	V-band 59–63 GHz
Circuit type	Lange coupler with varactor pairs back to back in each arm	Lange coupler with dual varactors in each arm (see Fig. 12.22)	Branchline hybrid with single varactor in each arm (see Fig. 12.21)	Hybrid branchline (two circuits in series)
Maximum phase shift	105°	160 ± 20°	109 ± 3° (360 ± 17° for 3 chips in series)	160° (variation <5°)
Insertion loss	2.5 ± 0.5 dB	2.7 ± 1.3 dB	1.8 ± 0.3 dB (4.2 ± 0.9 dB for 3 chips in series)	6 ± 2.5 dB
VSWR (max)	—	1.5	1.6	—
Input power level	10 dBm	—	20 dBm	—
Varactor bias range	0 to −10V	0 to −8V	0 to −16V	0 to −19V
Size	0.077 × 0.1 in^2 (two chips in series)	0.148 × 0.068 in^2	3.1 × 2.4 mm^2 (per chip)	1.5 × 2.7 mm^2

12.6 PERFORMANCE SUMMARY

Digital FET Phase Shifters

As in the case of *p-i-n* diode phase shifters, digital FET phase shifters make use of a cascade of binary phase bits and are controlled by the binary output logic of a digital computer. All four circuit techniques, namely, the switched line, hybrid coupled, loaded line, and high-pass low-pass, are adopted in FET phase shifters. As discussed in Chapter 8 (Section 8.8), the high-pass low-pass technique adopts switching between a high-pass filter and a low-pass filter to achieve the phase shift. The circuit provides constant phase shift over a wide bandwidth. It is particularly attractive for monolithic integration because of easy implementation of lumped inductors and capacitors.

Typical performance parameters reported for some of the practical monolithic digital FET phase shifters [53, 56–57, 74, 75] are listed in Table 12.3. An important feature of these phase shifters is the broad bandwidth [58, 75]. However, the insertion loss is high, typically in a range of about 5 to 12 dB from L- to V-bands. The switching time of GaAs FETs (<1 ns) is considerably less than that of *p-i-n* diodes and the dc power consumption is practically negligible. The peak power capability is also small, typically on the order of a few watts.

Analog Varactor Diode Phase Shifters

Table 12.4 lists the performance parameters of some of the practical varactor phase shifters reported up to V-band [41, 45, 57, 64]. All these phase shifters use a hybrid coupled reflection-type circuit. With the Lange coupler as the hybrid, bandwidth in excess of an octave has been achieved [64]. The insertion loss for a maximum phase shift variation of 360° is in the range of 3.5 to 10 dB for phase shifters reported at C-, X-, and Ku-bands [41, 45, 64]. At V-band, the loss is much higher: 9 to 19 dB for a 360° phase shifter [57].

REFERENCES

1. M.J. Howes and D. V. Morgan, *Gallium Arsenide Materials, Devices and Circuits,* John Wiley and Sons, New York, 1985.
2. R.E. Williams, *GaAs Processing Techniques,* Artech House, MA, 1984.
3. R.S. Pengelly, *Microwave Field Effect Transistors,* Research Studies Press, John Wiley and Sons, U.K., 1986.
4. R. Soares, *GaAs MESFET Circuit Design,* Artech House, MA, 1988.
5. R.A. Pucel (ed.), *Monolithic Microwave Integrated Circuits,* Selected Reprint Series, IEEE Press, New York, 1985.
6. H. Thomas *et al.* (eds.), *Gallium Arsenide for Devices and Integrated Circuits,* Proc. UWIST GaAs School, Peter Peregrinus, U.K., 1986.
7. J.V. Di Lorenzo and D.D. Khandelwal, *GaAs FET Principles and Technology,* Artech House, MA, 1982.
8. D. Laighton, J. Sasonoff, and J. Selin, "Silicon-on-Sapphire (SOS) Monolithic Transreceiver

Module Components for L- and S-Band," *Govt. Microcircuit Applications Conf., Digest Papers*, Vol. 8, 1980, pp. 299–302.

9. J.R. Knight, D. Effer, and P.R. Evans, "The Preparation of High Purity Gallium Arsenide by Vapour Phase Epitaxial Growth," *Solid State Electronics*, Vol. 8, 1965, pp. 178–180.

10. H.M. Hobgood *et al.*, "High Purity Semi-Insulating GaAs Material for Monolithic Microwave Integrated Circuits," *IEEE Trans. on Electron Devices*, Vol. ED-28, February 1981, pp. 140–149.

11. R.L. Van Tuyl *et al.*, "A Manufacturing Process for Analog and Digital Gallium Arsenide Integrated Circuits," *IEEE Trans. on Microwave Theory and Tech.*, Vol. MTT-30, July 1982, pp. 935–941.

12. W.R. Wisseman *et al.*, "GaAs Microwave Devices and Circuits with Sub-Micron Electron-Beam Defined Features," *Proc. IEEE*, Vol. 71, May 1983, pp. 667–675.

13. A. Gupta *et al.*, "Yield Considerations for Ion-Implanted GaAs MMICs," *IEEE Trans. on Electron Devices*, Vol. ED-30, January 1983, pp. 16–19.

14. A.Y. Cho *et al.*, "Low Noise and High Power GaAs Microwave Field Effect Transistors Prepared by Molecular Beam Epitaxy," *J. Applied Physics*, Vol. 48, 1977, pp. 346–349.

15. P.E. Luscher, "Crystal Growth by Molecular Beam Epitaxy," *Solid State Technology*, Vol. 20, December 1977, pp. 43–52.

16. H. Morkoc, J. Andrews, and V. Aeki, "GaAs MESFET Prepared by Organo-Metallic Chemical Vapour Deposition," *Electronics Letters*, Vol. 15, No. 4, February 1979, pp. 105–106.

17. S. Salimian, C.B. Cooper, and M.E. Day, "Dry Etching of Via Connections for GaAs MMIC Fabrication," *J. Vac. Sci. Technol.*, Vol. B5, No. 6, November/December 1987, pp. 1606.

18. D.C. A'Avanzo, "Proton Isolation for GaAs Integrated Circuits," *IEEE Trans. on Microwave Theory and Tech.*, Vol. MTT-30, July 1982, pp. 955–963.

19. A.K. Gupta, J.A. Higgins, and C.P. Lee, "High Electron Mobility Transistors for Millimeter Wave and High Speed Digital IC Applications," *Characterization of Very High Speed Semiconductor Devices and Integrated Circuits, SPIE*, Vol. 795, 1987, pp. 68–70.

20. F. Ali, I.J. Bahl, and A. Gupta (eds.), *Microwave and Millimeter Wave Heterostructure Transistors and Their Applications*, Artech House, MA, 1989.

21. R.A. Pucel, "Design Considerations for Monolithic Microwave Circuits," *IEEE Trans. on Microwave Theory and Tech.*, Vol. MTT-29, June 1981, pp. 513–534.

22. B. Bhat and S.K. Koul, *Stripline-Like Transmission Lines for Microwave Integrated Circuits*, Wiley Eastern Pvt. Ltd., India, 1989.

23. T.C. Edwards, *Foundation for Microstrip Circuit Design*, John Wiley and Sons, U.K., 1981.

24. K.C. Gupta, R. Garg, and I.J. Bahl, *Microstrip Lines and Slotlines*, Artech House, MA, 1979.

25. R.H. Jansen, "A Novel CAD Tool and Concept Compatible with the Requirements of Multilayer GaAs MMIC Technology," *IEEE Int. Microwave Symp. Digest*, 1985, pp. 711–714.

26. F.W. Grover, *Inductance Calculations*, Van Nostrand, NJ, 1946.

27. F.E. Terman, *Radio Engineering Handbook*, McGraw-Hill, 1943.

28. H.M. Greenhouse, "Design of Planar Rectangular Microelectronic Inductors," *IEEE Trans. on Parts, Hybrids and Packaging*, Vol. PHP-10, No. 2, June 1974, pp. 101–109.

29. D. Cahana, "A New Transmission Line Approach for Designing Spiral Microstrip Inductors for Microwave Integrated Circuits," *IEEE Int. Microwave Symp. Digest*, 1983, pp. 245–247.

30. M. Parisot *et al.*, "Highly Accurate Design of Spiral Inductors for MMICs with Small Size and High Cut-Off Frequency Characteristics," *IEEE Int. Microwave Symp. Digest*, 1984, pp. 106–111.

31. I. Wolff and H. Kapusta, "Modeling of Circular Spiral Inductors for MMICs," *IEEE Int. Microwave Symp. Digest*, 1987, pp. 123–126.

32. G.D. Alley, "Interdigital Capacitors and Their Applications to Lumped Element Microwave

Integrated Circuits," *IEEE Trans. on Microwave Theory and Tech.*, Vol. MTT-18, December 1970, pp. 102–103.

33. J.L. Hobdell, "Optimisation of Interdigital Capacitors," *IEEE Trans. on Microwave Theory and Tech.*, Vol. MTT-27, September 1979, pp. 788–791.

34. R. Esfandiari *et al.*, "Design of Interdigitated Capacitors and Their Application to GaAs Monolithic Filters," *IEEE Trans. on Microwave Theory and Tech.*, Vol. MTT-31, January 1983, pp. 57–64.

35. Y.C. Lim and R.A. Moore, "Properties of Alternately Charged Coplanar Strips by Conformal Mappings," *IEEE Trans. on Electron Devices*, Vol. ED-15, March 1968, pp. 173–180.

36. H.S. Veloric *et al.*, "Capacitors for Microwave Applications," *IEEE Trans. on Parts, Hybrids, and Packaging*, Vol. PHP-12, June 1976, pp. 83–89.

37. M.E. Elta *et al.*, "Tantalum Oxide Capacitors for GaAs Monolithic Integrated Circuits," *IEEE Electron Devices Letters*, Vol. EDL-3, May 1982, pp. 127–129.

38. M. Durschlag and J.L. Vorhaus, "A Tantalum Based Process for MMIC On-Chip Thin-Film Components," *IEEE GaAs IC Symp. Tech. Digest*, 1982, pp. 146–149.

39. D.R. Chen and D.R. Decker, "Monolithic Microwave Integrated Circuits (MMICs)—The Next Generation of Microwave Components," *Microwave J.*, Vol. 23, May 1980, pp. 67–78.

40. Y. Ayasli, "Microwave Switching with GaAs FETs," *Microwave J.*, Vol. 25, November 1982, pp. 61–74.

41. D.E. Dawson *et al.*, "An Analog X-Band Phase Shifter," *IEEE Microwave and Millimeter Wave Monolithic Circuits Symp. Digest*, 1984, pp. 6–10.

42. R.K. Mains, G.H. Haddard, and D.F. Peterson, "Investigation of Broadband, Linear Phase Shifters Using Optimum Varactor Diode Doping Profiles," *IEEE Trans. on Microwave Theory and Tech.*, Vol. MTT-29, November 1981, pp. 1158–1164.

43. R.V. Garver, "360 Degree Varactor Linear Phase Modulator," *IEEE Trans. on Microwave Theory and Tech.*, Vol. MTT-17, March 1969, pp. 137–147.

44. E.C. Niehenke, V.V. Dimarco, and A. Friedberg, "Linear Analog Hyperabrupt Varactor Diode Phase Shifters," *IEEE Int. Microwave Symp. Digest*, 1985, pp. 657–660.

45. C.L. Chen *et al.*, "A Low-Loss Ku-Band Monolithic Analog Phase Shifter," *IEEE Trans. on Microwave Theory and Tech.*, Vol. MTT-35, March 1987, pp. 315–320.

46. G.H. Nesbit *et al.*, "Monolithic Transmit/Receive Switch for Millimeter-Wave Application," *IEEE GaAs IC Symp. Digest*, October 1989, pp. 147–150.

47. K. Wilson and L.A. Hing, "Diode Based MMICs," *Military Microwaves Conf. Proc.*, July 1988, pp. 205–210.

48. K.H.G. Duh *et al.*, "Millimeter Wave Low Noise HEMT Amplifiers," *IEEE Int. Microwave Symp. Digest*, 1988, pp. 923–926.

49. K. Hikosaka *et al.*, "A 30 GHz-1W Power HEMT," *IEEE Electron Devices Letters*, Vol. EDL-8, November 1987, pp. 521–523.

50. E. Sovero *et al.*, "35 GHz Performance of Single and Quadrupole Channel Power Heterojunction HEMTs," *IEEE Trans. on Electron Devices*, Vol. ED-33, October 1986, pp. 1434–1438.

51. J.L. Cazaux *et al.*, "An Analytical Approach to the Capacitance Voltage Characteristics of Double-Heterojunction HEMTs," *IEEE Trans. on Electron Devices*, Vol. ED-35, August 1988, pp. 1223–1231.

52. J.L. Cazaux *et al.*, "The Use of Double Heterojunction Diodes in Monolithic Phase Shifters," *18th Eur. Microwave Conf. Digest*, 1988, pp. 1005–1010.

53. Y. Ayasli *et al.*, " A Monolithic Single Chip X-Band Four Bit Phase Shifter," *IEEE Trans. on Microwave Theory and Tech.*, Vol. MTT-30, December 1982, pp. 2201–2206.

54. V. Sokolov *et al.*, "A Ka-Band GaAs Monolithic Phase Shifter," *IEEE Trans. on Microwave Theory and Tech.*, Vol. MTT-31, December 1983, pp. 1077–1083.

55. A.A. Lane and F.A. Myers, "MMIC Phase Shifters," *Military Microwave Conf. Digest,* 1988, pp. 211–216.

56. C. Andricos, I.J. Bahl, and E.L. Griffin, "C-Band 6 Bit GaAs Monolithic Phase Shifter," *IEEE Trans. on Microwave Theory and Tech.,* Vol. MTT-33, December 1985, pp. 1591–1596.

57. C.K. Pao *et al.,* "V-Band Monolithic Phase Shifters," *10th Annual IEEE GaAs IC Symp. Tech. Digest,* November 1988, pp. 269–272.

58. M.J. Schindler and M.E. Miller, "A 3-Bit K/Ka Band MMIC Phase Shifter," *IEEE Microwave and Millimeter Wave Monolithic Circuits Symp. Digest,* 1988, pp. 95–98.

59. Y. Ayasli *et al.,* "Wideband Monolithic Phase Shifter," *IEEE Trans. on Microwave Theory and Tech.,* Vol. MTT-32, December 1984, pp. 1710–1714.

60. Y. Ayasli *et al.,* "Wideband S-C Band Monolithic Phase Shifter," *IEEE Microwave and Millimeter Wave Monolithic Circuits Symp. Digest,* 1984, pp. 11–13.

61. A.W. Jacomb-Hood, D. Seielstad, and J.D. Merrill, "A Three Bit Monolithic Phase Shifter at V-Band," *IEEE Microwave and Millimeter Wave Monolithic Circuits Symp. Digest,* 1987, pp. 81–84.

62. K. Wilson, J.M.C. Nicholas, and G. McDermoth, "A Novel MMIC Phase Shifter," *IEEE Microwave and Millimeter Wave Monolithic Circuits Symp. Digest,* 1985.

63. K. Wilson *et al.,* "A Novel MMIC X-Band Phase Shifter," *IEEE Trans. on Microwave Theory and Tech.,* Vol. MTT-33, December 1985, pp. 1572–1578.

64. D.M. Krafcsik *et al.,* "A Dual-Varactor Analog Phase Shifter Operating at 6 to 18 GHz," *IEEE Trans. on Microwave Theory and Tech.,* Vol. MTT-36, December 1988, pp. 1938–1941.

65. A. Chu *et al.,* "Monolithic Analog Phase Shifters and Frequency Multipliers for mm-Wave Phased Array Applications," *Microwave J.,* Vol. 29, December 1986, pp. 105–112.

66. L.C.T. Liu *et al.,* "A 30 GHz Monolithic Receiver," *IEEE Trans. on Microwave Theory and Tech.,* Vol. MTT-34, December 1986, pp. 1548–1552.

67. J.L. Vorhaus, R.A. Pucel, and Y. Tajima, "Monolithic Dual-Gate GaAs FET Digital Phase Shifter," *IEEE Trans. on Microwave Theory and Tech.,* Vol. MTT-30, July 1982, pp. 982–992.

68. R.A. Pucel *et al.,* "A Multi-Chip GaAs Monolithic Transmit/Receive Module for X-Band," *IEEE Int. Microwave Symp. Digest,* 1982, pp. 489–492.

69. A.A.A. Desalles, "Optical Control of GaAs MESFETs," *IEEE Trans. on Microwave Theory and Tech.,* Vol. MTT-31, October 1983, pp. 812–820.

70. J.L. Gautier, D. Pasquet, and P. Pouvil, "Optical Effects on the Static and Dynamic Characteristics of a GaAs MESFET," *IEEE Trans. on Microwave Theory and Tech.,* Vol. MTT-33, September 1985, pp. 819–822.

71. H. Mizuno, "Microwave Characteristics of an Optically Controlled GaAs MESFET," *IEEE Trans. on Microwave Theory and Tech.,* Vol. MTT-31, July 1983, pp. 596–600.

72. R.N. Simons and K.B. Bhasin, "Analysis of Optically Controlled Microwave/Millimeter Wave Device Structures," *IEEE Trans. on Microwave Theory and Tech.,* Vol. MTT-34, December 1986, pp. 1349–1355.

73. P.R. Herczfeld *et al.,* "Optical Phase and Gain Control of a GaAs MMIC Transmit-Receive Module," *18th Eur. Microwave Conf. Digest,* 1988, pp. 831–836.

74. P.R. Maloney, M.A. Mezger, and J.P. Sasonoff, "L-Band GaAs Transceiver Components," *IEEE GaAs IC Symp. Digest,* 1985, pp. 121–124.

75. M.J. Schindler *et al.,* "Monolithic 6 to 18 GHz 3 Bit Phase Shifter," *IEEE GaAs IC Symp. Digest,* 1985, pp. 129–132.

Chapter 13
Millimeter-Wave Phase Shifters

13.1 INTRODUCTION

Chapters 4–7 of Volume I present a detailed discussion on a wide variety of ferrite phase shifters, and Chapters 8–12 of this volume cover the semiconductor device phase shifters. It is apparent that while significant advances have taken place in the techniques and technology of both these categories of phase shifters, their practical realization has been confined mostly to microwave frequencies. Only a few specific phase-shifter types have been extended to millimeter-wave frequencies. For example, among the various ferrite phase shifters, the nonreciprocal twin-toroid type and the reciprocal dual-mode type have been demonstrated up to 94 GHz. The details of these phase shifters are presented in Chapters 4 and 5 of the companion volume. Among the semiconductor device phase shifters, the monolithic GaAs FET and varactor diode phase shifters have been reported up to V-band. The characteristics of these phase shifters are described in Chapter 12.

In this chapter, we discuss the general limitations of ferrite and semiconductor device phase shifters for operation at millimeter-wave frequencies. Other special phase-shifting techniques that are suitable particularly for millimeter-wave frequency range are described.

13.2 FERRITE PHASE SHIFTERS

At very high frequencies, away from the gyromagnetic resonance ($\omega >> \omega_0$), the elements μ and κ of the permeability tensor of ferrite can be approximated as $\mu \simeq \mu_0$, and $\kappa \simeq \mu_0 \omega_m / \omega$. Since the value of κ is inversely proportional to the frequency, for a material with a given saturation magnetization ($4\pi M_s$), the achievable differential phase shift per unit length decreases with an increase in frequency. Therefore, the higher the operating frequency is, the larger the saturation magnetization of the ferrite must be.

Ferrite phase shifters designed for normal operation satisfy the condition—namely, $\gamma 4\pi M_s / \omega \simeq 0.25$ to 0.6, or $4\pi M_s$ (kG)$/f$(GHz) $\simeq 0.1$ to 0.2. Accordingly, for operation at a typical frequency of 50 GHz, the ferrite material is required to have a $4\pi M_s$ value within the range of 5000 G to 10,000 G, and above 50 GHz, its value must be even higher. Commercially available high-magnetization spinel ferrites provide $4\pi M_s$ values on the order of 5000 G with square hysteresis loop and good dieletric properties. It is worthwhile to note that with the choice of ferrite having $4\pi M_s$ = 5000 G, if the operating frequency is increased beyond 50 GHz, the ratio $\gamma 4\pi M_s/\omega$ would fall below the minimum specified limit, with a consequent increase in conductor and dielectric loss in the device. The nonavailability of ferrite material with sufficiently high saturation magnetization is therefore a major factor in limiting the upper frequency of operation. Another important factor, which is perhaps even more crucial, is the difficulty in manufacturing very small size devices at the highest frequencies.

As described in Chapters 4 and 5 of the companion volume, the twin-toroid and the dual-mode phase shifters are two specific ferrite devices that have found potential for extension to millimeter-wave frequencies. The base loss levels of both these phase shifters at millimeter-wave frequencies are compared in Figure 5.40 [1], and their practical performance characteristics are compared in Table 5.4 [2] (see Volume I). It may be noted that at 60 GHz, the twin-toroid cross section is 0.058 × 0.038 in.², the dual-mode structure has an outer diameter of 0.25 in., and both offer an insertion loss of about 1.5 dB. The upper practical frequency limit for both these devices is reported to be 94 GHz [1, 2].

13.3 SEMICONDUCTOR DEVICE PHASE SHIFTERS

Semiconductor device hybrid microwave integrated circuit (MIC) phase shifters using discrete *p-i-n* diodes, metal semiconductor field-effect transistors (MESFET), and varactors are described in Chapters 9, 10, and 11, respectively. Hybrid MIC phase shifters are commonly realized in planar transmission line media of which the microstrip is the most popular. The microstrip permits easy mounting of active devices with the ground plane serving as a natural heat sink for heat dissipation. These phase shifters are quite popular at microwave frequencies. Above the microwave frequency range, the insertion loss increases beyond acceptable levels ($\gtrsim 2.5$ dB), thereby precluding their use in practice. This increase in insertion loss occurs due to the increase in propagation loss (primarily conductor loss) in the microstrip line as well as the semiconductor device loss. Furthermore, the need to define narrow linewidths and meet stringent dimensional tolerances increases fabricational complexity. It may be noted that there are other transmission structures—namely, the suspended stripline and the fin-line, which offer considerable fabricational ease and have much lower loss compared to the microstrip at millimeter-wave frequencies. However, because these structures have air gaps between the planar surfaces of the substrate and the metallic enclosure, mounting of active devices, particularly shunt mounting, and heat dissipation pose difficult problems. While these transmission lines are popularly used for realizing a variety of passive millimeter-wave components, including mixers and modulators up to about 120 GHz, they are not the preferred media for semiconductor device phase shifters.

Compared to the hybrid MIC, the monolithic technique has better potential for realizing millimeter-wave phase shifters. The inherent advantages of monolithic technology and the recent advances in monolithic phase shifters are reviewed in Chapter 12. Monolithic phase shifters are realized on GaAs substrates. The active devices commonly used are the FETs for digital phase shifting and varactors for analog phase shifting. Since these devices are grown *in situ* on the GaAs substrate, the problem of mounting and bonding and the associated parasitics commonly encountered in hybrid MICs are eliminated. Also, since GaAs has a much higher electron mobility than other semiconductors, the FETs and varactors fabricated in GaAs have lower loss and higher cut-off frequency, thereby enabling higher frequency of operation. As discussed in Chapter 12, monolithic FET and varactor diode phase shifters are reported up to V-band, but the insertion loss of these millimeter-wave devices with the present state of the art is too high (>6 dB) to be of practical use. Considerable scope exists, however, for improving the GaAs device fabrication techniques towards reducing the insertion loss and extending the frequency of operation.

13.4 DIELECTRIC-PLASMA GUIDE PHASE SHIFTERS

Techniques of realizing millimeter-wave phase shifters using high-purity semiconductors as the dielectric guide structure have been reported by several investigators [3–6]. The basic mechanism involves changing the phase of the RF propagating wave by creating free carriers (plasma) into the guide and controlling its dielectric constant. This technique of using semiconductor bulk phenomena for achieving phase shift is an attractive alternative to the expensive microstrip-based hybrid MIC or monolithic circuit techniques.

Two methods have been reported for the generation of plasma in a semiconductor dielectric waveguide: one requiring the injection of electrons into the semiconducting medium via contacts [3, 4], and the other through optical illumination [5, 6]. Both these methods are discussed below.

Contact Injection Technique

An example of a millimeter-wave phase shifter using the contact injection technique from Jacobs and Chrepta [3] is illustrated in Figure 1.12 of Chapter 1 of the companion volume. The device makes use of plasma injection into a silicon dielectric guide by means of a *p-i-n* diode structure placed on its broad wall. When a forward bias is applied to the *p-i-n* diode, mobile charge carriers are injected into the I-region.

The portion of the I-region at the upper edge of the triangular piece is set into conduction first, and as the current is increased, the conductivity increases in the downward direction. This results in a change in the effective height of the dielectric guide with a consequent decrease in β_y and an increase in β_z. The parameters β_y and β_z are the propagation constants of the dielectric guide in the y- and z-directions, respectively (refer to Figure 1.12 in Volume I). If l is the length of the p-i-n diode structure in the z-direction, and β_z and β_z' are the propagation constants corresponding to zero bias and a forward bias, respectively, the differential phase shift $\Delta\phi$ is given by $(\beta_z - \beta_z')l$. The experimental results of this phase shifter [3], as well as the measured propagation parameters of other plasma injection devices [7–8], however, show high losses (>6 dB) because of excess metallization for contact to the junction area. Secondly, large phase shifts have not been achieved, the reason for which is attributed to excess heating of the waveguide.

Optical Injection Technique

The problem of high loss encountered in the contact injection method is circumvented in the optical injection technique by creating a surface layer plasma rather than a bulk plasma. Optical control also offers other advantages, which include (1) near perfect isolation between the controlling and controlled devices, (2) immunity from electromagnetic interference, (3) ultrafast response (~1 ns), and (4) high power-handling capability.

Figure 13.1(a) shows the schematic of an optically controlled semiconductor guide phase shifter [5]. The semiconductor guide can be made of Si, GaAs, silicon-on-sapphire, or silicon-on-alumina. The ends of the guide are tapered to provide efficient transition to standard rectangular metal waveguides at the input and output ports. The guide is illuminated on its broad surface with above-bandgap radiation from a laser beam. The absorption of photon energy creates electron-hole pairs in a region adjacent to the air-semiconductor interface, thereby forming a surface plasma layer. The initial depth of plasma injection can be controlled by appropriately choosing the wavelength of optical radiation and the absorption properties of the semiconductor guide material. For example, the absorption coefficient of GaAs for a photon energy of 1.55 eV is approximately 10^4 cm^{-1}. The plasma layer thickness corresponding to $1/e$ absorption depth for light is less than 1 μm. Varying the optical illumination intensity changes the plasma density and, hence, the refractive index of the plasma-occupied volume of the dielectric guide. The optically induced phase shift $\Delta\phi$ over the section of guide of length l is given by $(\beta_z - \beta_z')l$, where β_z and β_z' are the propagation constants in the guide with and without the plasma, respectively.

The phase shift and attenuation characteristics of the optically controlled phase shifter have been studied by modeling the device as a dielectric-plasma waveguide [5, 6, 9]. Both transverse electric (TE) and transverse magnetic (TM) mode solutions have been reported. The model analyzed by Vaucher, Striffler, and Lee [6] is a

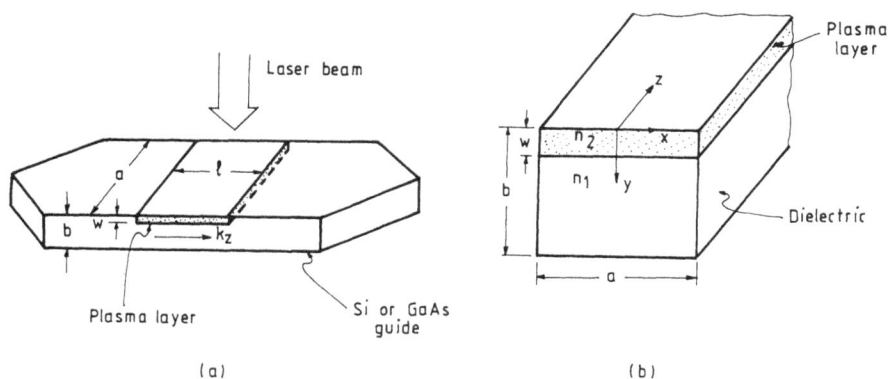

Figure 13.1 (a) Schematic of optically controlled phase shifter; (b) dielectric-plasma waveguide model. (After Lee, Mak, and DeFonza [5].)

rectangular dielectric guide with a thin surface plasma layer of uniform carrier concentration as shown in Figure 13.1(b). The plasma layer has a refractive index n_2 that is higher than the refractive index n_1 in the remaining volume of the guide. In the model considered by Butler, Wu, and Scott [9], the guiding structure extends to infinity in the lateral direction (x-direction in Figure 13.1(b)) and the plasma layer is assumed to have a nonuniform carrier concentration. The nonuniform distribution corresponding to an exponentially absorbed optical beam is chosen. Figure 13.2 shows the theoretical phase shift and attenuation characteristics for a TE mode in a silicon guide with an exponential plasma layer [9]. The plasma density profile is represented as $\bar{N}(y) = Ne^{-y/w}d$, where N is the plasma density at the surface ($y = 0$). Plots are shown as a function of N for three different values of the decay factor w_d. The frequency is 94 GHz. For low values of plasma density for which the plasma frequency is less than the frequency of the propagating millimeter wave, the plasma layer does not have much effect on the propagation. As the plasma density increases with an increase in the optical illumination intensity, the interaction between the wave and the plasma becomes stronger. In the density range where the plasma frequency crosses the operating frequency, the attenuation is large and the phase shift increases rapidly. As the density is increased further, the attenuation decreases, but the phase shift saturates. At large densities for which the skin depth of millimeter waves approaches the thickness of the plasma layer, the plasma region behaves like a metallic layer. The dielectric waveguide behaves essentially like an image guide. It can be seen from Figure 13.2 that the phase shifter can offer large phase shifts with low attenuation loss. Lee, Mak, and DeFonzo [10] and Lee [11] have reported measured phase shifts as high as 300°/cm in a Si waveguide having a cross section of 2.4×1.0 mm^2 and 1400°/cm in a GaAs waveguide having a cross section of 2.4×0.5 mm^2 at a frequency of 94 GHz.

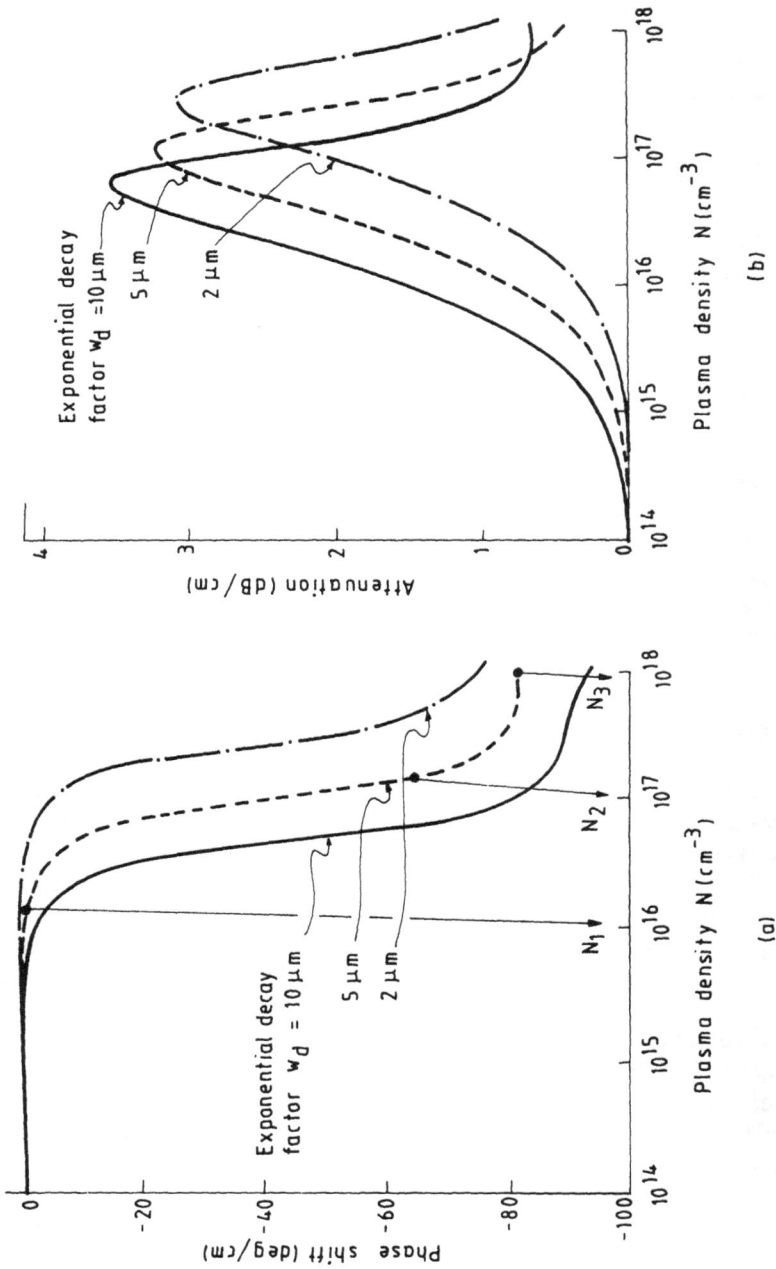

Figure 13.2 The propagation characteristics for a TE mode in a silicon waveguide with an exponential plasma layer of various decay constants: (a) phase shift and (b) attenuation. (From Butler, Wu, and Scott [9], copyright © 1986 IEEE, reprinted with permission.)

13.5 SEMICONDUCTOR-SLAB-LOADED WAVEGUIDE PHASE SHIFTER

The optically controlled dieletric guide millimeter-wave phase shifter shown in Figure 13.1(a) is an open structure. The feasibility of using an alternate closed structure in the form of a semiconductor-slab-loaded rectangular waveguide for millimeter-wave phase shifting has been investigated by Hadjicostas, Scott, and Butler [12] and

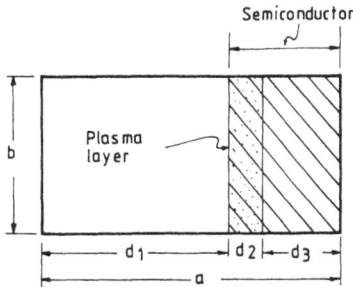

Figure 13.3 Rectangular waveguide with a semiconductor slab showing surface plasma layer created by optical injection. (After Hadjicostas, Scott, and Butler [12].)

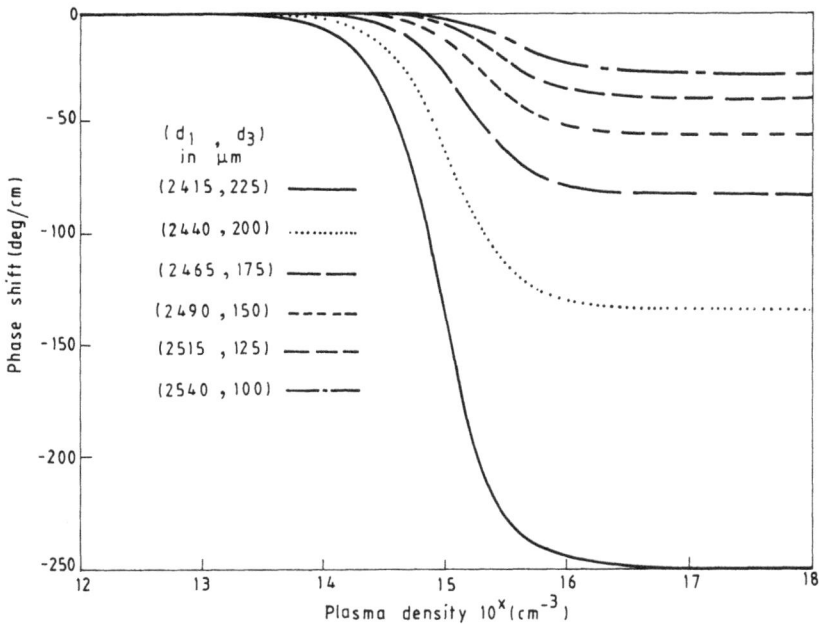

Figure 13.4 Phase shift *versus* plasma density of the guide shown in Figure 13.3. $a = 2.65$ mm, $b = 1.25$ mm, $d_2 = 10$ μm, semiconductor: GaAs. (From Wu, Butler, and Scott [13], copyright © IEEE 1987, reprinted with permission.)

Wu [13]. Figure 13.3 shows the cross-sectional view of the guide structure. A plasma layer is created on the surface of the semiconductor slab by optical injection through a small aperture in the left wall of the guide. The plasma region thus formed produces a change in the refractive index, thereby changing the propagation constant of the guide. Wu, Butler, and Scott [13] have analyzed the structure by assuming uniform charge density in the plasma layer. In Figure 13.3, this layer is shown to be of thickness d_2. Figures 13.4 and 13.5 illustrate the phase shift and attenuation characteristics of the guide for the dominant LSE (longitudinal section electric) mode at 94 GHz [13]. The waveguide cross section is 2.65×1.27 mm^2 and the semiconductor material is GaAs. Plots are shown as a function of plasma density for different values of GaAs slab thickness $(d_2 + d_3)$ by keeping the plasma layer thickness constant $(d_2 = 10 \ \mu m)$. It may be noted that the nature of variation is similar to that shown in Figure 13.2 for the optically controlled open dielectric guide. For high plasma densities at which the plasma frequency is much higher than the operating frequency, the millimeter-wave fields are concentrated essentially in the air region of the waveguide. The phase shift saturates to a high value and the attenuation is reduced substantially. It can also be seen from Figure 13.4 that high phase shifts are

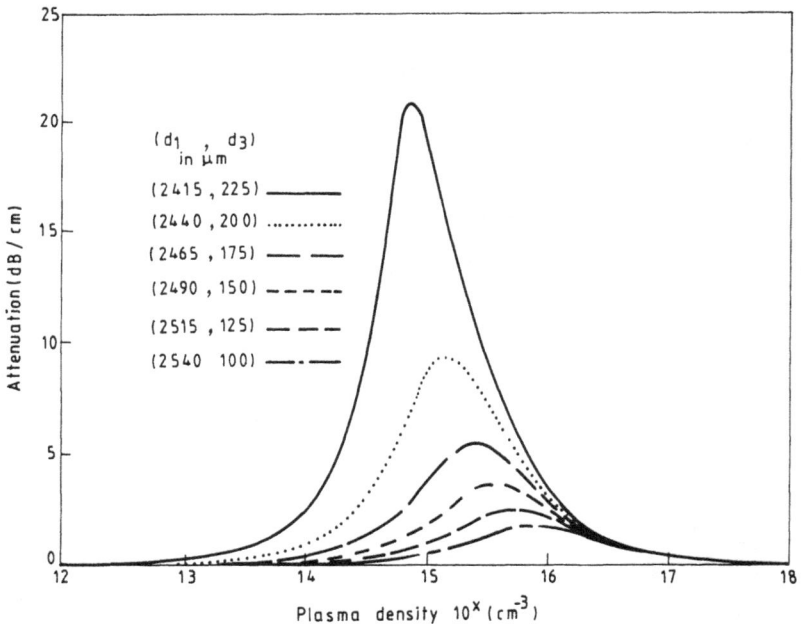

Figure 13.5 Attenuation characteristics of the guide shown in Figure 13.3. $a = 2.65$ mm, $b = 1.25$ mm, $d_2 = 10 \ \mu m$, semiconductor: GaAs. (From Wu, Butler, and Scott [13], copyright © IEEE 1987, reprinted with permission.)

attained when the slab thickness is increased to about 10% of the waveguide width $(d_2 + d_3 \simeq 0.1a)$.

13.6 ACTIVE PHASE SHIFTER USING VARACTOR-CONTROLLED GUNN

Cohen [14] has demonstrated analog active phase shifting using a varactor-controlled injection-locked Gunn oscillator at microwave as well as millimeter-wave frequencies. The phase shifter is basically a widely tunable Gunn VCO. Figure 13.6 shows the circuit schematic using lumped elements. The VCO is first varactor tuned to the desired operating frequency and injection locked to a stable frequency source. Any subsequent change in the varactor voltage changes the phase of the output signal relative to the locking signal. Thus, if $E \sin(\omega_L t)$ is the locking signal, the output of the varactor-controlled VCO is given by $E \sin(\omega_L t + \phi(V))$, where $\phi(V)$ is the voltage dependent phase. For changing the operating frequency, the VCO is to be unlocked, varactor tuned to a new frequency, and relocked. The phase of the output signal then follows the varactor voltage. The maximum theoretical phase change in a single-stage phase shifter of this type is reported to be $\pm 90°$. Higher phase shift can be obtained by feeding the output signal to a frequency multiplier. If $E \sin(\omega_L t + \phi(V))$ is the input to a frequency multiplier of order N, the output is given by $E \sin(N\omega_L t + N\phi(V))$. The phase gets multiplied by a factor N.

Figure 13.7 shows the measured phase shift *versus* varactor bias voltage for a millimeter-wave active phase shifter having a locking gain of 22 dB and a power output of 160 mW. The graph is reproduced from Cohen [14]. Plots are shown for three different injection-locked frequencies: 46.75, 47, and 47.25 GHz. The phase shift range is approximately 160° and the phase-voltage response is nearly linear. By combining this phase shifter with a frequency doubler, Cohen [14] has reported continuously variable active phase shifting up to 320° at 94 GHz with 40 mW power output.

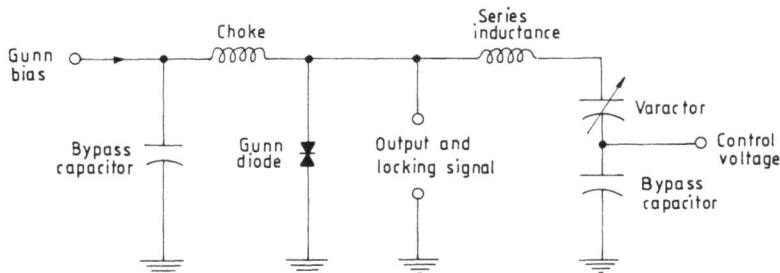

Figure 13.6 Schematic of a varactor-controlled Gunn VCO as active phase shifter. (From Cohen [14], copyright © IEEE 1984, reprinted with permission.)

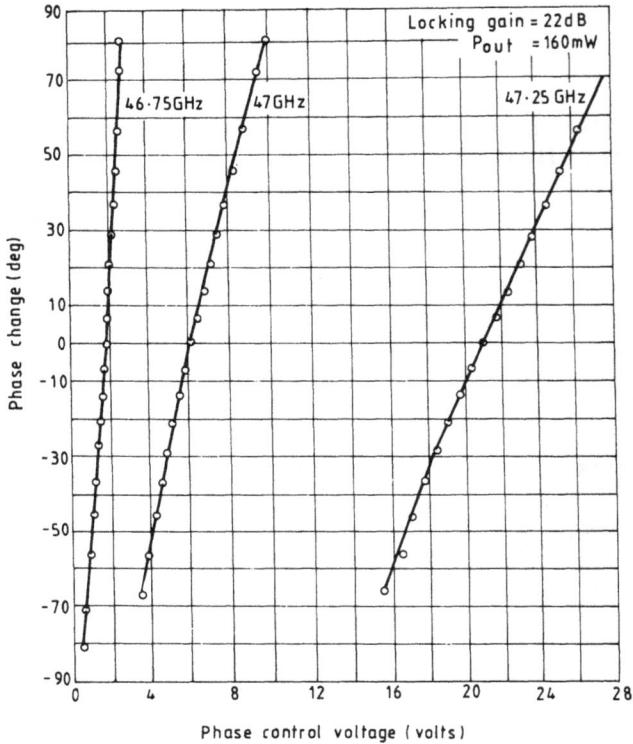

Figure 13.7 Measured phase change *versus* varactor control voltage for an active Gunn phase shifter. (From Cohen [14], copyright © IEEE 1984, reprinted with permission.)

13.7 SUSPENDED SLOT AND MICROSTRIP TRANSMISSION LINES WITH MAGNETIZED SEMICONDUCTOR SUBSTRATES

Among the variants of the basic slotline and the microstrip, the suspended slotline and the suspended microstrip are the most suitable for millimeter-wave integrated circuits. By employing magnetized semiconductor substrates in these structures, Krowne, Mostafa, and Zaki [15] have carried out a full-wave spectral domain analysis and studied the nonreciprocal propagation characteristics at millimeter-wave frequencies. Figure 13.8 shows the geometry of a typical slotline structure investigated by these authors. The slotline employs a dielectric substrate of thickness h_3 and relative dielectric constant ε_{rd}, and a semiconductor substrate of thickness h_2. The dc magnetic field is assumed to be oriented perpendicular to the direction of propagation, but is inclined at an angle ϕ with respect to the plane of the substrate. In the

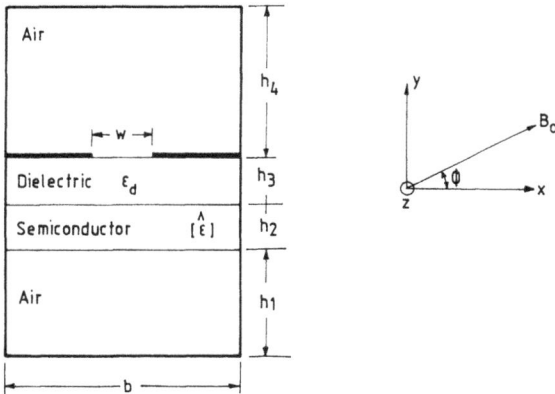

Figure 13.8 Cross section of a suspended slotline. Magnetic field bias B_0 is at inclination angle ϕ to the planar surface.

$$\varepsilon_d = \varepsilon_0 \varepsilon_{rd}, \quad [\hat{\varepsilon}] = \varepsilon_0 \varepsilon_{rs}[U] + \frac{1}{j\omega}[\rho]^{-1}$$

(From Krowne, Mostafa, and Zaki [15], copyright © IEEE 1988, reprinted with permission.)

magnetized semiconductor layer, Maxwell's equations can be written as

$$\nabla \times \vec{E} = -j\omega\mu_0\vec{H} \qquad (13.1)$$

$$\nabla \times \vec{H} = j\omega[\hat{\varepsilon}]\vec{E} \qquad (13.2)$$

where $[\hat{\varepsilon}]$ is the complex permittivity tensor of the semiconductor and is given by

$$[\hat{\varepsilon}] = \varepsilon[U] + \frac{1}{j\omega}[\rho]^{-1} \qquad (13.3)$$

$$\varepsilon = \varepsilon_0 \varepsilon_{rs} \qquad (13.4)$$

The parameters ε and $[\rho]$ denote the static permittivity and the macroscopic resistivity tensor, respectively, of the semiconductor, and $[U]$ is a unit matrix. The expression for $[\rho]$ is given by [15]

$$[\rho] = \begin{bmatrix} j\omega + \tau_p^{-1} & 0 & \omega_c \sin\phi \\ 0 & j\omega + \tau_p^{-1} & -\omega_c \cos\phi \\ -\omega_c \sin\phi & \omega_c \cos\phi & j\omega + \tau_p^{-1} \end{bmatrix} \frac{1}{\varepsilon\omega_p^2} \qquad (13.5)$$

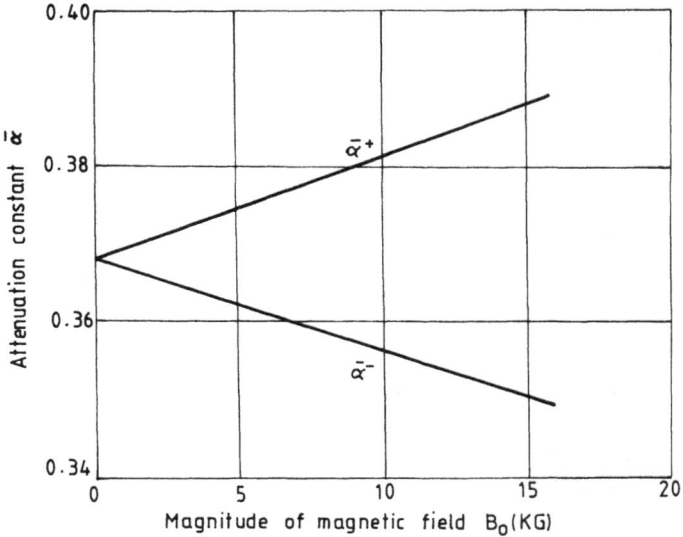

(a)

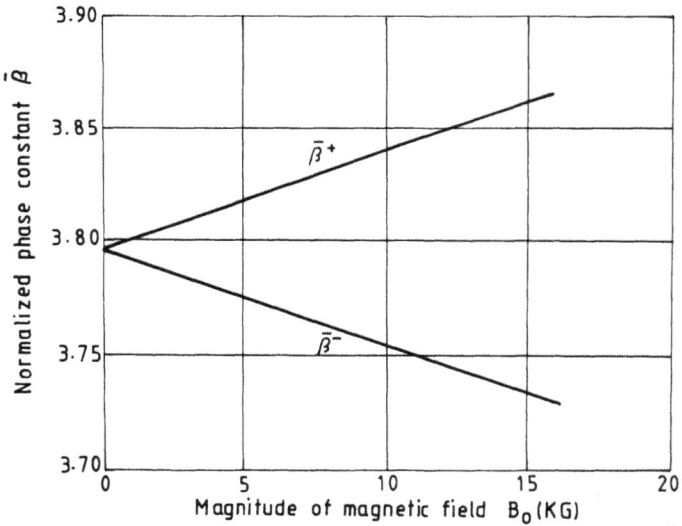

(b)

Figure 13.9 (a) Variation of $\bar{\alpha}^+$ and $\bar{\alpha}^-$ with magnetic field B_0; (b) Variation of $\bar{\beta}^+$ and $\bar{\beta}^-$ with magnetic field B_0. Structure at Figure 13.8: $b = 2.35$ mm, $h_1 = h_4 = 2.1$ mm, $h_2 = 0.25$ mm, $h_3 = 0$, $w = 1.0$ mm, $\varepsilon_{rd} = \varepsilon_{rs} = 12.5$, freq. $= 75$ GHz; $\phi = 0$, $\omega_p = 6.28 \times 10^{12}$ rad/s ($n_0 = 10^{16}$/cc), $\tau_p = 10^{-13}$ sec. (From Krowne, Mostafa, and Zaki [15], copyright © IEEE 1988, reprinted with permission.)

where ω_p, ω_c, and τ_p denote the angular plasma frequency, angular cyclotron frequency, and momentum relaxation time, respectively, of the semiconductor plasma. The expressions for ω_p and ω_c are given by

$$\omega_p = \frac{q^2 n_0}{m^* \varepsilon} \tag{13.6}$$

$$\omega_c = \frac{q B_0}{m^*} \tag{13.7}$$

where q, m^*, n_0 are the electron charge, its effective mass, and electron density, respectively, and B_0 is the applied dc magnetic field. Because of the anisotropy in the semiconductor layer and the field displacement effect [16], the structure exhibits different propagation constants for the forward ($+z$) and reverse ($-z$) directions of propagation. If β^+ and α^+ denote the phase constant and attenuation constant, respectively, for the wave propagating in the $+z$ direction, and β^- and α^- denote the corresponding quantities for the wave propagating in the $-z$ direction, then the difference in phase constants given by $\Delta\beta = \beta^+ - \beta^-$ and the difference in attenuation constants given by $\Delta\alpha = \alpha^+ - \alpha^-$ are both functions of B_0 and the angle ϕ. It has been shown by Krowne, Mostafa, and Zaki [15] that both $\Delta\beta$ and $\Delta\alpha$ decrease monotonically as ϕ is increased from $0°$ (B_0 along x-direction) to $90°$ (B_0 along y-direction).

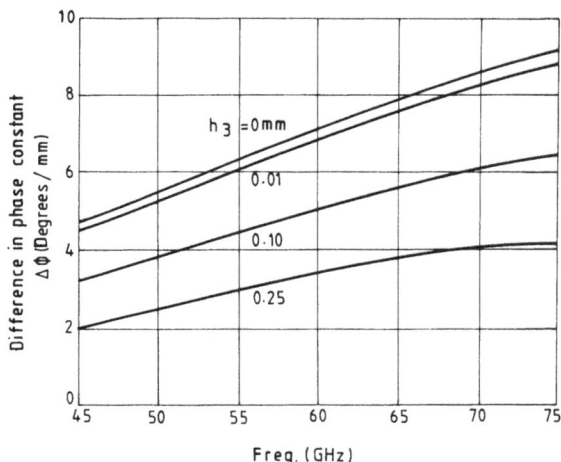

Figure 13.10 Dispersion behavior of $\Delta\beta = \beta^+ - \beta^-$ with h_3 as a parameter. Structure at Figure 13.8: $b = 2.35$ mm, $h_1 = h_4 = 2.1$ mm, $h_2 = 0.25$ mm, $w = 1.0$ mm, $\varepsilon_{rd} = \varepsilon_{rs} = 12.5$, $\phi = 0$, $\omega_p = 6.28 \times 10^{12}$ rad/s ($n_0 = 10^{16}$/cc), $\tau_p = 10^{-13}$ sec, $\omega_c = 3.14 \times 10^{12}$ rad/s ($B_0 = 12$ kG). (From Krowne, Mostafa, and Zaki [15], copyright © IEEE 1988, reprinted with permission.)

Thus, maximum phase change $\Delta\beta$ is achieved when the magnetic field is oriented parallel to the substrate, but perpendicular to the direction of propagation. Figure 13.9 illustrates typical variation in the normalized attenuation constants $\bar{\alpha}^+ = \alpha^+/\beta_0$ and $\bar{\alpha}^- = \alpha^-/\beta_0$, and the normalized phase constants $\bar{\beta}^+ = \beta^+/\beta_0$ and $\bar{\beta}^- = \beta^-/\beta_0$ of the slotline structure (Figure 13.8 with $h_3 = 0$ and $\phi = 0°$) as a function of the magnetic field B_0. The various fixed parameters are indicated in the figure. It can be seen that both $\Delta\beta$ and $\Delta\alpha$ increase with an increase in the strength of the magnetic field B_0. Figure 13.10 shows variation in $\Delta\beta$ as a function of frequency from 45 to 75 GHz, with B_0 fixed at 12 kG. Plots are shown for four different values of dielectric substrate thickness ($h_3 = 0$, 0.01, 0.1, and 0.25 mm). It can be seen that $\Delta\beta$ increases with a decrease in h_3 and reaches a maximum when $h_3 = 0$. This is an expected trend because of increased interaction between the slot fields and the magnetoplasma as h_3 is reduced.

REFERENCES

1.   C.R. Boyd, "A 60 GHz Dual-Mode Ferrite Phase Shifter," *IEEE Int. Microwave Symp. Digest*, 1982, pp. 257–259.

2.   W.E. Hord, "Microwave and Millimeter Wave Ferrite Phase Shifters," *Microwave J.*, Vol. 32, supplement to September 1989 issue, pp. 81–89.

3.   H. Jacobs and M.M. Chrepta, "Electronic Phase Shifter for Millimeter Wave Semiconductor Dielectric Integrated Circuits," *IEEE Trans. on Microwave Theory and Tech.*, Vol. MTT-22, April 1974, pp. 411–417.

4.   B.J. Levin and G.G. Weidner, "Millimeter Wave Phase Shifter," RCA Review, Vol. 34, 1973, pp. 489–505.

5.   C.H. Lee, P.S. Mak, and A.P. DeFonza, "Optical Control of Millimeter Wave Propagation in Dielectric Waveguides," *IEEE J. Quantum Electron.* Vol. QE-16, pp. 277–288, March 1980.

6.   A.M. Vaucher, C.D. Striffler, and C.H. Lee, "Theory of Optically Controlled Millimeter-Wave Phase Shifters," *IEEE Trans. on Microwave Theory and Tech.*, Vol. MTT-31, February 1983, pp. 209–216.

7.   B.J. Levin and G. Weidner, "A Distributed *p-i-n* Diode Phaser for Millimeter Wavelengths," *IEEE Int. Microwave Symp. Digest*, 1973, pp. 63–65.

8.   G.R. Vanier and R.M. Mindock, "Diode Structures for a Millimeter Wave Phase Shifter," *IEEE Int. Microwave Symp. Digest*, 1975, pp. 173–175.

9.   J.K. Butler, T.F. Wu, and M.W. Scott, "Non-Uniform Layer Model of a Millimeter-Wave Phase Shifter," *IEEE Trans. on Microwave Theory and Tech.*, Vol. MTT-34, January 1986, pp. 147–155.

10.   C.H. Lee, P.S. Mak, and A.P. DeFonzo, "Millimeter-Wave Switching by Optically Generated Plasma in Silicon," *Electronics Letters*, Vol. 14, 1978, pp. 733–734.

11.   C.H. Lee, "Optical Generation and Control of Microwaves and Millimeter Waves," *IEEE Int. Microwave Symp. Digest*, 1987, pp. 811–814.

12.   G. Hadjicostas, M.W. Scott, and J.K. Butler, "Optically Controlled Millimeter Wave Phase Shifter in a Metallic Waveguide," *IEEE Int. Microwave Symp. Digest*, 1987, pp. 657–660.

13.   T.F. Wu, J.K. Butler, and M.W. Scott, "Characteristics of Metallic Waveguides Inhomogeneously Filled With Dielectric Materials With Surface Plasma Layers," *IEEE Trans. on Microwave Theory and Tech.*, Vol. MTT-35, July 1987, pp. 609–614.

14.   L.D. Cohen, "Active Phase Shifters for the Millimeter and Microwave Bands," *IEEE Int. Microwave Symp. Digest,* 1984, pp. 397–399.
15.   C.M. Krowne, A.A. Mostafa, and K.A. Zaki, "Slot and Microstrip Guiding Structures Using Magnetoplasmons for Non-Reciprocal Millimeter-Wave Propagation," *IEEE Trans. on Microwave Theory and Tech.,* Vol. MTT-36, December 1988, pp. 1850–1860.
16.   B.R. McLeod and W.G. May, "A 35 GHz Isolator Using a Coaxial Solid-State Plasma in a Longitudinal Magnetic Field," *IEEE Trans. on Microwave Theory and Tech.,* Vol. MTT-19, June 1971, pp. 510–516.

Chapter 14
Surface Acoustic Wave and Magnetostatic Wave Delay-Line Phase Shifters

14.1 INTRODUCTION

Microwave acoustic delay lines based on *surface acoustic wave* (SAW) and *magnetostatic wave* (MSW) technologies offer phase shift in the form of true time delay. SAW technology is well developed and is widely applied to a variety of components, including nondispersive and dispersive delay lines. The literature is so extensive that several review articles and special issues are devoted to this subject [1–7]. The bulk of the literature, however, pertains to applications at low frequencies up to VHF. In the microwave frequency range, SAW technology is practical up to about 2.5 GHz. Compared to SAW technology, MSW technology is relatively new. It holds enormous potential for use in the microwave frequency range from 1 to about 20 GHz [8–11]. Both SAW and MSW devices are compatible with planar technology and both can be used as tapped delay lines. In terms of electronic tunability, SAW delay lines have limited tuning capability, when there is variation in applied voltage. On the other hand, MSW delay lines can be easily tuned by an externally applied bias magnetic field.

In this chapter, we review the principles and key operating features of SAW and MSW delay lines, with special emphasis on their use at microwave frequencies.

14.2 BASICS OF SURFACE ACOUSTIC WAVES

14.2.1 Piezoelectricity [12]

Surface acoustic wave technology makes use of the phenomenon of coupling between acoustic and electric fields that occurs in piezoelectric crystals. Piezoelectric crystals possess an asymmetry in the arrangement of atomic dipoles such that straining the material causes additional polarization charges to appear. These polarization charges give rise to electric fields. The material can be subjected to stress patterns by the application of a periodic field to a spatially periodic electrode printed on the plane surface of the substrate. Piezoelectric crystals essentially offer a convenient medium for coupling electric fields to acoustic fields and vice versa. The constitutive relations for the medium are given by

$$[T] = [t][S] - [e]^T[E] \tag{14.1}$$

$$[D] = [e][S] + [\varepsilon][E] \tag{14.2}$$

where $[T]$, $[S]$, $[D]$, and $[E]$ denote the stress, strain, electric displacement, and electric field, respectively. The parameters $[t]$, $[e]$, and $[\varepsilon]$ represent the stiffness constant (elasticity tensor), piezoelectric constant, and dielectric constant, respectively, of the material. These are, in general, tensor quantities. The superscript T indicates transposition of the matrix $[e]$. It may be noted that D and E are vectors, each of which have three components; T and S are vectors, with six independent components each; and $[t]$, $[e]$, and $[\varepsilon]$ are matrices on the order of 6×6, 3×6, and 3×3 respectively.

14.2.2 Properties of Surface Acoustic Waves [1–7, 12–15]

Surface acoustic waves, also known as *Raleigh waves,* propagate along the surface of a piezoelectric solid and decay with depth into the material. Most of the energy is confined within a depth of about one wavelength. Figure 14.1(a) shows a simple illustration of a surface wave in a semi-infinite piezoelectric solid. The wave is uniform in the x-direction and propagates in the z-direction according to $e^{j(\omega t - kz)}$. The surface wave comprises a compressional wave and a shear wave, and is accompanied by an electrostatic wave. Figure 14.1(b) shows particle displacements u_z and u_y corresponding to compressional and shear motions as a function of z at the surface. Both u_y and u_z decay as a function of y in the slab. The electrostatic field generated by the acoustic field can be described in terms of an electric potential ϕ. The electric field $\vec{E}$ can be determined from the potential using the relation

$$\vec{E} = -\nabla \phi \tag{14.3}$$

Piezoelectric Coupling Coefficient

Suppose a thin conducting layer is placed on the surface of a piezoelectric solid carrying surface waves. The potential ϕ produced because of the surface waves induces charges in the conducting layer. Using a transmission line analogy, this is equivalent to adding a capacitance, thus causing a reduction in the surface wave

(a)

(b)

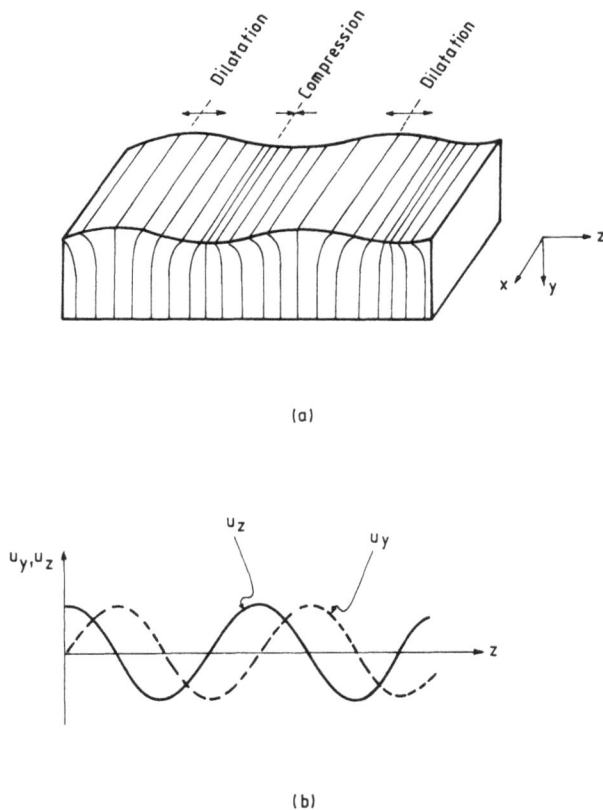

Figure 14.1 (a) Illustration of a surface wave in a semi-infinite piezoelectric solid; (b) particle displacements u_y and u_z as a function of z at the surface.

velocity v_s. The effective piezoelectric coupling coefficient K is defined in terms of this change in surface wave velocity as [1]

$$K^2 = 2 \left| \frac{\Delta v_s}{v_s} \right| \tag{14.4}$$

The coupling coefficient K is a significant parameter because it determines the efficiency of the transducer in converting the electrical signal to acoustic signal in practical SAW devices.

Piezoelectric Substrates

The commonly used substrates for SAW delay lines are lithium niobate (LiNbO$_3$), quartz, and bismuth germanium oxide (Bi$_{12}$GeO$_{12}$). Because of acoustic anisotropy, these materials exhibit beam steering effect. That is, the surface wave velocity depends on the direction of propagation, and the direction of power flow, in general, is not perpendicular to the phase front, but makes an angle α with respect to this perpendicular direction. This angle α is known as the beam steering angle. It is given by

$$\alpha = \frac{1}{v_s} \frac{dv_s}{d\theta} \qquad (14.5)$$

where θ is the angle between the direction of propagation and the z-axis. For piezoelectric crystals, therefore, it is common to specify the plane of the crystal surface and the desirable direction of propagation. For SAW devices, the angle θ is chosen at which the velocity is either a maximum or minimum (so that $\alpha = 0$). Another factor influencing the cut and orientation is the requirement for strong piezoelectric coupling. For example, in a YZ LiNbO$_3$ substrate, the effective coupling constant K and the surface wave velocity are both maximum in the z-direction ($\alpha = 0$ at $\theta = 0$). This is the preferred cut for most SAW devices. The terminology YZ is used to imply that the exposed surface of the LiNbO$_3$ substrate is in a plane normal to the y-axis, and the direction of surface wave propagation is in the z-direction.

Apart from the above main requirements, there are other considerations that may be significant for certain applications. Table 14.1 summarizes the preferred cuts and orientations, and also the basic SAW properties of various piezoelectric substrates as compiled from the available literature [3, 12, 14]. It may be noted that YZ LiNbO$_3$ has high coupling efficiency compared to other materials. On the other hand, ST-X quartz offers excellent temperature stability, although it is weak in piezoelectric coupling. Bi$_{12}$GeO$_{12}$ offers very low surface wave velocity, which is useful if long delays are required. Aluminum nitride film (AlN) on sapphire substrate has higher surface wave velocity compared to other materials. This property is useful for realizing devices at higher frequencies by enabling relaxed dimensional tolerances for the transducers.

Surface Wave Velocity and Attenuation

An important parameter of a SAW delay line is the surface acoustic wave velocity. In piezoelectric materials, the surface wave velocity is less than the slowest bulk wave velocity by about 10%, and is less than the velocity of electromagnetic waves in free space by about five orders of magnitude. As listed in Table 14.1, the surface

Table 14.1

Surface Acoustic Wave Properties of Piezoelectric Materials for Delay Lines [3, 12, 14]

Material	Cut	Propagation Direction	Velocity v_s m/s (approx.)	K^2 % (approx.)	Attenuation at 1 GHz $(dB/\mu s)$ (approx.)	Temp. Coeff. $ppm/°C$ (approx.)	Remarks
LiNbO$_3$	Y	Z	3.48×10^3	4.3	1.6	85	Efficient coupling, wide bandwidth
Quartz	ST	X	3.15×10^3	0.17	—	<3	Temperature stable
Bi$_{12}$GeO$_{20}$	(100)	(011)	1.68×10^3	1.2	1.5	−122	Long delay
LiTaO$_3$	Z	Y	3.33×10^3	0.93	—	67	—
GaAs	(100)	(011)	2.86×10^3	0.07	—	—	Compatible with MMIC
AlN/Al$_2$O$_3$ (film thickness 0.1 wavelength)	X	Z	6.17×10^3	0.63	—	—	High-frequency application

wave velocity is typically in the range of 2.0×10^3 to 6.0×10^3 m/s. Thus, several microseconds of delay is achieved over a crystal length of a few centimeters.

The attenuation of surface waves is higher than that of bulk waves and increases with an increase in frequency. Figure 14.2 illustrates typical variation in the attenuation of surface waves as a function of frequency for YZ-LiNbO$_3$ substrate.

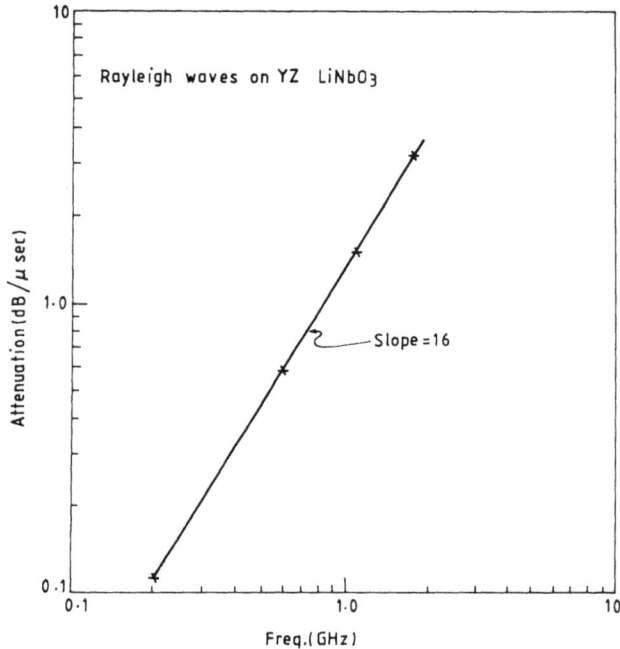

Figure 14.2 Attenuation of Rayleigh waves on a YZ LiNbO$_3$ substrate as a function of frequency. (From Lean and Broers [13], reprinted with permission of *Microwave Journal*, copyright © 1970 Horizon House—Microwave Inc.)

14.3 SAW DELAY LINES

14.3.1 Interdigital Transducer [1, 14–24]

Of the various techniques available for generating and detecting surface acoustic waves [1], the interdigital transducer (IDT) technique has become the most popular for SAW devices [16–20]. The IDT consists of a series of interleaved, thin conducting electrodes deposited on a piezoelectric surface. Figure 14.3 shows the schematic of uniform IDT, in which the width w of the electrodes is equal to the interelectrode gap width s. The structure has a periodicity given by $p = 2(s + w)$. When

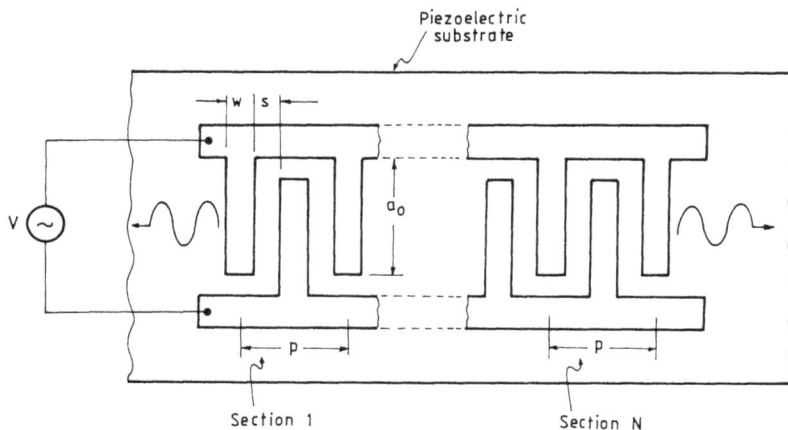

Figure 14.3 Schematic diagram of interdigital transducer.

a sinusoidal voltage signal is applied to the transducer, a spatially periodic electric field is generated by the electrodes. This causes a strain pattern of periodicity p in the piezoelectric crystal. If the frequency of the applied signal is such that the wavelength λ_s of the surface wave is equal to p, then there is strong coupling to acoustic surface wave energy. By symmetry, these surface waves are launched in both directions.

Analysis

The performance of transducers is vital for determining the response of SAW delay lines. Extensive theoretical studies have been reported on IDT using different theoretical models [18–24]. Here we follow the simple cross-field model [18] and summarize the essential characteristics of the transducer. The transducer is divided into N identical sections, each of the length p, as shown in Figure 14.3. The total number of electrodes is $(2N + 1)$. At the synchronous frequency f_0, the length p is equal to the wavelength of surface waves. If C_s is the static capacitance per section, then the total capacitance of the IDT is $C_T = NC_s$. The electrical equivalent circuit of the IDT can be represented as a series or parallel combination of C_T and radiation immittance representing acoustic wave excitation. Figure 14.4 shows the series and shunt representations for the IDT input immittance. These equivalent circuits are applicable up to S-band frequencies. The expression for the input impedance $Z(f)$ is given by

$$Z(f) = R_a(f) + jX_a(f) + (1/j\omega C_T)$$

$$= [G_a(f) + jB_a(f) + j\omega C_T]^{-1}, \quad \omega = 2\pi f \qquad (14.6)$$

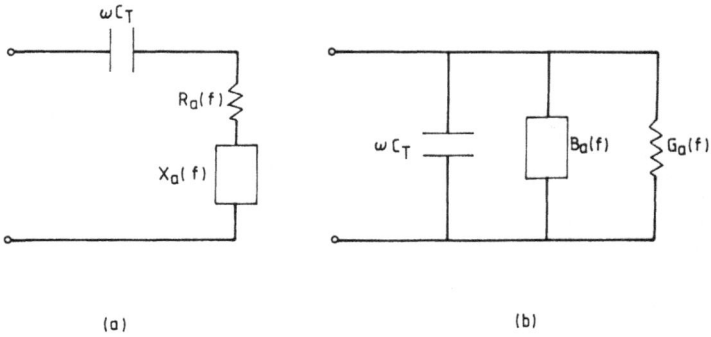

Figure 14.4 Equivalent circuit representation of IDT: (a) series circuit and (b) parallel circuit.

From the crossed-field model, the expressions for $G_a(f)$ and $B_a(f)$ are given by

$$G_a(f) = 2G_0 \left[\tan\left(\frac{\theta}{4}\right) \sin\left(\frac{N\theta}{2}\right) \right]^2 \qquad (14.7a)$$

$$B_a(f) = G_0 \tan\left(\frac{\theta}{4}\right) \left[4N + \tan\left(\frac{\theta}{4}\right) \sin(N\theta) \right] \qquad (14.7b)$$

where

$$G_0 = \frac{\omega_0 C_s K^2}{2} \qquad (14.8a)$$

$$\theta = 2\pi \left(\frac{\omega}{\omega_0}\right) \qquad (14.8b)$$

$$\omega_0 = 2\pi f_0 = \frac{2\pi v_s}{L} \qquad (14.8c)$$

For frequencies near acoustic synchronism, the expressions for $G_a(f)$ and $B_a(f)$ in (14.7) are given approximately by [18]

$$G_a(f) \simeq \hat{G}_a \left(\frac{\sin x}{x}\right)^2 \qquad (14.9a)$$

$$B_a(f) \simeq \hat{G}_a \left(\frac{\sin 2x - 2x}{2x^2} \right) \tag{14.9b}$$

$$x = N\pi \frac{(\omega - \omega_0)}{\omega_0} \tag{14.10}$$

These approximate relations are reported to be accurate within 10% for $|(\omega - \omega_0)/\omega_0| < 0.2$. At the synchronous frequency ω_0, $G_a(f_0) = \hat{G}_a$ and $B_a(f_0) = 0$. $\hat{G}_a$ is the maximum value of $G_a(f)$. It is known as the synchronous conductance and is given by

$$\hat{G}_a = \frac{4}{\pi} K^2 (\omega_0 C_s) N^2 \tag{14.11}$$

It can be seen that $\hat{G}_a$ is proportional to the aperture a_0 (length of the electrode) and N^2, and its value is much smaller than $\omega_0 C_T$. If ohmic losses are neglected, then the power dissipated in the resistive part of the equivalent circuit accounts for radiated acoustic power. Figure 14.5 shows typical variation of $G_a(f)$ and $B_a(f)$ as a function of frequency for an IDT having fifteen electrodes ($N = 15$) on a LiNbO$_3$ substrate [18]. The bandwidth and conversion loss of the IDT can be obtained from the plot of $G_a(f)$ versus frequency.

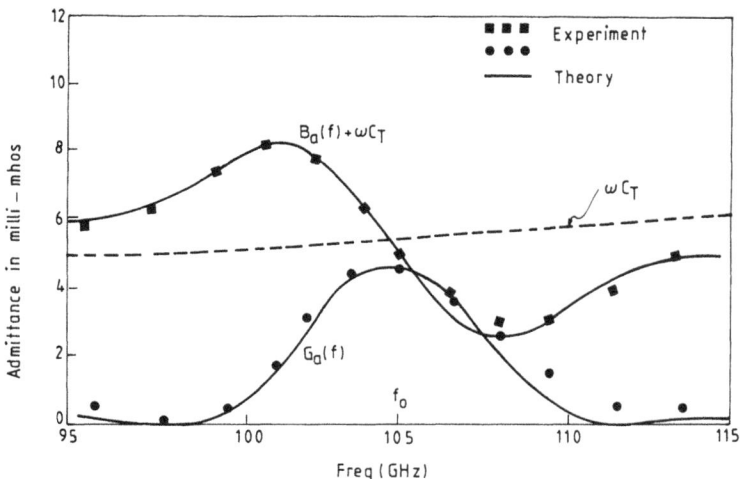

Figure 14.5 Measured radiation admittance for an $N = 15$ transducer on YZ LiNbO$_3$ compared with theoretical curves calculated from the "crossed-field" model. (From Smith *et al.* [18], copyright © 1969 IEEE, reprinted with permission.)

The conversion loss of the IDT can be minimized by resonating the capacitance C_T with an inductor at f_0. Figure 14.6 shows the equivalent circuit of an IDT connected to a generator having a resistance R_g through a matching network which, in this case, consists of a simple inductor. For the range $K^2 N \ll 1$, which normally holds in practice, we obtain from (14.6), (14.7), and (14.11):

$$R_a(f_0) \simeq \hat{R}_a = \frac{\hat{G}_a}{(\omega_0 C_s N)^2} \tag{14.12}$$

It may be noted that $\hat{R}_a$ is independent of N but varies inversely as a_0.

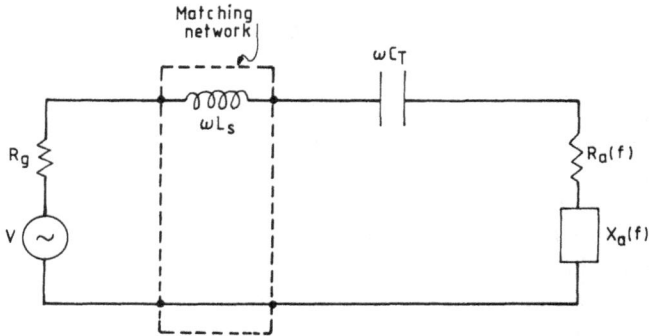

Figure 14.6 Equivalent circuit of IDT with a tuning inductor.

With an appropriate choice of a_0, $\hat{R}_a$ can be made equal to the generator resistance R_g. Thus, at $f = f_0$, all the available electric energy is converted to acoustic energy. However, since the energy flow is bidirectional, only one-half of the radiated energy is directed towards the output transducer.

Bandwidth

The bandwidth of the circuit with series tuning is a function of N and K^2. For large values of N, the fractional bandwidth $(\Delta f / f_0)$ is approximately equal to $1/N$, and for small values of N, it is governed by the electrical Q of the series resonant circuit. The maximum bandwidth is found to occur for an intermediate optimum value of N given by [14]

$$N_{\text{opt}} \simeq \frac{\sqrt{\pi}}{2K} \tag{14.13}$$

The maximum fractional bandwidth is then given by

$$\left(\frac{\Delta f}{f_0}\right)_{max} \simeq \frac{1}{N_{opt}} \tag{14.14}$$

Figure 14.7 shows the bandwidth as a function of N for a series-tuned transducer with reference to 1.5-dB points on the conversion loss curve. It can be seen that an IDT on LiNbO$_3$ can offer bandwidth in excess of 20%, which is much higher than that offered by quartz ($\simeq$5%).

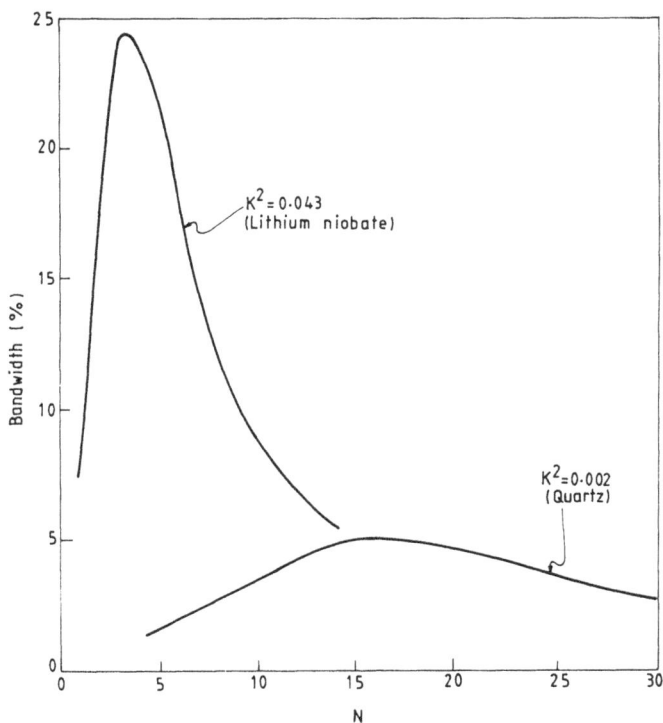

Figure 14.7 Bandwidth of a series-tuned transducer to the 1.5-dB points of the conversion loss (crossed-field model, $R_g = \hat{R}_a$). (From Morgan [14], copyright © 1973 Ultrasonics, reprinted with permission.)

Frequency Response

When a voltage impulse is applied to an IDT, the mechanical deformation that is produced in the substrate corresponds to the electrode geometry. The time (impulse) response that propagates parallel to the transducer axis bears the image of the electrode structure. For example, for a uniform IDT, the impulse "image" is sinusoidal with the period and length of the transducer. The frequency response $H(f)$ can be obtained from the impulse response $h(t)$ by taking its Fourier transform:

$$H(f) = \int_{-\infty}^{\infty} h(t)e^{-j2\pi ft}dt \qquad (14.15)$$

In order to find the frequency response of a uniform IDT, we may consider the structure as having $(N + 1)$ taps (N periodic sections) and equal weights separated by a delay of T between successive taps. The impulse response $h_0(t)$ of an equally weighted N-tap structure is shown in Figure 14.8(a). This can be considered as a continuous response $h(t)$ over the period NT (Figure 14.8(b)) sampled at frequency intervals $f_0 = 1/T$. The frequency response is obtained from (14.15) by setting $h(t)$

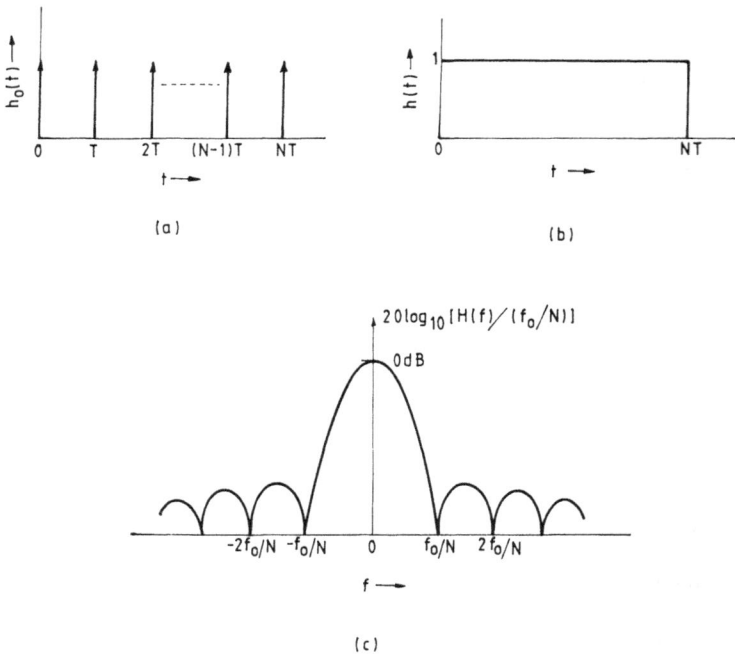

Figure 14.8 IDT as an equal weighted $(N + 1)$ tap structure: (a) impulse response, (b) corresponding continuous response, and (c) frequency response.

= 1 and integrating over the period 0 to NT. Thus

$$H(f) = \int_0^{NT} e^{-j2\pi ft} dt$$

$$= e^{-j\pi Nf/f_0} \left(\frac{N}{f_0}\right) \frac{\sin(\pi Nf/f_0)}{(\pi Nf/f_0)} \tag{14.16}$$

where

$$f_0 = \frac{1}{T} \tag{14.17}$$

The relative magnitude of $H(f)$ in dB is shown as a function of frequency in Figure 14.8(c). It may be mentioned that a symmetric impulse response $h(t)$ gives rise to a frequency response $H(f)$ whose phase varies linearly with frequency. Since the impulse response considered above is symmetric, about $t = NT/2$, the phase of $H(f)$ varies linearly with frequency with phase reversal at points given by integral multiples of (f_0/N). The 3-dB bandwidth is given by

$$(\Delta f)_{3\ dB} = \frac{0.9 f_0}{N} \tag{14.18}$$

The actual impulse response of a uniform IDT is a truncated sinusoid. Figure 14.9 shows the impulse response of a five-electrode ($N = 2$) structure and its frequency response.

The impulse response and hence the frequency response of an IDT can be controlled by employing a nonuniform geometry, that is, by varying the aperture a_0

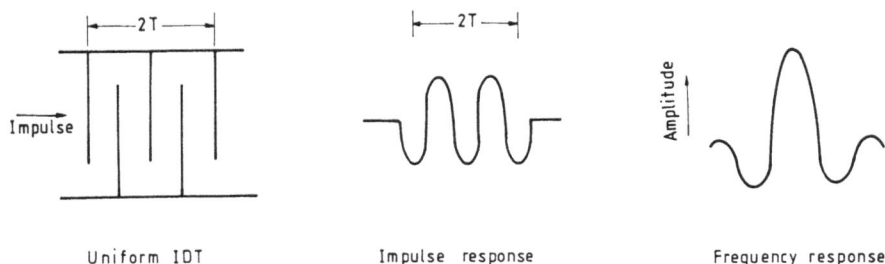

Figure 14.9 Impulse and frequency response of a uniform IDT.

or the periodicity p or both, along the structure. Nonuniform IDTs have been described extensively in the literature [19–24]. These have been used as the basis of dispersive delay lines [25].

14.3.2 Characteristics of SAW Delay Lines

A delay line basically consists of two IDTs separated by a distance L on a piezoelectric substrate, as illustrated in Figure 14.10. Surface waves generated by the input IDT travel in both directions. Waves propagating to the right reach the output IDT after a delay where they are detected by the inverse piezoelectric effect. Waves reaching the two sides of the substrate are absorbed by means of acoustic absorbers. Because of bidirectionality, each IDT introduces an inherent loss of 3 dB. Thus, the minimum theoretical insertion loss of a delay line is 6 dB. The time delay is given by $t = L/v_s$, where v_s is the surface wave velocity.

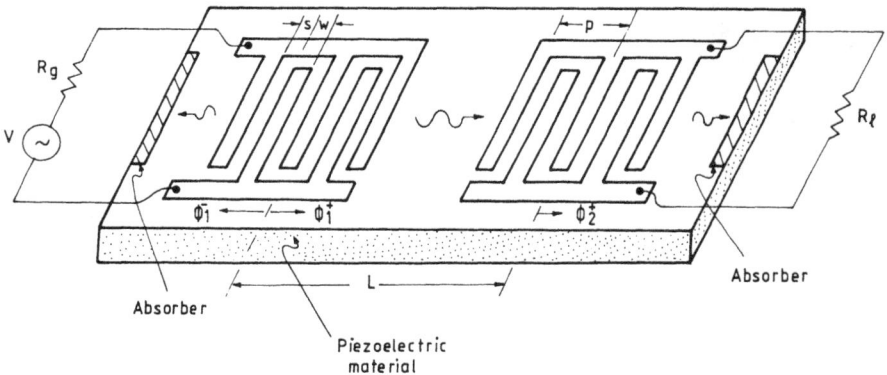

Figure 14.10 Schematic of SAW delay line.

The overall frequency response of a SAW delay line is the product of the frequency responses of the input and output IDTs modified by the interface circuitry. The propagation region between the two transducers introduces additional time delay or linear phase shift. In the following, we first consider the response and insertion loss of the input (or transmitting) IDT with the generator circuit, and then those of the output (or receiving) IDT with the load circuit.

Input IDT

The surface acoustic wave generated by the input IDT has an amplitude proportional to the voltage appearing at the terminals of the IDT. Let ϕ_1^+ be the amplitude of the surface wave propagating to the right, as shown in Figure 14.10. Referring to the shunt equivalent circuit shown in Figure 14.11(a) for the input IDT driven by a generator of impedance R_g, let V_1 be the actual voltage appearing across the input IDT. Then ϕ_1^+ and V_1 are related by the frequency response $H_1(f)$ of the transmitting IDT. We can write [12]

$$\phi_1^+ = H_1(f)V_1 \tag{14.19}$$

From the equivalent circuit (Figure 14.11(a)), we can relate V_1 to the generator voltage as

$$V_1 = \frac{V}{(1 + Y_a(f)R_g)} \tag{14.20}$$

where

$$Y_a(f) = G_a(f) + jB_a(f) + j\omega C_T \tag{14.21}$$

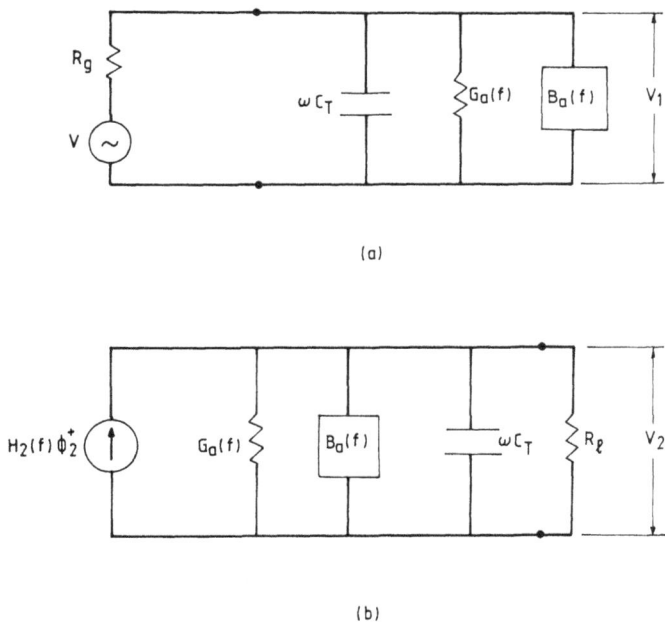

(a)

(b)

Figure 14.11 Equivalent circuits of (a) input IDT driven by generator and (b) output IDT connected to load.

Substituting (14.20) in (14.19), we have

$$\phi_1^+ = H_1(f)\, P_1(f)V \tag{14.22}$$

where

$$P_1(f) = \frac{1}{(1 + Y_a(f)R_g)} \tag{14.23}$$

Thus, the overall response on the input side is a product of two terms: $H_1(f)$, which is obtained from the Fourier transform of the electrode voltages, and $P_1(f)$, which is determined by the external circuit conditions.

We now evaluate the insertion loss due to the transmitting IDT. Because of the bidirectionality, the power delivered to the forward or reverse traveling surface wave is $G_a(f)\, V_1^2/2$. The maximum available power from the source is given by $V^2/4R_g$. The insertion loss α_1 of the bidirectional IDT is given by

$$\alpha_1(\text{dB}) = 10 \log_{10} \left| \frac{V^2/4R_g}{G_a(f)\, V_1^2/2} \right| \tag{14.24}$$

Using (14.20) and (14.21) in (14.24), we obtain

$$\alpha_1(\text{dB}) = 10 \log_{10} \left| \frac{(1 + G_a(f)R_g)^2 + \{R_g(\omega C_T + B_a(f))\}^2}{2G_a(f)R_g} \right| \tag{14.25}$$

It may be noted that at the synchronous frequency f_0 and under matched conditions, the insertion loss reduces to 3 dB.

Output IDT

We now consider the equivalent circuit of the output IDT connected to a load. This is shown in Figure 14.11(b). The output IDT is assumed to be identical to the input IDT. The forward-traveling surface wave reaching the output IDT induces a current $H_2(f)\phi_2^+$, which is indicated by the current generator in Figure 14.11(b). $H_2(f)$ is the frequency response of the IDT as a receiver, and ϕ_2^+ is given by

$$\phi_2^+ = \phi_1^+ \, e^{-j2\pi f t_0}, \quad t_0 = L/v_s \tag{14.26}$$

The current through the load resistor R_l is given by

$$I_2 = H_2(f)P_2(f)\phi_2^+ \tag{14.27}$$

where $P_2(f)$ is the response due to the external circuit and is given by

$$P_2(f) = \frac{1}{[(1 + G_a(f)R_l) + j\omega C_T R_l]} \tag{14.28}$$

The insertion loss expression for the output IDT is the same as that given by (14.25), with R_g replaced by R_l.

Delay Line Response

Using (14.22) and (14.26) in (14.27), we can obtain the following expression for the transconductance transfer function:

$$G_{12}(f) = \frac{I_2}{V} = H_1(f)P_1(f)H_2(f)P_2(f)e^{-j2\pi f t_0} \tag{14.29}$$

If the roles of the two IDTs as transmitter and receiver are interchanged, then by the principle of reciprocity, the transconductance transfer function remains the same. As evident from (14.29), the overall response of the delay line is obtained as the product of the frequency responses of the two IDTs modified by the external circuits and the propagation time delay. Under matched conditions, the minimum possible insertion loss of the delay line employing two bidirectional IDTs is 6 dB.

The frequency response of a delay line can be distorted by several second-order effects. These include the internal reflections of the IDT, diffraction and propagation losses, triple transit echo, and spurious losses. These aspects have been studied by several investigators [25–35] and are considered briefly below:

Internal Reflections

The frequency response of an IDT can be distorted by the multiple reflections of the surface waves at the electrodes. Two types of reflections occur: one is the regeneration reflection, and the second is mechanical electrical loading. An incident SAW induces a voltage at the IDT terminals, which in turn produces surface waves in both directions. The backward-traveling wave is called the regenerated wave. The second effect is due to the partial reflection of the wave at each electrode. The presence of the electrodes on the substrate loads the surface wave electrically by virtue of the velocity change and loads the wave mechanically by virtue of their mass. Both these effects are manifested as periodic changes in the wave impedance along the transducer.

Several techniques are available to help minimize the acoustic wave distortion due to these internal reflections. One simple technique is to introduce dummy electrodes [29, 30]. For eliminating the acoustic reflections due to the wave impedance discontinuities, a split-electrode technique is commonly used [31, 32]. That is, each electrode (in a conventional IDT) is replaced by a pair to double the periodicity of the structure. The center-to-center spacing between successive electrodes is then equal to a quarter wavelength at the synchronous frequency, thereby canceling the reflections. This technique shifts the surface wave stop band and the synchronous bulk wave frequencies to twice their original values while retaining the periodicity of surface waves. Thus, any bulk wave scattering is also eliminated.

Diffraction and Propagation Effects

The propagation of surface waves from the input IDT to the output IDT should ideally introduce only a time delay. That means that in the frequency domain, the amplitude must be constant and the phase response linear. In practice, diffraction causes the acoustic beam to widen, thus producing both amplitude and phase errors [33]. Furthermore, because propagation distances between different pairs of electrodes (a pair constituted of one electrode from input IDT and another from output IDT) are different, the corresponding phase delays differ. This causes an error in the frequency response. Both these errors, however, can be compensated for by suitably adjusting the electrode structure.

Triple Transit Signal

In a delay line, part of the signal arriving at the output IDT is the twice reflected signal, first from the output IDT and then from the input IDT. This is known as the triple transit signal and has a delay equal to three times the delay of the main signal. If the IDTs are bidirectional and matched, the magnitude of this signal is 12 dB below that of the main signal. The triple transit signal can introduce significant ripples in the amplitude and phase response. Several techniques have been reported in the literature [34, 35] for suppressing the triple transit signal. These include making one of the transducers unidirectional and using dummy reflecting transducers for canceling the reflections.

Spurious Losses

There are two main sources of spurious loss. One is the ohmic loss due to the finite conductivity of the electrodes. This can be reduced by depositing thicker metallic films for the electrodes. The second source is the generation of bulk waves. Bulk waves generated at the input IDT travel away from the surface, and hence do not, in general, affect the frequency response of the delay line. The loss due to bulk waves is frequency dependent and normally occurs on the high-frequency side as these waves travel faster than surface waves.

Upper Frequency Limit

The upper frequency of operation of SAW delay line is limited mainly by two major factors. One is the practical resolution achievable in the IDT pattern, and the second is the propagation loss [36].

As an illustration of the order of dimensions involved in an IDT for a microwave delay line, we consider a simple design on a LiNbO₃ substrate at L-band. We choose a uniform IDT so that the frequency response is of the form $(\sin x)/x$ centered around 1 GHz. At $f_0 = 1$ GHz, the time delay T from one positive electrode to the next is

$$T = \frac{1}{f_0} = 10^{-3} \ \mu s \qquad (14.30)$$

For LiNbO₃, $v_s = 3.5 \times 10^3$ m/s and $K^2 = 4.3\%$. The periodicity p of the IDT is given by

$$p = T v_s = 3.5 \ \mu m \qquad (14.31)$$

For uniform periodic IDT, we have

$$w = s = p/4 = 0.875 \ \mu m \qquad (14.32)$$

The optimum number of periodic sections N_{opt} and the 3-dB bandwidth can be calculated using (14.13) and (14.18), respectively. Thus

$$N_{opt} \simeq \frac{\sqrt{\pi}}{2K} \simeq 4 \qquad (14.33)$$

$$(\Delta f)_{3\,dB} = \frac{0.9 f_0}{N_{opt}} \simeq 225 \ MHz \qquad (14.34)$$

With the determination of N_{opt}, the optimum value of acoustic aperture (finger overlap) depends on the loss due to transducer, tuning elements, parasitic elements, beam steering, and diffraction [37]. To reduce the beam steering and diffraction losses, it is desirable to keep the acoustic aperture as wide as possible. However, the transducer loss and matching considerations limit the extent to which the aperture can be increased. Slobodnik and Szabo [37] have shown that the overall insertion loss reaches a minimum for an aperture of about one hundred wavelengths. Thus, in the above example, the aperture a_0 of the IDT is 350 μm.

From the above calculations, it is seen that the finger width w and the spacing s between fingers are generally less than 1 μm for frequencies above 1 GHz. For achieving such high pattern resolution, both scanning beam electron systems [38] and optical projection systems [39, 40] are used; the former for linewidths greater than about 0.1 μm and the latter for linewidths greater than about 0.5 μm.

The surface wave propagation loss depends strongly on the quality of surface finish. Besides surface imperfections such as scratches and damaged layers, surface waves suffer attenuation due to scattering by crystal imperfections and nonlinearly generated harmonics [13, 41]. It can be seen from Figure 14.2 that for LiNbO$_3$, the propagation loss is about 1.6 dB/μs at 1 GHz, which increases to about 6 dB/μs at 2.5 GHz [13]. If two identical IDTs are spaced by a distance L = 2.5 cm on LiNbO$_3$, the propagation delay is $t_0 = L/v_s \simeq 7.1$ μs, and the propagation loss at 1 GHz is approximately 11.4 dB.

Another undesirable effect of surface irregularities is dispersion at higher frequencies. For example, surface irregularities with a peak value of 35Å are reported to cause a dispersion of about 10^{-3}/GHz [36]. Dispersion is also introduced through mass loading of IDT. In order to keep the dispersion small, the metal film must be made thinner at higher frequencies. This requirement, however, conflicts with the need to keep the electrodes thick to minimize the resistive losses.

14.3.3 Microwave SAW Delay Lines

Because of the various practical difficulties enumerated above, the realization of SAW devices has been confined mainly to the MHz region (below 1 GHz). However, using sophisticated technologies for the preparation of supersmooth substrates [42, 43] and for realizing high resolution IDTs [39, 40], SAW delay lines have been realized up to about 2.5 GHz. For example, Lean and Broers [13] have reported a 1.5-μs SAW delay line at 2.5 GHz using the electron-beam technique. The delay line consists of two IDTs separated by 200 mil on a YZ LiNbO$_3$ substrate. Each IDT consists of twelve pairs of interleaved aluminum electrodes having a length of 150 μm, a linewidth of 0.4 μm, a spacing of 0.3 μm, and a thickness of 700Å. Without any external matching network, the delay line is reported to offer an insertion loss of 29 dB and a bandwidth of 210 MHz centered around 2.5 GHz [13]. This insertion loss includes the 6-dB loss due to the bidirectionality of the two IDTs, the propagation loss over 1.5-μs delay, the beam steering loss, and the conversion loss from RF to acoustic energy or vice versa at the IDTs.

Silva $et\ al.$ [40] have reported a 1.25-μs delay line at 2.2 GHz fabricated on a 37.95°, X-cut LiNbO$_3$ substrate using direct optical projection. In this, the IDTs have ten pairs of electrodes with a linewidth of 0.42 μm and acoustic aperture (length) of 100 μm. The delay line is reported to offer an insertion loss of 19 dB and a 3-dB bandwidth of 170 MHz.

Hartemann and Arnodo [44] have reported a 1.25-μs delay line on YZ-LiNbO$_3$ using two grating array transducers at 2.55 GHz. Each transducer makes use of two grating arrays of eight fingers as illustrated in Figure 14.12. The finger width and

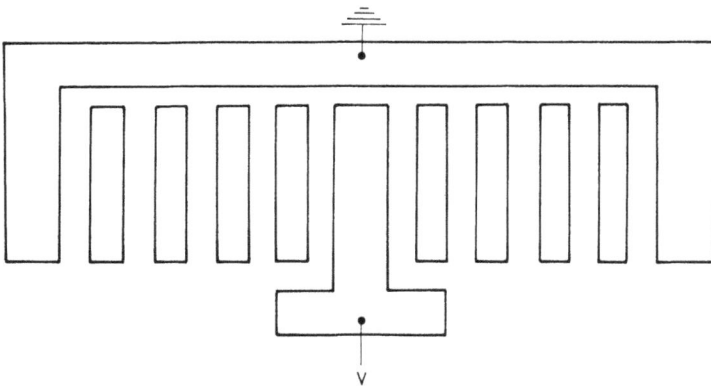

Figure 14.12 Schematic of a grating array transducer. (After Hartemann and Arnodo [44].)

the spacing between adjacent fingers are equal to $\lambda_s/2$ instead of $\lambda_s/4$, as in a standard uniform IDT. The grating transducer therefore offers the advantage of fabricational ease. As shown in Figure 14.12, the gratings are placed between two electrodes and the electrode width is twice the finger width. The surface waves launched by both the gratings are in phase at the center frequency. The amplitude of the surface wave is independent of the number of electrodes, whereas in a standard uniform IDT, the amplitude is proportional to the number of interelectrode spacings. With a finger width of 0.66 μm and an active length of fingers equal to 142 μm for the grating transducers, a 1.5-μs delay line is reported to offer an insertion loss of 39 dB at 2.55 GHz, and a 3-dB bandwidth of 87 MHz [44].

For integration with monolithic microwave integrated circuit (MMIC) components, GaAs is a preferred substrate over LiNbO$_3$. Webster [45] has reported a 1.5-GHz SAW delay line using (100) cut, (110) propagation GaAs substrate. "Three-halves" (3/2) electrode interdigital transducers [46] operating at the second overtone are used. Figure 14.13 shows the schematic of the IDT. The transducer consists of three interdigital electrodes per wavelength at the fundamental frequency. The electrode width and the interelectrode spacing are each equal to $\lambda_s/6$. This structure, like the double-electrode IDT [32], overcomes the problem of mechanical reflections in the passband. In the delay line reported by Webster [45], each IDT consists of 173 electrodes with electrode width and interelectrode spacing each equal to 0.62 μm, and acoustic aperture equal to 350 μm. Transducer metallization consists of 60Å Cr and 600Å Al. With two such IDTs spaced by a distance of 3 mm, the delay line (without external tuning) is reported to offer an insertion loss of 33 dB at the fundamental frequency of 0.76 GHz and 55 dB at the second overtone of 1.52 GHz.

SAW delay lines can be used as variable time-delay devices by applying a variable dc bias voltage to a pair of surface electrodes [47–49]. Figure 14.14 shows

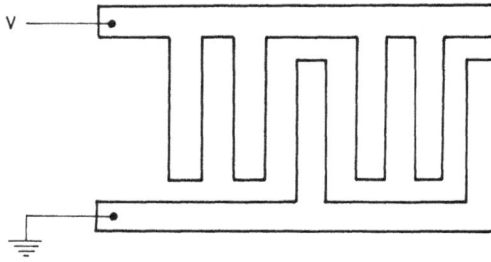

Figure 14.13 "Three-halves" electrode interdigital transducer.

the schematic of a delay line employing a 38° rotated Y-cut, X-propagating LiNbO$_3$ (38 YX LNO) [49]. The cut and orientation provide for high acoustic velocity (4800 m/s), as well as large electroacoustic effect, thus permitting high-frequency operation as well as optimum time-delay tuning sensitivity. Two parallel strip surface electrodes are printed on either side of the acoustic path. A dc voltage applied between the two strips produces an in-plane E-field. With this arrangement, Thaxter, Carr, and Silva [49] have reported a fractional change in time delay of 0.3×10^{-6} per volt in a 1-GHz SAW delay line. It may be noted that a bias of a few hundred volts is required for achieving small variation in time delay.

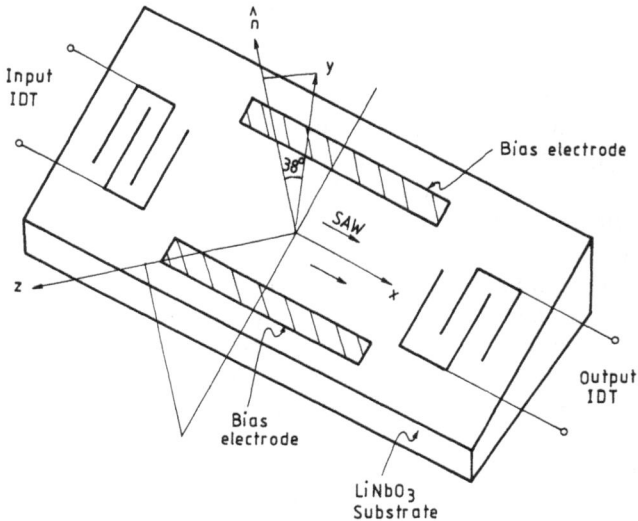

Figure 14.14 Schematic of a voltage-controlled SAW delay line on 38 YX LNO. (After Thaxter, Carr, and Silva [49].)

Surface acoustic waves are basically nondispersive. However, it is possible to make dispersive delay lines by introducing thin layers of different materials on the substrate. Dispersion depends on the thickness of the layer and the velocity difference between the substrate and the layered medium [50]. The delayed signals can be tapped electronically using transducers along the propagation path [51].

14.4 BASICS OF MAGNETOSTATIC WAVES

14.4.1 Materials

Magnetostatic waves (MSW) are slow, dispersive electromagnetic waves that propagate in a magnetically biased ferrimagnetic medium. The most commonly used ferrimagnetic material for MSW device applications is the *single-crystal yttrium iron garnet* (YIG). High-quality YIG films have low ferromagnetic resonance linewidths ($\Delta H < 0.5$ Oe), thus qualifying for low-loss microwave devices [52]. YIG films are generally grown using the *liquid phase epitaxy* (LPE) technique [53]. An ideal substrate for growing YIG film is *gadolinium gallium garnet* (GGG). The GGG substrate is chemically inert and is compatible with YIG from the point of view of lattice match and thermal expansion. Using the LPE process, film thicknesses of more than 100 μm can be grown without any detectable composition variations on the GGG substrate. The value of ΔH is typically 0.3 Oe at 9 GHz. The saturation magnetization $4\pi M_s$ of pure YIG film is around 1750G and can be varied over a range of 100G to 1800G by using gallium and tantalum doping. Other materials for MSW devices include lithium ferrite films, which exhibit both high magnetization and high Curie temperature. Using the LPE technique, film thicknesses up to about 20 μm have been reported on MgO substrates [54]. The value of ΔH is about 10 Oe.

14.4.2 Fundamental MSW Modes

Studies on MSW propagation in unbounded YIG film are well documented in the literature [55–57]. YIG film supports three pure propagating modes depending on the orientation of the bias magnetic field relative to the direction of propagation. The propagating modes are (1) *magnetostatic surface waves* (MSSW), (2) *magnetostatic forward volume waves* (MSFVW), and (3) *magnetostatic backward volume waves* (MSBVW). Figure 14.15 illustrates the three specific orientations of the bias magnetic field H_0 and the nature of dispersion in an unbound YIG medium.

In the MSSW mode, the bias magnetic field H_0 is in the plane of the YIG film but perpendicular to the wave vector $\vec{k}$ (Figure 14.15(a)). It has anisotropic propagation in the plane of the film. The wave energy is confined to a single surface for a given direction of magnetic field and propagation. For a reversal in either the bias field or propagation direction, the waves switch over to the opposite surface of the YIG film. This property makes the MSSW transducers unidirectional.

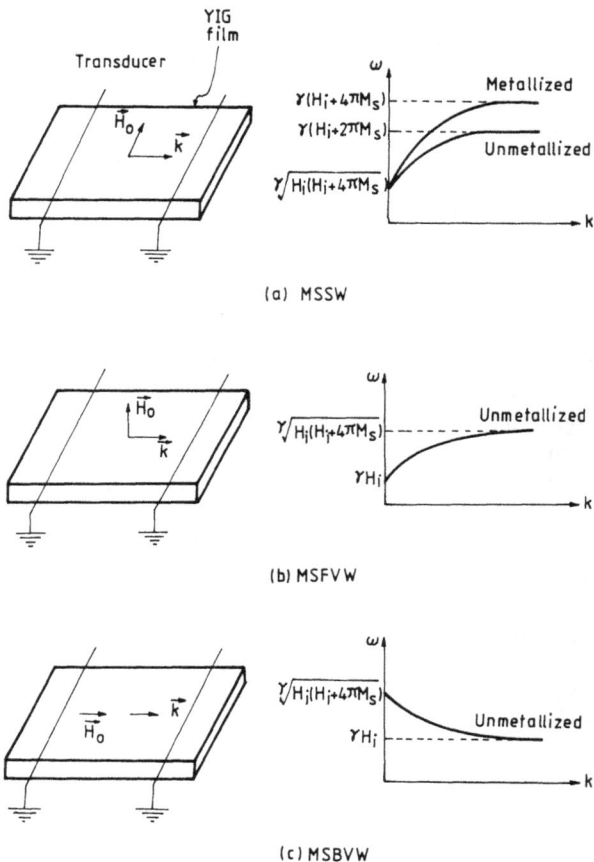

Figure 14.15 Fundamental MSW propagating modes. H_i: internal magnetic field, $\vec{k}$: wave vector.

In the MSFVW mode, H_0 is perpendicular to the film surface (Figure 14.15(b)). This mode is quasi-isotropic in the plane of the film. The mode energy is distributed over the volume of the film similar to that in a rectangular metal waveguide.

In the MSBVW mode, H_0 is parallel to the wave vector $\vec{k}$ (Figure 14.15(c)). This mode is characterized by anisotropic propagation and its mode energy is distributed across the film. The phase and group velocities are oppositely directed.

Figure 14.16 illustrates the potential bandwidth limits of MSW devices as a function of internal magnetic field [54]. The bandwidth of MSSW is large at lower frequencies and decreases at higher frequencies. For example, at 2 GHz, a bandwidth of 1.5 GHz is possible, whereas at 10 GHz it reduces to about 300 MHz. Volume waves, on the other hand, allow wide band operation at higher frequencies in the

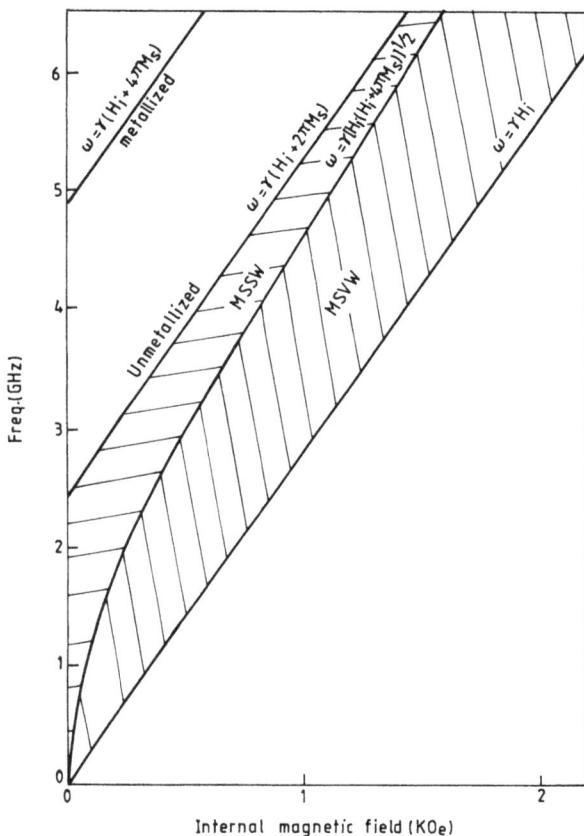

Figure 14.16 Frequency limits of magnetostatic waves. (From Castera [54], copyright © 1984, American Institute of Physics, reprinted with permission.)

microwave range. It can be seen from Figure 14.16 that metallizing the YIG surface increases the bandwidth.

The propagation loss of magnetostatic waves in YIG film is approximately given by 76.4 ΔH dB/μs [58], where ΔH is in oersteds. Thus, for a commercially available YIG film with a typical value of ΔH equal to 0.3 Oe at 9 GHz, the propagation loss is about 23 dB/μs. In contrast, for a SAW device on LiNbO$_3$, typical propagation loss is more than 100 dB/μs at 9 GHz.

Magnetostatic waves travel with velocities that are two to four orders slower than electromagnetic waves ($v_s \simeq 3 \times 10^6$–3×10^8 cm/s). Linewidths are typically in the range of 10 to 500 μm at microwave frequencies up to about 20 GHz. This is a distinct advantage over the SAW devices from the point of view of fabrication.

More importantly, MSWs permit electronically tunable time delays by controlling the bias magnetic field. Typical time delays achievable are in the range of 10 to 500 ns.

14.5 MSW DELAY LINES

14.5.1 MSW Guiding Structures

Several guiding structures for MSW propagation have been analyzed in the literature. These include a YIG film deposited on a substrate [56], a coplanar waveguide using YIG film [59], a metallic trough partially filled with YIG material [60], a rectangular rod with metallic boundaries [61], a YIG-slab-loaded rectangular waveguide [62–64], a YIG film with suitably spaced parallel ground planes [59, 65–67], and a multilayer planar structure with ground plane [9, 68–73]. In the following, we summarize the dispersion relations and salient characteristics of a few typical structures.

YIG-Slab-Loaded Rectangular Waveguide [62]

Figure 14.17 shows the cross section of a rectangular waveguide loaded with a YIG slab in the H-plane. In the magnetostatic limit, we have $\nabla \times \vec{H} = 0$ so that $\vec{H}$ can be expressed as

$$\vec{H} = \nabla \phi \tag{14.35}$$

where ϕ is the scalar magnetic potential. We assume propagation in the z-direction according to $e^{j(\omega t - kz)}$, and the direction of dc magnetic field along the x-direction, which is conducive to MSSW propagation.

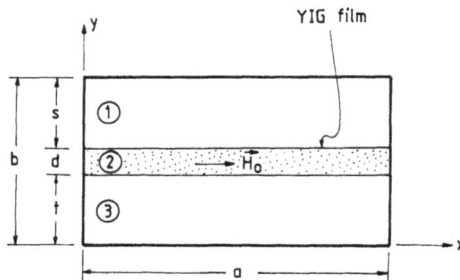

Figure 14.17 YIG-slab-loaded rectangular waveguide.

Using (14.35) in $\nabla \cdot [\mu]\vec{H} = 0$, we obtain the following relation in ϕ for the YIG region:

$$\mu_0 \frac{d^2\phi}{dx^2} + \mu \left(\frac{d^2\phi}{dy^2} + \frac{d^2\phi}{dz^2} \right) = 0 \tag{14.36}$$

where μ_0 is the permeability of free space and μ is the diagonal element of the permeability tensor of YIG. For the air region (14.36) reduces to

$$\frac{d^2\phi}{dx^2} + \frac{d^2\phi}{dy^2} + \frac{d^2\phi}{dz^2} = 0 \tag{14.37}$$

Referring to the structure shown in Figure 14.17, the solutions in the three regions can be expressed as [62]

$$\phi_1 = \sum_{n=0}^{\infty} A_n \cos\left(\frac{n\pi x}{a} \right) \cosh[q_n(b - y)] e^{-jkz} \tag{14.38a}$$

$$\phi_2 = \sum_{n=0}^{\infty} \cos\left(\frac{n\pi x}{a} \right) [B_n \cosh(p_n y) + C_n \sinh(p_n y)] e^{-jkz} \tag{14.38b}$$

$$\phi_3 = \sum_{n=0}^{\infty} D_n \cos\left(\frac{n\pi x}{a} \right) \cosh(q_n y) e^{-jkz} \tag{14.38c}$$

where

$$p_n^2 = \frac{\mu_0}{\mu} \left(\frac{n\pi}{a} \right)^2 + k^2 \tag{14.39a}$$

$$q_n^2 = \left(\frac{n\pi}{a} \right)^2 + k^2 \tag{14.39b}$$

At the interfaces $y = t$ and $(t + d)$, both ϕ and B_y are continuous. Using these boundary conditions in (14.38) and eliminating the unknown coefficients, A_n, B_n, C_n, and D_n yield the desired dispersion relation. It is given by [62]

$$e^{2p_n d} = \left[\frac{\mu p_n - \kappa k - \mu_0 q_n \tanh(q_n s)}{\mu p_n + \kappa k + \mu_0 q_n \tanh(q_n s)} \right] \left[\frac{\mu p_n + \kappa k - \mu_0 q_n \tanh(q_n t)}{\mu p_n - \kappa k + \mu_0 q_n \tanh(q_n t)} \right] \tag{14.40}$$

where κ is the off-diagonal component of the permeability tensor. The expressions for μ and κ are as given in Chapter 3 (in Volume I).

$$\mu = \mu_0 \left[1 + \frac{\omega_0 \omega_m}{\omega_0^2 - \omega^2} \right] \tag{14.41}$$

$$\kappa = \mu_0 \frac{\omega \omega_m}{\omega_0^2 - \omega^2} \tag{14.42}$$

where

$$\omega_0 = \gamma H_0 \tag{14.43a}$$

$$\omega_m = \gamma 4\pi M_s \tag{14.43b}$$

Finite-Width YIG Film Between Ground Planes [10, 65–67]

Figure 14.18 shows the cross section of a YIG film situated between two ground planes. The propagation is along the z-direction and is perpendicular to the width of the film. Ishak [10] has reported magnetostatic dispersion relations for the lowest order width modes corresponding to the three orientations of the dc magnetic field along the coordinate axes. The expressions are as follows:
MSSW mode (H_0 along x):

$$e^{2p_1 d} = \left[\frac{\bar{\mu} p_1 - \bar{\kappa} k - q_1 \tanh(q_1 s)}{\bar{\mu} p_1 + \bar{\kappa} k + q_1 \tanh(q_1 s)} \right] \left[\frac{\bar{\mu} p_1 + \bar{\kappa} k - q_1 \tanh(q_1 t)}{\bar{\mu} p_1 - \bar{\kappa} k + q_1 \tanh(q_1 t)} \right] \tag{14.44}$$

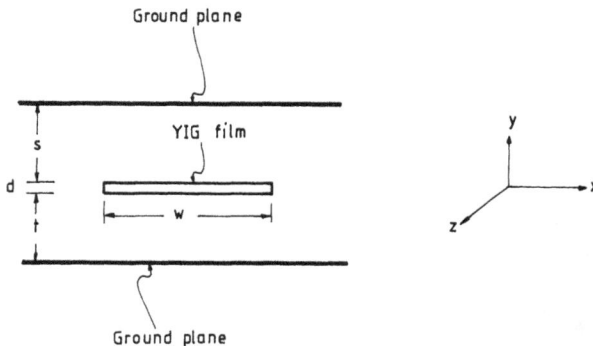

Figure 14.18 Geometry of finite-width YIG film between ground planes.

where

$$p_1^2 = \frac{1}{\bar{\mu}} \left(\frac{\pi}{w}\right)^2 + k^2 \qquad (14.45a)$$

$$q_1^2 = \left(\frac{\pi}{w}\right)^2 + k^2 \qquad (14.45b)$$

$$\bar{\mu} = \frac{\mu}{\mu_0}, \quad \bar{\kappa} = \frac{\kappa}{\mu_0} \qquad (14.45c)$$

It may be noted that (14.44) is the same as (14.40), with $n = 1$ and $a = w$.
MSFVW mode (H_0 along y):

$$\tanh(\eta k_1 d) = \eta \left[\frac{\tanh(k_1 t) + \tanh(k_1 s)}{\eta^2 - \tanh(k_1 t)\tanh(k_1 s)}\right] \qquad (14.46)$$

where

$$k_1^2 = \left(\frac{\pi}{w}\right)^2 + k^2 \qquad (14.47a)$$

$$\eta^2 = -\bar{\mu} = \left|\frac{\omega_0 \omega_m}{\omega^2 - \omega_0^2} - 1\right| \qquad (14.47b)$$

MSBVW mode (H_0 along z):

$$e^{2p_1'd} = \left[\frac{\bar{\mu}p_1' - \bar{\kappa}k_z - q_1'\tanh(q_1's)}{\bar{\mu}p_1' + \bar{\kappa}k_z + q_1'\tanh(q_1's)}\right]\left[\frac{\bar{\mu}p_1' + \bar{\kappa}k_z - q_1'\tanh(q_1't)}{\bar{\mu}p_1' - \bar{\kappa}k_z + q_1'\tanh(q_1't)}\right] \qquad (14.48)$$

where

$$p_1'^2 = \frac{k^2}{\bar{\mu}} + k_z^2 \qquad (14.49a)$$

$$q_1'^2 = k_z^2 + k^2 \qquad (14.49b)$$

$$k_z = \frac{\pi}{w} \qquad (14.49c)$$

The above dispersion relations can be used to compute the time-delay characteristics of several practical structures. In Figure 14.18, the portion below the YIG film may consist of a grounded dielectric substrate with microstrip transducers, and the upper portion can be a GGG substrate with the top surface metallized. Figure 14.19 illustrates the arrangement with the GGG substrate and YIG film raised. If in (14.44) through (14.48) $s \rightarrow \infty$, the structure corresponds to YIG film on a dielectric substrate with ground plane (Figure 14.20).

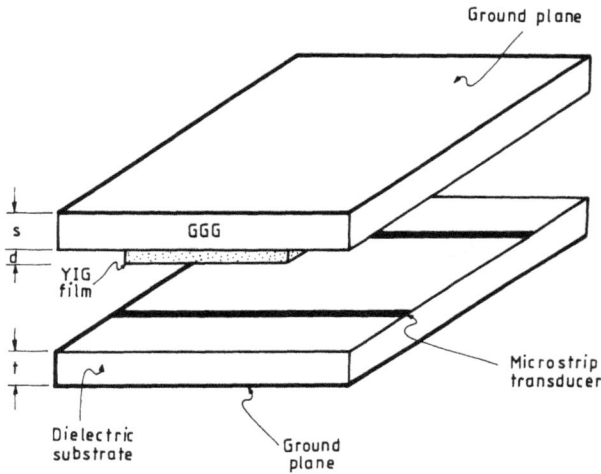

Figure 14.19 Schematic of a delay-line configuration (GGG substrate raised).

Multilayer YIG Film on Ground Plane

MSW dispersion can be controlled in a multilayer structure by introducing desirable thicknesses and saturation magnetization levels of films during the process of film growth [9]. Adkins and Glass [9, 69, 70] have derived dispersion relations for a four-layer structure of the type shown in Figure 14.21. The four layers can have arbitrary thicknesses (t_n, $n = 1$ to 4) and magnetizations ($4\pi M_n$, $n = 1$ to 4). Adkins [9] has summarized the dispersion relations for all three MSW modes, which are reproduced below (with slight modification in notation for the permeability tensor components):

MSSW mode (H_0 along x):

$$|\phi_2\theta_1 - \phi_1\theta_2\alpha_3\beta_3||e^{-|k|t_3} - e^{|k|t_3}| + \phi_1\theta_1|\beta_3 e^{-|k|t_3} + \alpha_3 e^{|k|t_3}|$$
$$- \phi_2\theta_2|\beta_3 e^{|k|t_3} + \alpha_3 e^{-|k|t_3}| = 0 \tag{14.50}$$

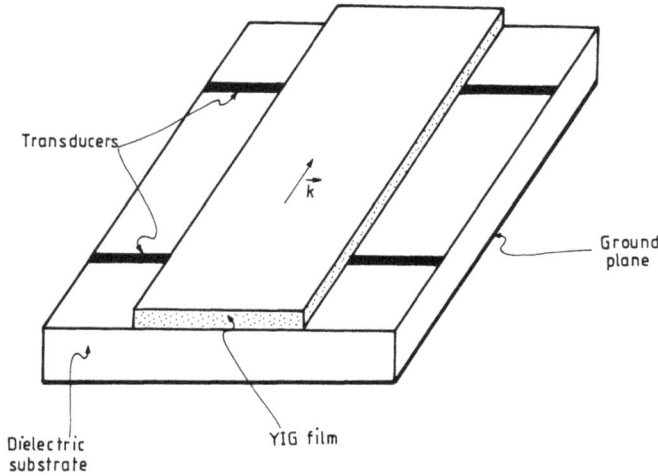

Figure 14.20 Geometry of a YIG film on a dielectric substrate with ground plane.

where

$$\phi_1 = 1 + \left(\frac{\bar{\mu}_2 - \bar{\chi}_2 S - \gamma}{\bar{\mu}_2 + \bar{\chi}_2 S + \gamma}\right) e^{2|k|t_1} \tag{14.51a}$$

$$\phi_2 = \bar{\mu}_2 - \bar{\chi}_2 S - (\bar{\mu}_2 + \bar{\chi}_2 S) \left|\frac{\bar{\mu}_2 - \bar{\chi}_2 S - \gamma}{\bar{\mu}_2 + \bar{\chi}_2 S + \gamma}\right| e^{2|k|t_2} \tag{14.51b}$$

$$\theta_1 = (\bar{\mu}_4 - \bar{\chi}_4 S) e^{|k|t_4} - \left(\frac{\bar{\mu}_4 - \bar{\chi}_4 S - 1}{\bar{\mu}_4 + \bar{\chi}_4 S + 1}\right)(\bar{\mu}_4 + \bar{\chi}_4 S) e^{-|k|t_4} \tag{14.51c}$$

$$\theta_2 = e^{|k|t_4} + \left(\frac{\bar{\mu}_4 - \bar{\chi}_4 S - 1}{\bar{\mu}_4 + \bar{\chi}_4 S + 1}\right) e^{-|k|t_4} \tag{14.51d}$$

$$\gamma = \frac{(\bar{\mu}_1 - \bar{\chi}_1 S)(1 - e^{-2|k|t_1})}{1 + \{(\bar{\mu}_1 - \bar{\chi}_1 S)/(\bar{\mu}_1 + \bar{\chi}_1 S)\} e^{2|k|t_1}} \tag{14.51e}$$

$$\alpha_n = \bar{\mu}_n - \bar{\chi}_n S \tag{14.51f}$$

$$\beta_n = \bar{\mu}_n + \bar{\chi}_n S \tag{14.51g}$$

$$\bar{\mu}_n = \frac{\mu_n}{\mu_0}, \quad \bar{\chi} = \frac{\kappa_n}{\mu_0} \tag{14.51h}$$

$$S = \pm 1 \tag{14.51i}$$

The parameters μ_n and κ_n are the diagonal and off-diagonal components of the permeability tensor for the nth layer. The expressions for μ and κ are given in (14.41) and (14.42), respectively.

Adkins [9] has pointed out that the most appropriate geometry for volume wave consists of two magnetic layers separated by one nonmagnetic dielectric layer. The dispersion relations provided below for MSFVW and MSBVW modes are for the structure with layers 1 and 3 nonmagnetic and layers 2 and 4 magnetic [9]:

MSFVW mode (H_0 along y):

$$\{e^{-2|k|t_1}\,(\alpha_2^2 + 1) + (\alpha_2^2 - 1) - 2\alpha_2 \cot(\alpha_2|k|t_2)\}\{(1 - \alpha_1^2)$$

$$+ 2\alpha_1 \cot(\alpha_1|k|t_4)\} + e^{-2|k|t_1}\{e^{2|k|t_1}(\alpha_2^2 + 1)$$

$$+ (\alpha_2^2 - 1) + 2\alpha_2 \cot(\alpha_2|k|t_2)\}(1 + \alpha_1^2)\,e^{-2|k|t_3} = 0 \tag{14.52}$$

where

$$\alpha_n^2 = -\bar{\mu}_n \tag{14.53}$$

MSBVW mode (H_0 along z):

$$\{e^{-2|k|t_1}((\bar{\mu}_2\alpha_2)^2 + 1) + (\bar{\mu}_2\alpha_2)^2 - 1 - 2(\bar{\mu}_2\alpha_2) \cot(\alpha_2|k|t_2)\}$$

$$\cdot\{1 - (\bar{\mu}_1\alpha_1)^2 + 2(\bar{\mu}_1\alpha_1) \cot(\alpha_2|k|t_2)\}$$

$$+ e^{-2|k|t_1}\{e^{2|k|t_1}((\bar{\mu}_2\alpha_2)^2 + 1) + (\bar{\mu}_2\alpha_2)^2 - 1$$

$$+ 2(\bar{\mu}_2\alpha_2) \cot(\alpha_2|k|t_2)\}\{1 + (\bar{\mu}_1\alpha_1)^2\}\,e^{-2|k|t_3} = 0 \tag{14.54}$$

where

$$\alpha_n^2 = -\frac{1}{\bar{\mu}_n} \tag{14.55}$$

In (14.52) to (14.55), α_n and $\bar{\mu}_n$ refer to the nth magnetic layer; that is, α_1 and $\bar{\mu}_1$ refer to layer 2, and α_2 and $\bar{\mu}_2$ refer to layer 4 in Figure 14.21.

Time-Delay Characteristics

YIG-Slab-Loaded Rectangular Waveguide—Figures 14.22 to 14.24 show typical time-delay characteristics of MSSW and MSFVW modes in a YIG-slab-loaded guide (Figure 14.17), as reported by Radmanesh, Chu, and Haddad [62]. All three graphs are

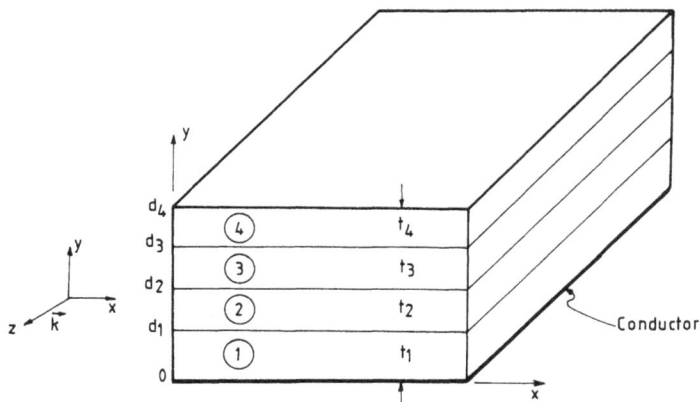

Figure 14.21 Geometry of multilayer YIG film on ground plane. (After Adkins [9].)

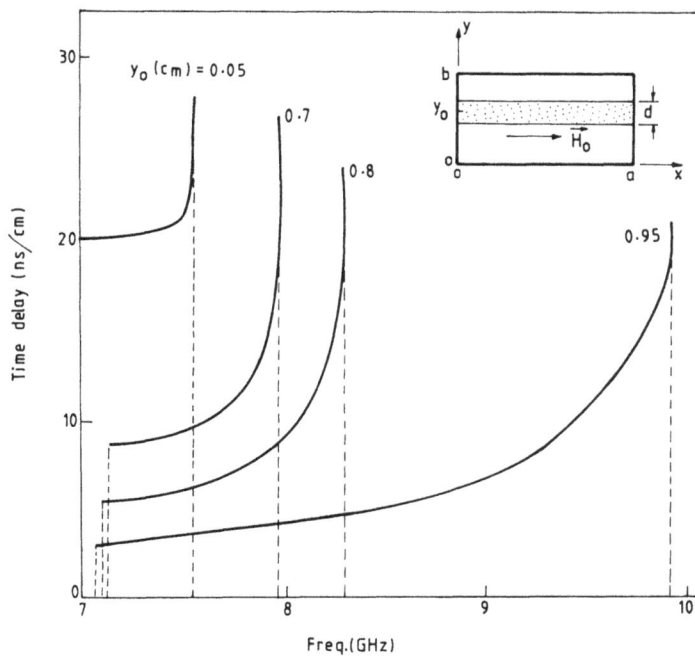

Figure 14.22 Time delay *versus* frequency of first MSSW mode in YIG-slab-loaded rectangular guide for various slab positions. $a = 2$ cm, $b = 1$ cm, $d = 0.1$ cm. $4\pi M_s = 1750$G, $H_0 = 1800$ Oe. (From Radmanesh, Chu, and Haddad [62], reprinted with permission of *Microwave Journal,* copyright © 1986 Horizon House—Microwave Inc.)

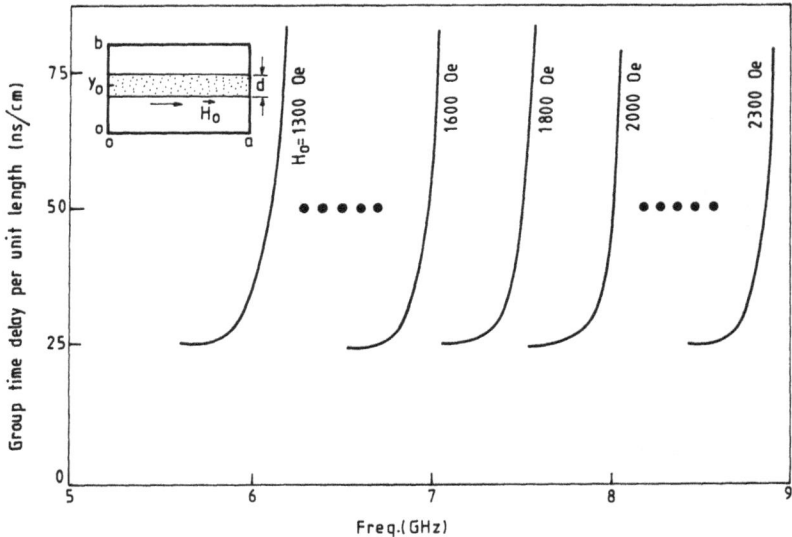

Figure 14.23 Effect of magnetic field on the group time-delay characteristics of first MSSW mode in a YIG-slab-loaded rectangular guide. $a = 2$ cm, $b = 1$ cm, $d = 0.1$ cm, $y_0 = 0.5$ cm, $4\pi M_s = 1750$G. (From Radmanesh, Chu, and Haddad [62], reprinted with permission of *Microwave Journal*, copyright © 1986 Horizon House—Microwave Inc.)

for a rectangular waveguide 2×1 cm^2 in size loaded with a YIG film with a thickness of $d = 0.1$ cm and saturation magnetization $4\pi M_s = 1750$G. Time delay per unit length is calculated using the relation $t_0 = (\partial\omega/\partial k)^{-1}$. For the lowest order width MSSW mode ($n = 1$ in (14.40)), the effect of varying the film position on the time-delay characteristic is shown in Figure 14.22 and that of varying the dc magnetic field is shown in Figure 14.23. It can be seen from Figure 14.22 that the structure offers large time delay when the film is close to the bottom broad wall of the guide, and large bandwidth when the film is close to the top wall. The characteristics are nonreciprocal. Figure 14.23 shows the tuning capability by varying the bias magnetic field. The time-delay curves are shifted to higher ranges of frequencies for higher values of bias field. For the same guide structure, Figure 14.24 shows the variation in group time delay *versus* frequency for the MSFVW mode. Plots are shown for the zero- ($n = 0$) and first-order width ($n = 1$) modes for three different positions of the YIG film. It can be seen that the time delay increases as the slab is moved towards the middle of the waveguide. The characteristics are reciprocal and symmetrical around the center of the guide. For large time delay, the slab must be positioned at the center of the guide, whereas for large bandwidths it should be placed near the top or bottom broad wall of the guide. The time-delay characteristics as a function of the bias magnetic field are similar to those shown in Figure 14.23 for the MSSW mode.

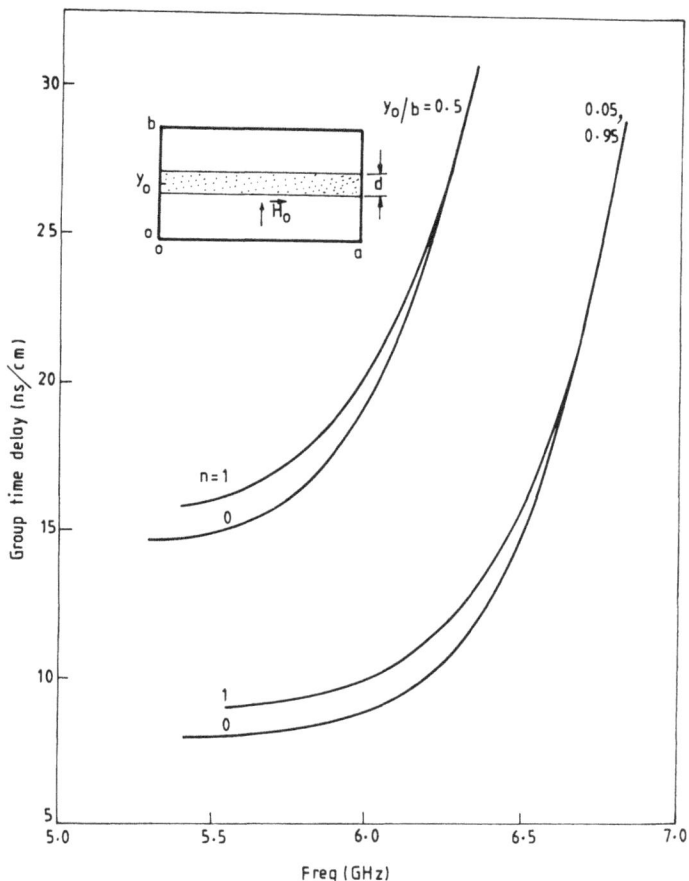

Figure 14.24 Group time-delay *versus* frequency for zero- and first-order MSFVW modes in YIG-slab-loaded waveguide. $a = 2$ cm, $b = 1$ cm, $d = 0.1$ cm, $4\pi M_s = 1750$G, $H_0 = 1800$ Oe. (From Radmanesh, Chu, and Haddad [64], copyright © 1987 IEEE, reprinted with permission.)

YIG Film Above a Ground Plane—Figure 14.25 shows a typical time-delay characteristic reported by Daniel, Adam, and O'Keeffe [74] for the MSFVW mode in a YIG film spaced by a distance equal to its thickness above a ground plane. Plots are shown for three different values of spacing d (10, 20, and 50 μm). The characteristics clearly illustrate the dispersion control that can be achieved by varying the ground plane spacing and film thickness. For the parameters considered in Figure 14.25, the structure offers linearly dispersive time delay over nearly 1 GHz at X-band. Furthermore, the differential time delay over the approximately linear region varies inversely with the thickness of the film.

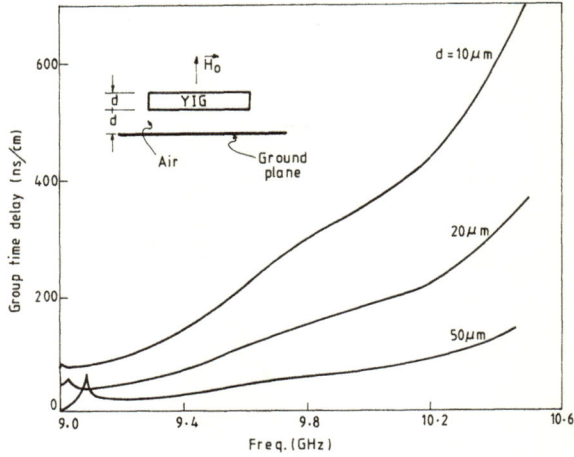

Figure 14.25 Group time-delay *versus* frequency for MSFVW mode in a YIG film of thickness d spaced by a distance d above a ground plane. Internal magnetic bias field $H_i = 3200\text{G}$. (From Daniel, Adam, and O'Keeffe [74], copyright © 1979 IEEE, reprinted with permission.)

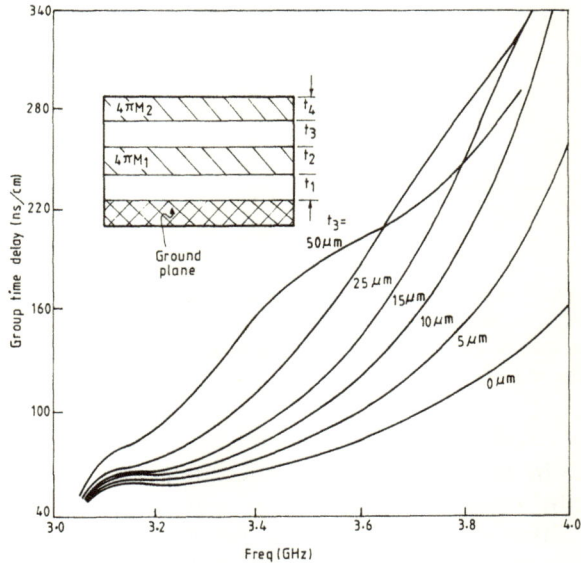

Figure 14.26 Dispersion for composite MSFVW mode of triple-layer geometry as a function of layer spacing. $4\pi M_1 = 4\pi M_2 = 1643\text{G}$, $H_0 = 2700$ Oe, $t_1 = 254$ μm, $t_2 = 15$ μm, $t_4 = 15$ μm. (From Adkins [9], reproduced with permission of Birkhauser Boston.)

Multilayer—YIG Film Structure—Dispersion control in a multilayer structure (Figure 14.21) by varying the thicknesses and magnetization levels of film layers has been discussed by Adkins [9]. A structure that uses two magnetic layers separated by a nonmagnetic layer is reported to be particularly useful for producing linear dispersion in MSFVW and MSBVW modes. Two typical graphs from Adkins [9] that illustrate this feature are reproduced in Figures 14.26 and 14.27. Linearization can be achieved either by varying the spacing between the two films having the same magnetizations (Figure 14.26) or by varying the magnetization of one of the films while keeping the thicknesses of layers fixed (Figure 14.27).

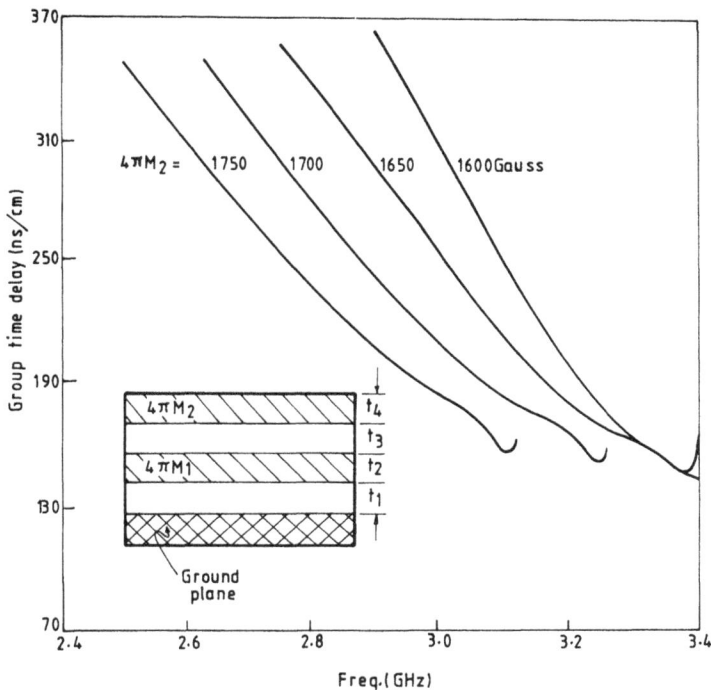

Figure 14.27 Dispersion for composite MSBVW mode of triple-layer geometry as a function of magnetization. $4\pi M_1 = 1600G$, $t_1 = 25$ μm, $t_2 = 15$ μm, $t_3 = 25$ μm, $t_4 = 15$ μm, $H_0 = 700$ Oe. (From Adkins [9]. Reproduced with permission of Birkhauser Boston.)

14.5.2 Transducers

Magnetostatic waves can be excited in YIG films by means of transducers. Transducing these waves is easy because of the strong coupling between the RF currents in the transducer and the MSW magnetic field components. Several types of transducers have been reported in the literature that are useful for MSW excitation and detection. These include the microstrip [75], meanderlines [76–79], gratings [76, 78], and interdigital transducers [80, 81].

The microstrip offers a simple and convenient means of exciting MSW. The transducer is realized in the form of a short-circuited or open-circuited microstrip, as illustrated in Figure 14.28. Transducers based on a short-circuited microstrip operate efficiently at frequencies for which the width b of the YIG film is less than $\lambda/4$, where λ is the electromagnetic guide wavelength in the microstrip. In the case of an open-circuited microstrip, efficient transduction is achieved when the film width b is approximately $\lambda/2$ long. Ganguly and Webb [75] have analyzed a microstrip line on a dielectric substrate with a YIG film overlay for MSSW propagation, and have derived an expression for the equivalent radiation resistance (R_m) per unit length of the line. For a shorted microstrip transducer of length b (see Figure 14.28), the input impedance Z_i is given by

$$Z_i = R_i + jX_i = Z_f \tanh(\alpha_f + j\beta_f)b \tag{14.56}$$

where Z_f, α_f, and β_f are the characteristic impedance, attenuation constant, and phase constant, respectively, of the YIG-film-loaded microstrip. If a standard dielectric microstrip of characteristic impedance Z_0 is used to feed this YIG-loaded transducer section, then the ratio of the power P_m converted to MSSW (assuming no ohmic losses) to the power P_{in} delivered by the source is given by

$$\frac{P_m}{P_{in}} = \frac{4Z_0 R_i}{[(R_i + Z_0)^2 + X_i^2]} \tag{14.57}$$

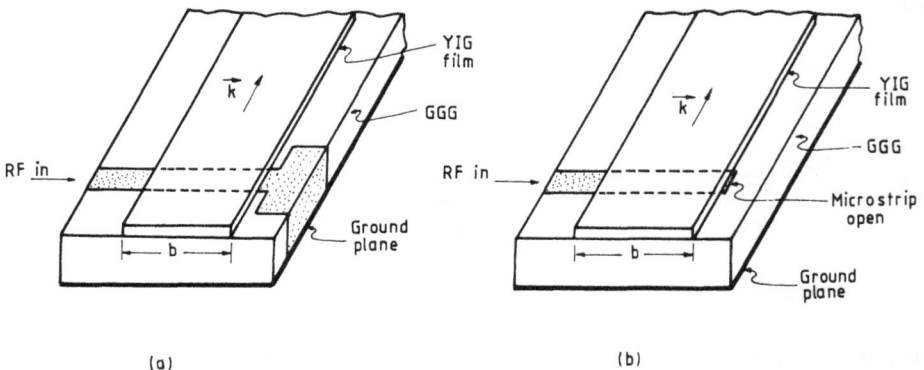

Figure 14.28 MSW transducers: (a) shorted microstrip and (b) open-circuited microstrip.

Ganguly and Webb [75] have shown that for $\alpha_f b$, $\beta_f b << 1$; $R_i \simeq R_m b/2$ and $X_i \simeq 0$, and thus (14.57) reduces to

$$\frac{P_m}{P_{in}} \simeq \frac{2Z_0 R_m b}{(Z_0 + R_m b/2)^2} \tag{14.58}$$

Since MSSW excitation is unidirectional, it is possible to achieve almost total conversion from electromagnetic energy to MSSW energy by making $R_m b/2$ as close to Z_0 as possible. The transducer is capable of wideband operation for narrow strip widths. A bandwidth of more than 500 MHz is achievable for a strip width approximately five times the film thickness or smaller [75].

Besides MSSW mode, the two volume modes (MSFVW and MSBVW) can also be excited by a microstrip transducer. Both short-circuited and open-circuited microstrip sections of narrow width are useful.

Multielement periodic transducers in the form of meander lines and gratings are useful for narrowband devices. Figure 14.29 shows the schematics of these transducers. The overall frequency response of a multielement transducer can be expressed in the form [76, 79]

$$R(f) = E(f)A_1^2(f)\, A_2^2(f) \tag{14.59}$$

where $E(f)$ is the frequency response of an infinitesimally narrow single-line electrode, and $A_1(f)$ and $A_2(f)$ are array factors. $A_1(f)$ represents the array factor of a single microstrip conductor carrying uniform current. It is given by

$$A_1(f) = \left[\frac{\sin(kw/2)}{(kw/2)} \right] \tag{14.60}$$

where w is the width of the electrode and k is the wave number. $A_2(f)$ represents the array factor of a line current multielement structure. For a meander line structure (Figure 14.29(a)),

$$A_2(f) = \frac{\sin\left(\dfrac{kcN}{2}\right)}{\cos\left(\dfrac{kc}{2}\right)} \quad \text{for } N \text{ even} \tag{14.61}$$

$$= \left[1 - \frac{\sin\left\{\dfrac{(N+1)kc}{2}\right\}}{\cos\left(\dfrac{kc}{2}\right)} \right] \left\{ 2\sin\left(\frac{Nkc}{2}\right) \right.$$

$$\left. + \frac{\sin\left\{\dfrac{(N+1)kc}{2}\right\}}{\cos\left(\dfrac{kc}{2}\right)} \right\} \right]^{1/2} \quad \text{for } N \text{ odd} \qquad (14.62)$$

where c is the center-to-center electrode spacing, that is, one-half of the meander line period, and N is the number of fingers in the meander line. For an N-element grating structure (Figure 14.29(b)), $A_2(f)$ is given by

$$A_2(f) = \frac{\sin\left(\dfrac{Nkc}{2}\right)}{\sin\left(\dfrac{kc}{2}\right)} \qquad (14.63)$$

Figure 14.30 shows the schematic of an interdigital transducer for MSWs. The transducer is fed by a microstrip power divider. The fingers are $\lambda/2$ long so that the RF currents in adjacent fingers are in antiphase. For a transducer with N fingers, the frequency response is given by [80]

$$A(f) = \frac{\sin\left(\dfrac{N\pi x}{2}\right)}{\sin\left(\dfrac{\pi x}{2}\right)} \qquad (14.64)$$

where

$$x = \left(\frac{\pi}{c} - k\right) \Big/ \left(\frac{\pi}{c}\right) \qquad (14.65)$$

The above formula assumes that the finger width is negligible, and coupling of each finger is independent of the wave number k.

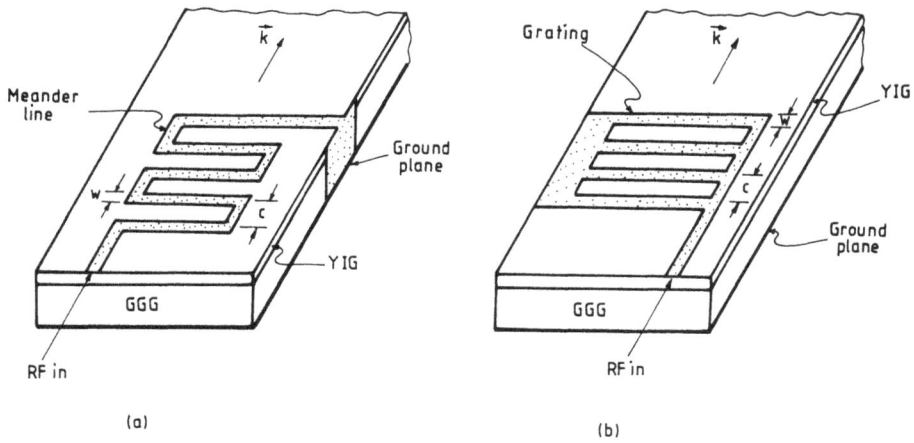

Figure 14.29 Schematic of (a) meanderline transducers and (b) grating transducer.

In practical devices, terminations are required to be introduced to suppress unwanted reflected waves from YIG sample ends and from the shorted transducers. Several techniques for MSW terminations have been reported in the literature. Some of the effective techniques include the use of ferrite powders [82] or GaAs thin films [83] to absorb MSW at the YIG edges, and cutting the edges of the YIG film at an angle ($\simeq 90°$) relative to the wave propagation direction [84]. Another effective method reported for suppression of spurious reflections is to use bias magnets to change the bias field in the region of the film ends [85]. The magnets help to increase the propagation delay and hence the propagation losses outside the device area.

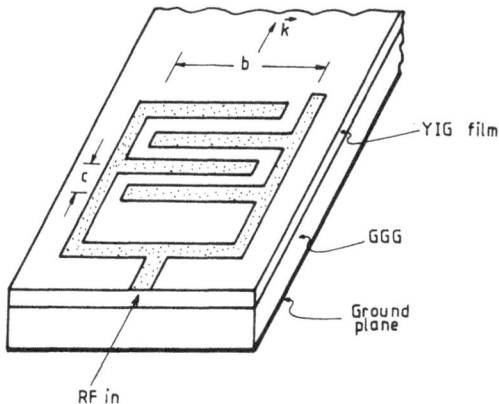

Figure 14.30 Schematic of an interdigital transducer.

14.5.3 Practical MSW Delay Lines

MSW delay lines can be conveniently fabricated in planar form. For example, Figure 14.31 shows a sandwich structure. The transducer patterns are realized on a metallized dielectric substrate using the standard photolithographic technique. The dielectric substrate can be alumina, RT-duroid, fused silica, or sapphire. YIG film is grown on a GGG substrate using the LPE technique and placed with a dielectric spacer in between. The entire sandwich structure is placed between the poles of biasing magnets.

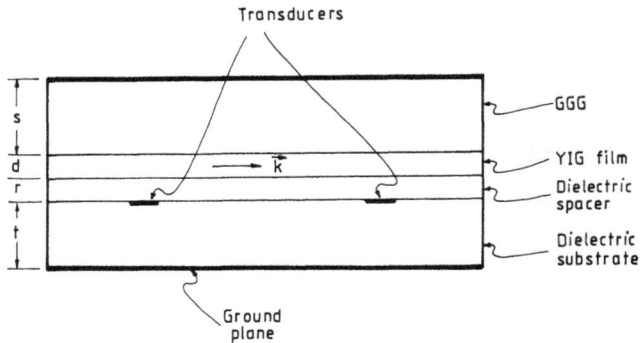

Figure 14.31 Cross-sectional side view of an MSW delay line.

As discussed in the preceding sections, MSW delay lines are nonlinearly dispersive. Most practical applications require either nondispersive or linearly dispersive delay lines. These characteristics can be achieved by controlling the magnetic parameters of the YIG film and the dimensional parameters of the delay-line structure. Furthermore, by making provision to change the bias magnetic field, the delay line can be made electronically tunable.

Nondispersive Delay Line

One convenient technique for achieving broadband nondispersive delay is to adjust the spacing between the YIG film and ground plane. Its effectiveness is illlustrated in Figure 14.22 for MSSW propagation in a YIG-slab-loaded rectangular waveguide [62]. Adam, O'Keeffe, and Daniel [86] have demonstrated a nondispersive MSFVW delay line employing a YIG film with suitably spaced ground planes on either side. The delay line is reported to have a constant time delay of 90 ns within ±5% over

400 MHz, with a 20-dB insertion loss at X-band. Other techniques of producing nondispersive delays include using a layered YIG film structure with different saturation magnetization for each layer [71], and graded-periodicity metal strip reflective arrays [87].

An electronically tunable nondispersive delay line that employs a cascade of MSSW and MSBVW delay lines has been demonstrated by Sethares, Owens, and Smith [88]. The slope of the delay *versus* frequency curve for the MSSW delay line is positive, whereas that of the MSBVW delay line is negative. The constant time delay *versus* frequency characteristic is achieved by linearizing the dispersion curves and making the two slopes equal in magnitude. The value of the time delay over the same frequency band can be changed by keeping the bias field of the MSSW line fixed and changing the bias field of the MSBVW line. An improved version of this cascaded delay line was later reported by Adkins *et al.* [11, 89]. Figure 14.32 shows the delay line schematic that was reported. In this device, the MSSW delay-line section makes use of a YIG film with a thickness of 25 μm with stepped ground plane spacing, and the MSBVW delay-line section uses a YIG film with a thickness of 50 μm with a ground plane spacing of 250 μm. The ends of the YIG wafers are

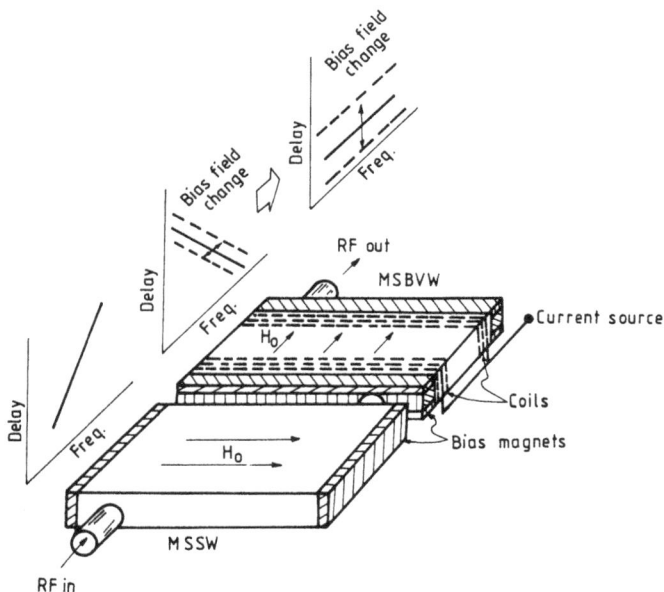

Figure 14.32 Electronically tunable MSW cascaded delay line. (From Adkins *et al.* [89], copyright © 1984, American Institute of Physics, reprinted with permission.)

beveled to reduce reflections. The device is reported to offer a maximum delay differential of 47 ns for a fixed bias of 410G on the MSSW delay line, and a variable bias in the range of 550G to 670G on the MSBVW line. The bandwidth of the device is approximately 250 MHz in the S-band [89].

Adam, Collins, and Owens [52] have reported a tapped nondispersive microwave delay line with a total delay of 250 ns and a tapped delay of 125 ns. Both tapped and full delays are constant within ±5% tolerance over a bandwidth of 225 MHz. The corresponding insertion losses are 9 dB and 14 dB, respectively.

Linearly Dispersive Delay Line

Linearly dispersive delay lines can be realized by adopting techniques similar to those used for nondispersive delay lines. Linearly dispersive delay lines have been demonstrated by (1) suitable positioning of the ground plane with respect to the YIG film [67, 74], (2) varying the ground plane spacing along the direction of propagation [90], and (3) using a multilayer YIG film structure [70].

Temperature Sensitivity

The main limitations of MSW delay lines are its low power-handling capability and sensitivity to temperature. The low power-handling capability is intrinsic to the magnetostatic interaction process and is in the range of milliwatts. Temperature sensitivity is a function of the doping material as well as its level. It arises due to the change of YIG magnetization with temperature. Castera [91] has investigated the temperature coefficient of frequency for all three MSW modes. It is defined as [91]

$$C = \frac{\Delta f}{f \Delta T} \tag{14.66}$$

where Δf and ΔT are the changes in frequency and temperature, respectively. For MSSW and MSBVW devices, C is negative, whereas for MSFVW it is positive. For all three modes, the magnitude of C decreases with an increase in frequency. For example, the value of C for MSSW delay lines made of pure YIG ($4\pi M_s = 1760G$) is reported to be -550 ppm/°C at 6 GHz and -210 ppm/°C at 17 GHz [92]. Frequency stability of MSW devices can, however, be improved by incorporating suitable temperature compensation schemes. One effective method is to sense the YIG temperature and compensate for temperature changes by changing the current through the coil of the electromagnet appropriately.

14.6 COMPARISON OF SAW AND MSW DELAY LINES

Both SAW and MSW delay lines, being true time-delay elements, belong to a separate class of phase shifters. Important features of these delay lines are compared in Table 14.2. The application of SAW delay lines is limited to about 2 GHz, whereas that of MSW delay lines extends from 1 to 20 GHz. Comparison between these delay lines is meaningful only in the overlapping frequency range of application, that is, at L-band. The advantages of SAW over MSW at L-band are its inherent nondispersive operation and higher power-handling capability. The advantages of the MSW delay line are its tunability and simple and invariant transducer geometry. Furthermore, although MSWs are basically dispersive, with suitable design modification it is possible to realize either a constant delay or a linearly varying delay with frequency.

Several experimental MSW delay lines are reported in the literature. Performance parameters of some of the typical nondispersive and dispersive delay lines at C- and X-bands [52, 88, 89, 74, 90] are compiled in Table 14.3. Bandwidth in excess of 1 GHz and time delays on the order of a few hundred nanoseconds have been reported. The insertion loss of these devices, however, is very high, typically in the range of 10 to 30 dB. Other drawbacks of MSW delay lines are its low power-handling capability (mW range) and sensitivity to temperature change. While the low power-handling capability is intrinsic to MSW interaction, considerable scope exists for improving the insertion loss and temperature stability.

Table 14.2
Comparison of Microwave SAW and MSW Delay Lines

	SAW	*MSW*
Frequency	1–2 GHz	1–20 GHz
Operation	Reciprocal	Can be reciprocal or nonreciprocal
Delay tunability	Impractical to tune	Easily tunable
Tapped delay	Yes	Yes
Dispersion	No	Yes
Time delay	1–5 μs/cm	10–500 ns/cm
Propagation loss	1 dB/μs at 1 GHz	0.005 dB/ns at 1 GHz
	60 dB/μs at 10 GHz	0.02 dB/ns at 10 GHz
Power handling	1W	1 mW
Bandwidth	—	>1 GHz
Transducer	Complex geometry, dimensions decrease with increase in frequency	Simple geometry, dimensions invariant with frequency
Temperature sensitive	—	Yes

Table 14.3

Performance Parameters of MSW Delay Lines

	Frequency	Bandwidth (MHz)	Time Delay (ns)	Insertion Loss (dB)	Propagating Mode
1. Nondispersive delay lines					
• Fixed delay					
Adam et al. [52]	S-band	200	70	8	MSSW
	X-band	400	90 ± 5%	20	MSFVW
	X-band	400	102 ± 5%	30	MSBVW
• Tapped delay					
Adam et al. [52]	S-band (f_0 = 2.7 GHz)	225 ± 5%	250 (total) 125 (tapped)	14 (total) 9 (tap delay)	—
• Tunable delay					
Sethares et al. [88]	S-band (f_0 = 2.86 GHz)	250	165–185 20G adjustment for down chirp, bias for tuning, rms phase error <20°	—	MSSW and MSFVW lines in cascade
Adkins et al. [89]	S-band (f_0 = 3.055 GHz)	250	146–193 MSSW line: 415G MSBVW line: 550–670G Rms phase error <12.2°	—	MSSW and MSBVW lines in cascade
2. Linearly dispersive delay lines					
Daniel et al. [74]	X-band (8.86 –10.08 GHz)	1220	50–230 ns/cm, linearity ± 3 ns/cm	27–45	MSFVW
Chang et al. [90]	S-band (f_0 = 3 GHz)	500	80 (differential delay)	20 (unmatched)	—

REFERENCES

1. R.M. White, "Surface Elastic Waves," *Proc. IEEE,* Vol. 58, 1970, pp. 1238–1276.
2. J.D. Maines and E.G.S. Paige, "Surface Acoustic Wave Components, Devices and Applications," *IEE Proc.,* Vol. 120, No. 10R, October 1973, (IEE Reviews), pp. 1078–1110.
3. M.G. Holland and L.T. Claiborne, "Practical Surface Acoustic Wave Devices," *Proc. IEEE,* Vol. 62, No. 5, May 1974, pp. 582–611.
4. *IEEE Trans. on Microwave Theory and Tech.,* Special Issue on Microwave Acoustic Signal Processing, Vol. MTT-21, April 1973. Also, *IEEE Trans. on Sonics and Ultrasonics,* Vol. SU-20, 1973.
5. *Proc. IEEE,* Special Issue on Surface Acoustic Wave Devices and Applications, Vol. 64, No. 5, 1976.
6. M. Lewis *et al.,* "Recent Developments in SAW Devices," *IEE Proc.,* Vol. 131, Pt. A, No. 4, June 1984, pp. 186–215.
7. D.P. Morgan (ed.), *Surface Acoustic-Wave Passive Interdigital Devices,* IEE Reprint Series 2, Peter Peregrinus Ltd., 1976.
8. J.C. Sethares and R. Floyd, "MSW Applications for Phased Array Antennas," *Circuits, Systems Signal Process,* Vol. 4, No. 1–2, 1985, pp. 334–350.
9. L.R. Adkins, "Dispersion Control in Magnetostatic Wave Delay Lines," *Circuits, Systems Signal Process,* Vol. 4,. No. 1–2, 1985, pp. 137–156.
10. W.S. Ishak, "Magnetostatic Wave Technology: A Review," *Proc. IEEE,* Vol. 76, No. 2, February 1988, pp. 171–187.
11. L.R. Adkins *et al.,* "Electronically Variable Time Delays Using Magnetostatic Wave Technology," *Microwave J.,* Vol. 29, March 1986, pp. 109–120.
12. S. Datta, *Surface Acoustic Wave Devices,* Prentice-Hall, 1986.
13. E.G. Lean and A.N. Broers, "Microwave Surface Acoustic Delay Lines," *Microwave J.,* Vol. 13, March 1970, pp. 97–101.
14. D.P. Morgan, "Surface Acoustic Wave Devices and Applications," *Ultrasonics,* Vol. 11, 1973, pp. 121–131.
15. M.B. Schulz and M.G. Holland, "Materials for Surface Acoustic Wave Components," *IEE Conf. Publ.,* Vol. 109, 1973, pp. 1–10.
16. R.M. White and F.W. Voltmer, "Direct Piezoelectric Coupling to Surface Elastic Waves," *Applied Physics Letters,* Vol. 7, 1965, pp. 314–316.
17. J.H. Collins, H.M. Gerard, and H.J. Shaw, "High Performance Lithium Niobate Acoustic-Surface-Wave Transducers and Delay Lines," *Applied Physics Letters,* Vol. 13, 1968, pp. 312–315.
18. W.R. Smith *et al.,* "Analysis of Interdigital Surface Wave Transducers by Use of an Equivalent Circuit Model," *IEEE Trans. on Microwave Theory and Tech.,* Vol. MTT-17, November 1969, pp. 856–864.
19. W.R. Smith *et al.,* "Design of Surface Wave Delay Lines with Interdigital Transducers," *IEEE Trans. on Microwave Theory and Tech.,* Vol. MTT-17, November 1969, pp. 865–873.
20. W.R. Smith *et al.,* "Analysis and Design of Dispersive Interdigital Surface Wave Transducers," *IEEE Trans. on Microwave Theory and Tech.,* Vol. MTT-20, July 1972, pp. 458–471.
21. G.L. Matthaei, D.Y. Wong, and B.P. Oshaughnessy, "Simplifications for the Analysis of Interdigital Surface Wave Devices," *IEEE Trans. on Sonics and Ultrasonics,* Vol. SU-22, 1975, pp. 105–114.
22. A.J. Bahr and R.E. Lee, "Equivalent Circuit Model for Interdigital Transducers With Varying Electrode Widths," *Electronics Letters,* Vol. 9, No. 9, 1973, pp. 281–282.
23. R.F. Milsom and M. Redwood, "Piezoelectric Generation of Surface Waves by Interdigital Array," *IEE Proc.,* Vol. 118, 1971, pp. 831–840.

24. P.R. Emtage, "Self-consistent Theory of Interdigital Transducers," *IEEE Ultrasonics Symp. Digest,* 1972, pp. 397–402.

25. R.H. Tancrell and M.G. Holland, "Acoustic Surface Wave Filters," *Proc. IEEE,* Vol. 59, 1971, pp. 393–409.

26. A.K. Ganguly and M.O. Vassell, "Frequency Response of Acoustic Surface Wave Filters," *J. Applied Physics,* Vol. 44, No. 3, 1973, pp. 1072–1085.

27. S.G. Joshi and R.M. White, "Dispersion of Surface Elastic Waves Produced by a Conducting Grating on a Piezoelectric Crystal," *J. Applied Physics,* Vol. 39, December 1968, pp. 5819–5827.

28. C.C. Tseng, "Frequency Response of an Interdigital Transducer for Excitation of Surface Elastic Waves," *IEEE Trans. on Electron Devices,* Vol. ED-15, August 1968, pp. 586–594.

29. R.H. Tancrell and R.C. Williamson, "Wavefront Distortion of Acoustic Surface Waves From Apodized Interdigital Transducers," *Applied Physics Letters,* Vol. 19, December 1971, pp. 456–459.

30. H.M. Gerard, G.W. Judd, and M.E. Pedinoff, "Phase Corrections for Weighted Acoustic Surface-Wave Dispersive Filters," *IEEE Trans. on Microwave Theory and Tech.* (corresp.), Vol. MTT-20, February 1972, pp. 188–192.

31. R.F. Mitchell and D.W. Parker, "Synthesis of Acoustic-Surface Wave Filters Using Double-Electrodes," *Electronics Letters,* Vol. 10, 1974, pp. 512.

32. T.W. Bristol *et al.,* "Applications of Double-Electrodes in Acoustic Surface Wave Design," *IEEE Ultrasonics Symp. Digest,* 1972, pp. 343–345.

33. T.L. Szabo and A.J. Slobodnik, Jr., "The Effect of Diffraction on the Design of Acoustic Surface Wave Devices," *IEEE Trans. on Sonics and Ultrasonics,* Vol. SU-20, 1973, pp. 240–251.

34. F.G. Marshall, "New Technique for the Suppression of Triple-Transit Signals in Surface-Acoustic-Wave Delay Lines," *Electronics Letters,* Vol. 8, 1972, pp. 311–312.

35. M.F. Lewis, "Triple Transit Suppression in Surface Acoustic Wave Devices," *Electronics Letters,* Vol. 8, 1972, pp. 553–554.

36. R.C. Williamson, "Problems Encountered in High Frequency Surface Wave Devices," *IEEE Ultrasonics Symp. Digest,* 1974, pp. 321–328.

37. A.J. Slobodnik, Jr., and T.L. Szabo, "Design of Optimum Acoustic Surface Wave Delay Lines at Microwave Frequencies," *IEEE Trans. on Microwave Theory and Tech.,* Vol. MTT-22, April 1974, pp. 458–462.

38. A.N. Broers, E.G. Lean, and M. Hatzakis, "1.75 GHz Acoustic-Surface Wave Transducer Fabrication by an Electron Beam," *Applied Physics Letters,* Vol. 15, 1969, pp. 98–101.

39. H.I. Smith, "Fabrication Technique for Surface Acoustic-Wave and Thin-Film Optical Devices," *Proc. IEEE,* Vol. 62, October 1974, pp. 1361–1387.

40. J.H. Silva *et al.,* "A 2.2 GHz SAW Delay Line Fabricated by Direct Optical Projection," *IEEE Trans. on Sonics and Ultrasonics,* Vol. SU-26, July 1979, pp. 312–313.

41. A.J. Slobodnik, Jr., "Surface-Quality Effects on SAW Attenuation on Y-Z LiNbO$_3$," *Electronics Letters,* Vol. 10, 1974, pp. 233–234.

42. J.M. Bennett and R.J. King, "Effect of Polishing Technique on the Roughness and Residual Surface Film on Fused Quartz Optical Flats," *Applied Optics,* Vol. 9, 1970, pp. 236–238.

43. R.W. Dietz and J.M. Bennett, "Bowl Feed Technique for Producing Supersmooth Optical Surfaces," *Applied Optics,* Vol. 5, 1966, pp. 881–882.

44. P. Hartemann and C. Arnodo, "Raleigh-Wave Delay Line Using Two Grating-Array Transducers at 2.55 GHz," *Electronics Letters,* Vol. 8, 1972, pp. 265–267.

45. R.T. Webster, "1.5 GHz GaAs Surface Acoustic Wave Delay Lines," *IEEE Trans. on Microwave Theory Tech.,* Vol. MTT-33, September 1985, pp. 824–827.

46. A.J. Slobodnik, Jr., J.H. Silva, and G.A. Roberts, "SAW Filters at 1 GHz Fabricated by Direct Step on the Wafer," *IEEE Trans. on Sonics and Ultrasonics,* Vol. SU-28, March 1981, pp. 105–106.

47. S.G. Joshi and B.B. Dasgupta, "Electrically Variable Surface Acoustic Waves Time Delay Using a Biasing Electric Field," *IEEE Ultrasonic Symp. Proc.*, 1981, pp. 319–323.

48. A.J. Budreau *et al.*, "Electrostatically Variable SAW Delay Lines—Theory and Experiment," *IEEE Trans. on Sonics and Ultrasonics*, Vol. SU-31, No. 6, November 1984, pp. 646–651.

49. J.B. Thaxter, P.H. Carr, and J.H. Silva, "Propagation of Transverse Bulk and Surface Acoustic Waves in LiNbO₃ Variable Time-Delay Devices," *IEEE Trans. on Ultrasonics, Ferroelectric, Frequency Control.*, Vol. 35, No. 5, September 1988, pp. 525–530.

50. D. Morgan and E.A. Ash, "Acoustic Surface Wave Dispersive Delay Lines," *IEE Proc.*, Vol. 116, July 1969, pp. 1125–1134.

51. S.T. Costanza, P.J. Hagon, and L.A. MacNevin, "Analog Matched Filter Using Tapped Acoustic Surface Wave Delay Line," *IEEE Trans. on Microwave Theory and Tech.*, Vol. MTT-17, November 1969, pp. 1042–1043.

52. J.D. Adam, J.H. Collins, and J.M. Owens, "Microwave Device Applications of Epitaxial Magnetic Garnets," *The Radio and Electronics Engr.*, Vol. 45, December 1978, pp. 738–748.

53. M. Shone, "The Technology of YIG Film Growth," *Circuits Syst. Sign. Processing*, Vol. 4, 1985, pp. 89–103.

54. J.P. Castera, "State of the Art in Design and Technology of MSW Devices," *J. Appl. Phys.*, Vol. 55, No. 6, March 1984, pp. 2506–2511.

55. M.S. Sodha and N.C. Srivastava, *Microwave Propagation in Ferrimagnetics*, Plenum, 1981.

56. R.W. Damon and J.R. Eshbach, "Magnetostatic Modes of a Ferromagnetic Slab," *J. Phys. Chem. Solids*, Vol. 19, 1961, pp. 308–320.

57. W.L. Bongianni, "Magnetostatic Propagation in a Dielectric Layered Structure," *J. Applied Physics*, Vol. 43, 1972, pp. 2541–2544.

58. C. Vittoria and N.D. Wilsey, "Magnetostatic Wave Propagation Losses in an Anisotropic Insulator," *J. Appl. Phys.*, Vol. 45, January 1974, pp. 414–420.

59. J.D. Adam, M.R. Daniel, and D.K. Schroder, "Magnetostatic-Wave Devices Move Microwave Design into Gigahertz Realm," *Electronics*, May 1980, pp. 123–128.

60. P. Young, "Effect of Boundary Conditions on the Propagation of Surface Magnetostatic Waves in a Transversely Magnetized Thin YIG Slab," *Electronics Letters*, Vol. 5, No. 18, September 1969, pp. 429–431.

61. B.A. Auld and K.B. Mehta, "Magnetostatic Waves in a Transversely Magnetized Rectangular Rod," *J. Applied Physics*, Vol. 38, No. 10, September 1967, pp. 4081–4082.

62. M. Radmanesh, C.M. Chu, and G.I. Haddad, "Magnetostatic Wave Propagation in a Yttrium Iron Garnet (YIG)-Loaded Waveguide," *Microwave J.*, Vol. 29, July 1986, pp. 135–140.

63. M. Radmanesh, C.M. Chu, and G.I. Haddad, "Magnetostatic Wave Propagation in a Finite YIG-Loaded Rectangular Waveguide," *IEEE Trans. on Microwave Theory and Tech.*, Vol. MTT-34, December 1986, pp. 1377–1382.

64. M. Radmanesh, C.M. Chu, and G.I. Haddad, "Magnetostatic Waves in a Normally Magnetized Waveguide Structure," *IEEE Int. Microwave Symp. Digest*, 1987, pp. 997–1000.

65. T.W. O'Keeffe and R.W. Patterson, "Magnetostatic Surface Wave Propagation in Finite Samples," *J. Applied Physics*, Vol. 49, September 1978, pp. 4886–4895.

66. J.D. Adam and S.N. Bajpai, "Magnetostatic Forward Volume Wave Propagation in YIG Strips," *IEEE Trans. on Magnetics*, Vol. MAG-18, September 1982, pp. 1598–1600.

67. M.R. Daniel, J.D. Adam, and T.W. O'Keeffe, "Linearly Dispersive Delay Lines at Microwave Frequencies Using Magnetostatic Waves," *IEEE Ultrasonic Symp. Proc.*, 1979, pp. 806–809.

68. N.S. Chang and Y. Matsuo, "Numerical Analysis of MSSW Delay Line Using Layered Magnetic Thin Slabs," *Proc. IEEE*, Vol. 66, 1978, pp. 1577–1578.

69. L.R. Adkins and H.L. Glass, "Propagation of Magnetostatic Surface Waves in Multiple Magnetic Layer Structures," *Electronics Letters*, Vol. 16, 1980, pp. 590–592.

70. L.R. Adkins and H.L. Glass, "Magnetostatic Volume Wave Propagation in Multiple Ferrite Layers," *J. Applied Physics*, Vol. 53, 1982, pp. 8928–8933.

71. J.P. Parekh and K.W. Chang, "MSFVW Dispersion Control Utilizing a Layered YIG Film Structure," *IEEE Trans. on Magnetics*, Vol. MAG-18, 1982, pp. 1610–1612.

72. M.R. Daniel and P.R. Emtage, "Magnetostatic Volume Wave Propagation in a Ferrimagnetic Double Layer," *J. Applied Physics*, Vol. 53, 1982, pp. 3723–3729.

73. A.K. Ganguly and C. Vittoria, "Magnetostatic Wave Propagation in Double Layers of Magnetically Anisotropic Slabs," *J. Applied Physics*, Vol. 45, 1974, pp. 4665–4667.

74. M.R. Daniel, J.D. Adam and T.W. O'Keeffe, "A Linearly Dispersive Magnetostatic Delay Line at X-Band," *IEEE Trans. on Magnetics*, Vol. MAG-15, No. 6, November 1979, pp. 1735–1737.

75. A.K. Ganguly and D.C. Webb, "Microstrip Excitation of Magnetostatic Surface Waves, Theory and Experiment," *IEEE Trans. on Microwave Theory and Tech.*, Vol. MTT-23, December 1975, pp. 998–1006.

76. H.J. Wu, C.V. Smith, Jr., and J.M. Owens, "Bandpass Filtering With Multibar Magnetostatic Surface Wave Microstrip Transducers," *Electronics Letters*, Vol. 13, September 1977, pp. 610–611.

77. P.R. Emtage, "Interaction of Magnetostatic Waves With a Current," *J. Applied Physics*, Vol. 49, August 1978, pp. 4475–4484.

78. J.C. Sethares, "Magnetostatic Surface Wave Transducers," *IEEE Trans. on Microwave Theory and Tech.*, Vol. MTT-27, November 1979, pp. 902–911.

79. J.D. Parekh and H.S. Tuan, "Meander Line Excitation of Magnetostatic Surface Waves," *Proc. IEEE*, Vol. 67, 1979, pp. 182–183.

80. J.D. Adams, R.W. Patterson, and T.W. O'Keeffe, "Magnetostatic Wave Interdigital Transducers," *J. Applied Physics*, Vol. 49, 1978, pp. 1797–1799.

81. J.C. Sethares and I.J. Weinberg, "Apodization of Variable Coupling Magnetostatic Surface Wave Transducers," *J. Applied Physics*, Vol. 50, No. 3, March 1979, pp. 2458–2460.

82. J.H. Collins, D.M. Hastie, J.M. Owens, and C.V. Smith, Jr., "Magnetostatic Wave Terminations," *J. Applied Physics*, Vol. 49, 1978, pp. 1800–1802.

83. J. Krug and P. Edenhofer, "Broadband Termination for Magnetostatic Surface Waves," *Electronics Letters*, Vol. 19, 1983, pp. 971–972.

84. W.S. Ishak, "Magnetostatic Surface Wave Devices for UHF and L-Band Applications," *IEEE Trans. on Magnetics*, Vol. MAG-19, September 1983, pp. 1880–1882.

85. V.L. Taylor, J.C. Sethares, and C.V. Smith, Jr., "MSW Terminations," *IEEE Ultrasonics Symp. Digest*, 1980, pp. 562–566.

86. J.D. Adam, T.W. O'Keeffe, and M.R. Daniel, "Magnetostatic Devices for Microwave Signal Processing," *Proc. Sov. Photo-Opt. Instrum. Eng. (USA)*, Vol. 241, 1980, pp. 96–103.

87. J.M. Owens, C.V. Smith, Jr., and T.J. Mears, "Magnetostatic Wave Reflective Array Filter," *IEEE Microwave Symp. Digest*, 1979, pp. 154–156.

88. J.C. Sethares, J.M. Owens, and C.V. Smith, Jr., "MSW Nondispersive Electronically Tunable Time Delay Elements," *Electronics Letters*, Vol. 16, No. 22, October 1980, pp. 825–826.

89. L. Adkins et al., "Electronically Variable Time Delays Using Cascaded Magnetostatic Delay Lines," *J. Applied Physics*, Vol. 55, No. 6, March 1984, pp. 2518–2520.

90. K.W. Chang, J.M. Owens, and R.L. Carter, "Linearly Dispersive Time Delay Control of MSSW by Variable Ground Plane Spacing," *Electronics Letters*, Vol. 19, July 1983, pp. 546–547.

91. J.P. Castera, "Magnetostatic Wave Temperature Coefficients," *Proc. RADC Microwave Magnetics Workshop*, 1981, pp. 178–186.

92. W.S. Ishak et al., "Tunable Magnetostatic Wave Oscillators Using Pure and Doped YIG Films," *IEEE Trans. on Magnetics*, Vol. MAG-20, September 1984, pp. 1229–1231.

Chapter 15
Selection Criteria for Phase Shifters in Systems

15.1 INTRODUCTION

Chapters 4–7 of Volume I and Chapters 8–14 of Volume II describe a wide variety of electronically variable phase shifters that can cater to a diverse range of specifications at microwave and millimeter-wave frequencies. The commonly specified electrical parameters of an electronic phase shifter are

- Center frequency of operation
- Bandwidth
- Insertion loss (for 360° phase shift)
- VSWR or return loss
- Switching time (for digital operation) or time required for 360° phase change (for analog operation)
- Switching power or energy (for digital operation) or dc holding power (for analog operation)
- Phase error
- Power-handling capability (peak and average)

It may be noted that in the case of active phase shifters, insertion gain is to be specified in place of insertion loss. For phased-array applications, it is important to specify the mode of operation—namely, digital or analog, reciprocal or nonreciprocal. Other factors include, size, weight, and geometry of the phase shifter, temperature dependence of phase shift and insertion loss, and radiation hardness.

The phase shifter specifications for a given application are governed by the system requirements. Phased-array radars employing thousands of elements particularly call for a careful selection of phase shifters, because the performance and cost of each phase shifter directly reflect on the overall performance and cost of the system. The choice of the phase shifter is governed by the per-element specifications of operating frequency, RF power handling, and switching speed. In this chapter, we provide broad selection criteria for phase shifters considering the present state-of-the-art performance.

15.2 SELECTION CRITERIA FOR DIGITAL PHASE SHIFTERS

Most of the practical phase shifters, both the ferrite and semiconductor types, are digital in nature [1–8]. Because the bulk requirement of phase shifters is for phased-array radars, digital types are preferred over the analog ones in view of the flexibility they offer for interfacing with the digital computer for control signals. Table 15.1 provides a classification of the various digital phase shifters. The figure number from an earlier chapter of this book and its companion volume provided with each phase-shifter type may be referred to for a typical circuit configuration.

With the present state of the art, the nonreciprocal twin-toroid ferrite phase shifter, the reciprocal dual-mode ferrite phase shifter, and the *p-i-n* diode (reciprocal) phase shifter are the major contenders for digital phase shifting applications. Performance characteristics of each of these phase-shifter types are presented in earlier chapters of this book and its companion volume: the twin-toroid phase shifter in Table 4.3, the dual-mode phase shifter in Table 5.2, and the *p-i-n* diode phase shifters in Table 9.1. A detailed comparison between the practical performance of the twin-toroid and the dual-mode phase shifters is provided in Table 5.4. Table 15.2 gives a consolidated summary and comparison of performance features of all three types. A comparison with the present performance states of GaAs *field-effect transistor* (FET) phase shifters is also provided because this technology holds considerable promise for future monolithic phased-array systems.

An important desirable property of a phase shifter for any application is that its insertion loss should be as low as possible. At lower microwave frequencies (L-, S-, and C-bands) both ferrite and *p-i-n* diode phase shifters have equally low insertion loss; in X- and Ku-bands the insertion loss of ferrites (expressed in dB) is nearly half that of *p-i-n* diodes; and above Ku-band, ferrites are far superior. The *p-i-n* diode phase shifters offer superior switching speeds compared to ferrites. It may be noted that the twin-toroid phase shifter offers much shorter switching time compared to dual-mode and approaches that of the *p-i-n* diode above Ka-band. The switching power required for *p-i-n* diode phase shifters is much smaller and the driver circuit is also simpler and cheaper than the ferrite counterparts. The power-handling capability of ferrites is much higher than that of diodes. For *p-i-n* diode phase shifters to handle high power levels, multiple diodes must be mounted to share the RF stress. This results in higher cost and also affects the reliability. Ferrites are temperature sensitive. However, this problem is partially circumvented by the use of a flux drive control circuit. Ferrite phase shifters are bulkier and heavier than *p-i-n* diode phase shifters at lower microwave frequencies, but become comparable at X-band and above.

From the point of view of applications, L-, S-, and C-bands are in general useful for ground-based systems; for example, L- and S-bands for surveillance type radars, S-band for ground and ship search and tracking radars, and C-band for microwave landing systems. Low loss and low cost are of primary importance, whereas size and weight are of secondary importance in these applications. Long-range radars must, in addition, have high power-handling capability. Frequencies at X-band and above are useful for airborne system applications, such as satellites and missiles. Small size, light weight, fast switching speed, low drive power, low RF power dissipation, and low cost become important considerations. The peak power requirement for airborne applications is generally small (below 100W).

For ground-based phased-array radars, both ferrite and p-i-n diode phase shifters may be used. Since size and weight are not major considerations for transmitting systems, ferrite phase shifters may be preferred because of their larger power-handling capability. The longer switching time of ferrites is generally acceptable for long-range radars. For aircraft landing systems and other short-range radars, fast switching speeds (~ 1 μs) afforded by the diode phase shifters may be required. Where nonreciprocal operation and intermediate switching speeds on the order of 5 μs are acceptable, the twin-toroid phase shifters with their low loss and high-power capability offer the best choice.

For airborne applications at X- and Ku-bands, both twin-toroid and dual-mode phase shifters match p-i-n diode phase shifters in size and weight, and their insertion loss is much lower. The switching time for ferrites at these frequencies is nearly five times shorter than at S-band. In particular, for the twin-toroid, the switching time (~ 1 μs) approaches that of the p-i-n diode.

At millimeter-wave frequencies, ferrite phase shifters enjoy uncontested superiority. Because of the reduction in the volume of ferrite to be switched, both switching time and switching energy are reduced. At V-band, the twin-toroid phase shifter has an insertion loss of 1.6 dB, a switching time of 1 μs, and requires a switching energy of 25 μJ. The dual-mode phase shifter has nearly the same insertion loss (~ 1.5 dB), has a switching time of 20 μs, and requires a switching energy of 100 μJ [2]. A twin-toroid phase shifter is suitable for rapid switching of multiple beams in a satellite communication system. The nonreciprocal feature of the twin-toroid phase shifter is not a problem in satellite system applications because separate antennas are used for transmission and reception. Moreover, in beam-forming networks comprising variable power dividers and phase shifters, the nonreciprocity in phase shift is an advantage in providing built-in isolation for the reflected signals in the power divider tree.

A variety of practical radar systems that use ferrite and p-i-n diode phase shifters have been reported in the literature [9, 10]. Examples of phased-array radar systems that use digital ferrite phase shifters are the Patriot multifunction array radar [9], the 3D AN/TPQ-36 X-band artillery locating radar [11], the AN/SPY-1 tactical multifunction array radar [12], and the 3D TRMS (Telefunken Radar Mobile Search) C-band search radar [13]. P-I-N diode phase shifters are also used in a number of practical systems, such as the AN/TPQ-37 S-band artillery locating radar [11] and the EMPAR (European Multifunction Phased-Array Radar) for shipboard use [14].

Monolithic GaAs FET phase shifters have been reported up to V-band (Table 12.3). The distinguishing features of this phase shifter compared to ferrite and

Table 15.1
Classification of Digital Phase Shifters

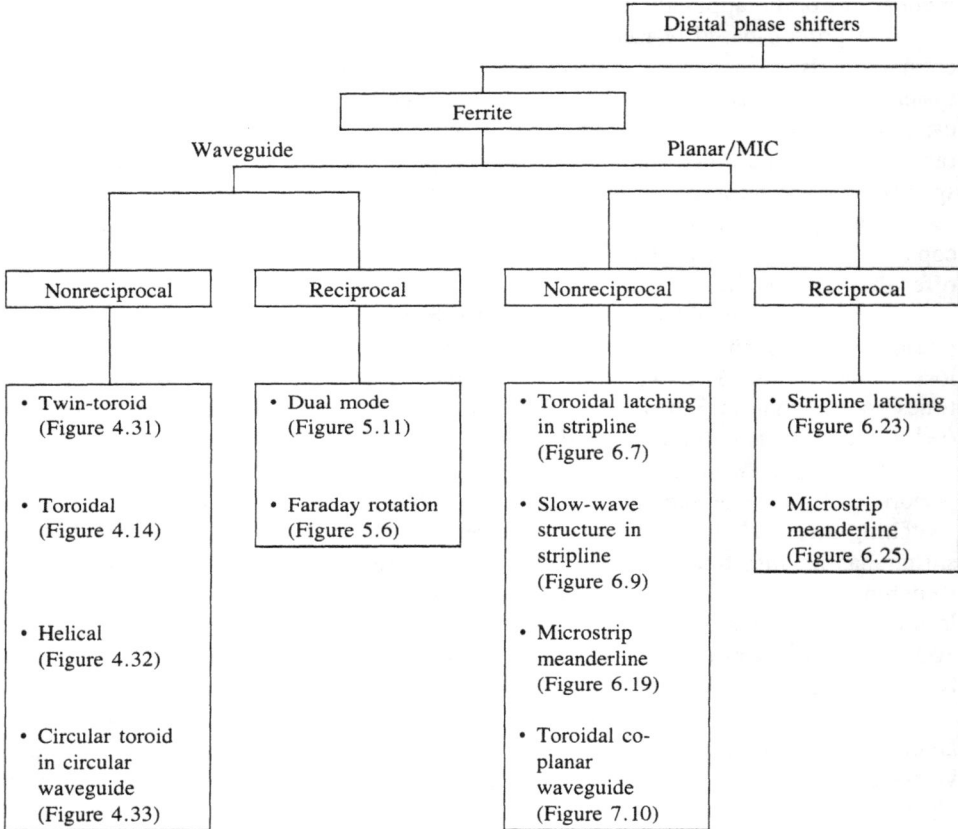

Digital phase shifters

Ferrite

Waveguide — Planar/MIC

Nonreciprocal	Reciprocal	Nonreciprocal	Reciprocal
• Twin-toroid (Figure 4.31)	• Dual mode (Figure 5.11)	• Toroidal latching in stripline (Figure 6.7)	• Stripline latching (Figure 6.23)
• Toroidal (Figure 4.14)	• Faraday rotation (Figure 5.6)	• Slow-wave structure in stripline (Figure 6.9)	• Microstrip meanderline (Figure 6.25)
• Helical (Figure 4.32)		• Microstrip meanderline (Figure 6.19)	
• Circular toroid in circular waveguide (Figure 4.33)		• Toroidal co-planar waveguide (Figure 7.10)	

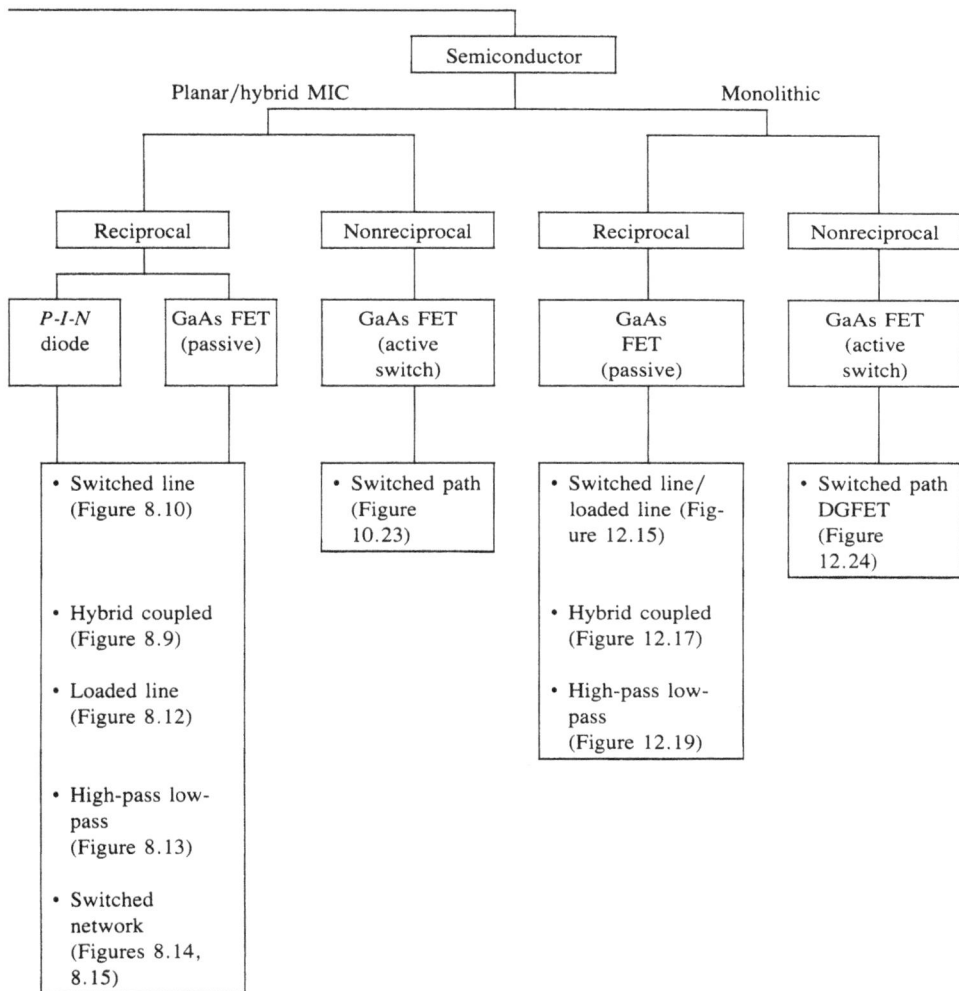

```
                          ┌─────────────────┐
                          │  Semiconductor  │
                          └─────────────────┘

     Planar/hybrid MIC                              Monolithic

  ┌──────────────┐  ┌──────────────┐     ┌──────────────┐  ┌──────────────┐
  │  Reciprocal  │  │ Nonreciprocal│     │  Reciprocal  │  │ Nonreciprocal│
  └──────────────┘  └──────────────┘     └──────────────┘  └──────────────┘

┌─────────┐ ┌─────────┐  ┌─────────┐     ┌─────────┐        ┌─────────┐
│ P-I-N   │ │GaAs FET │  │GaAs FET │     │ GaAs    │        │GaAs FET │
│ diode   │ │(passive)│  │(active  │     │ FET     │        │(active  │
│         │ │         │  │ switch) │     │(passive)│        │ switch) │
└─────────┘ └─────────┘  └─────────┘     └─────────┘        └─────────┘
```

Reciprocal (Planar/hybrid MIC)	Nonreciprocal (Planar/hybrid MIC)	Reciprocal (Monolithic)	Nonreciprocal (Monolithic)
• Switched line (Figure 8.10)	• Switched path (Figure 10.23)	• Switched line/ loaded line (Figure 12.15)	• Switched path DGFET (Figure 12.24)
• Hybrid coupled (Figure 8.9)		• Hybrid coupled (Figure 12.17)	
• Loaded line (Figure 8.12)		• High-pass low-pass (Figure 12.19)	
• High-pass low-pass (Figure 8.13)			
• Switched network (Figures 8.14, 8.15)			

Table 15.2
Selection Criteria for Digital Phase Shifters (L-V Bands)

	Ferrite Phase Shifters		Semiconductor Phase Shifters	
	Twin-Toroid (S-V Bands)	Dual-Mode (S-V Bands)	P-I-N Diode (L-Ku Bands)	GaAs FET (L-V Bands)
Geometry	Waveguide	Waveguide	Hybrid MIC	Monolithic
Reciprocal/ nonreciprocal	Nonreciprocal	Reciprocal	Reciprocal	Reciprocal
Bandwidth	5%–30%	5%–10%	10%–20%	Octave
Insertion loss in dB for 360° phase shift	0.5–1.0–1.6 (S–Ka–V)	0.5–0.9–1.5 (S–Ka–V)	0.5–2 (L–Ku)	5–12
Switching time (μs)	1–1–5 (V–Ka–S)	20–30–150 (V–Ka–S)	<1	0.001
Switching power/energy	25–30–150 μJ (V–Ka–S)	100–150–1000 μJ (V–Ka–S)	0.1–5W	Negligible
Peak power	1–100 kW (Ka–S)	1–40 kW (Ka–S)	kW (pulse width dependent)	W
Average power	1 kW (S-band)	500W (S-band)	W	mW
Temperature sensitivity	0.5°–3°/°C typical	0.5°–3°/°C typical	Negligible	—
Insertion phase trimming for ±10° tolerance	Usually required	Usually required	Not necessary	Not necessary
Radiation hardness	Excellent	Excellent	Poor	—
Weight	1–4 oz (Ka–L) <1 oz (>Ka)	1–4 oz (Ka–L) <1 oz (>Ka)	0.5–1 oz (Ku–L)	<0.1 oz

p-i-n diode phase shifters are its small size and weight, fast switching speed, broad bandwidth, negligible drive power, and higher reliability and reproducibility. The main drawbacks are its small power-handling capability and high insertion loss (5–12 dB). These problems are overcome in dual-gate active FET phase shifters. Vorhaus, Pucel, and Tajima [6] have reported an active phase-shifter bit having a gain of 3 dB over 10% bandwidth centered at 9.5 GHz. The current trend is to use

GaAs monolithic microwave integrated circuit (MMIC) chips for transceiver module functions, which include phase shifting, low-noise amplification, power amplification, and transmit-receive switch. MMIC technology offers the best potential for volume production of transceiver modules with integrated phase shifters [7, 8]. These advantages facilitate their application in electronically agile active phased-array radars. The PAVE PAWS system (by Raytheon) reported by Laighton [15] is an example of an active-aperture solid-state phased-array radar. The transceiver module that forms a key element of this system uses a 4-bit phase shifter, a two-stage low noise amplifier, and a transmit-receive switch.

15.3 SELECTION CRITERIA FOR ANALOG PHASE SHIFTERS

There are several applications in which only a few degrees of coverage, such as ±10°, are needed in elevation. Limited scan techniques enable considerable reduction in the number of phase shifters required in the system. For example, precision-approach radars that guide aircraft during final approach require only a limited coverage, on the order of 15° in azimuth as well as elevation. Such limited angle coverage can be achieved by using a small phased-array feed to illuminate a reflector.

Small phased arrays normally require phase-shifter circuits that have higher power-handling capability and that can increment phase shift over small intervals with high accuracy. Analog phase shifters are advantageous for such applications. With the facility for continuously variable phase control, the array can be set to any number of phase states.

Table 15.3 presents a classification of various analog phase shifters. Compared to the digital types (Table 15.1), the analog-type phase shifters reported are few in number. The analog ferrite phase shifters—namely, the *Reggia-Spencer* and the *rotary field,* are both reciprocal in nature. Among the two semiconductor types, the varactor diode phase shifter is reciprocal, and the vector summation FET phase shifters employing FETs as variable gain amplifiers are nonreciprocal.

The most prominent among the analog phase shifters is the rotary-field ferrite phase shifter [16, 17]. The detailed performance characteristics of this phase shifter are given in Table 5.3 of the companion volume. This phase shifter has uniformly low loss (≈0.6 dB) up to Ku-band. Its high power capability (≈40 kW at S-band, ≈4 kW at X-band) and high phase shift accuracy (≈1° rms phase error) make it ideally suited for small array applications. The phase shifter is also well suited for single-axis, electronically scanned large arrays. Its use for two-dimensional scanning is limited because of the large size and high power consumption. The AWACS (Airborne Warning and Control System) ultra-low-sidelobe antenna array (by Westinghouse) reported by Schrank [18] incorporates rotary-field ferrite phase shifters for beam scanning in elevation. This phase shifter has rms phase accuracies of less than 1° and is essentially frequency independent.

For low-power analog applications, varactor diode phase shifters offer considerable potential [19]. A small-bit varactor phase shifter is particularly useful as a

Table 15.3
Classification of Analog Phase Shifters

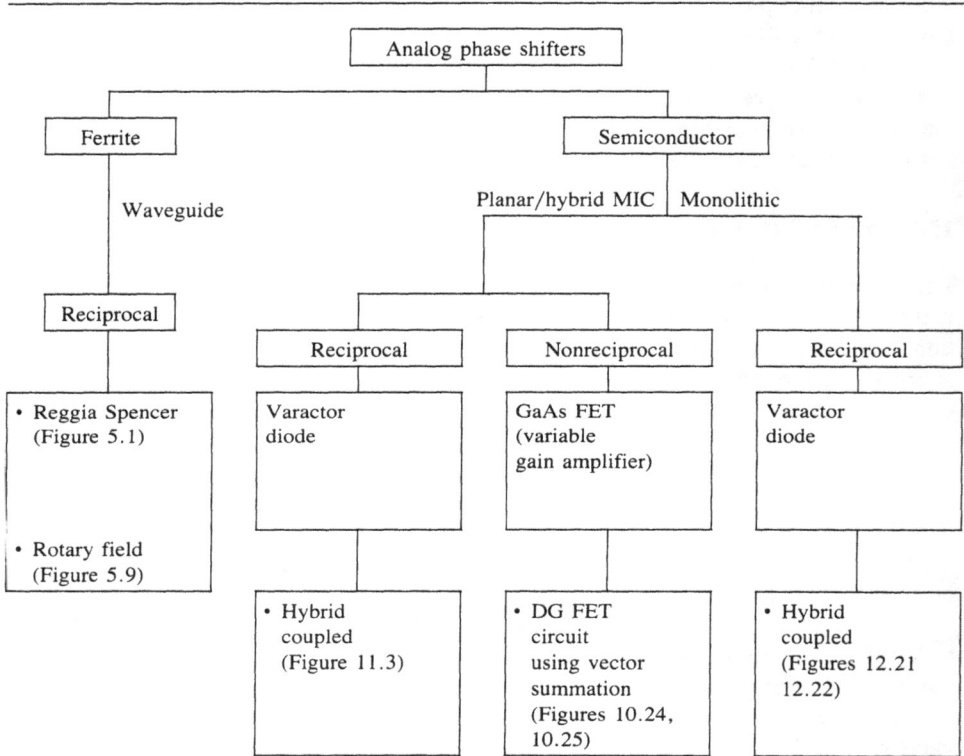

		Analog phase shifters		

Ferrite		Semiconductor		
Waveguide		Planar/hybrid MIC	Monolithic	

Reciprocal		Reciprocal	Nonreciprocal	Reciprocal
• Reggia Spencer (Figure 5.1)		Varactor diode	GaAs FET (variable gain amplifier)	Varactor diode
• Rotary field (Figure 5.9)		• Hybrid coupled (Figure 11.3)	• DG FET circuit using vector summation (Figures 10.24, 10.25)	• Hybrid coupled (Figures 12.21 12.22)

phase trimmer in a multibit phase-shifter module or as a phase adjuster in the reference arm of a nulling circuit of a microwave bridge. In the monolithic version, varactor diode phase shifters can be integrated with frequency multipliers to realize millimeter-wave transmitter modules. From the performance parameters listed in Table 12.2, it can be seen that monolithic varactor phase shifters offer large bandwidths, but their insertion loss is quite high even at microwave frequencies. Their practical utility as 360° analog phase shifters would require considerable improvement in insertion loss.

Nondispersive tunable magnetostatic wave (MSW) delay lines have potential applications in broadband phased arrays at microwave frequencies up to 20 GHz [20, 21]. Other applications include delay-line discriminators, target simulation, and electronic timing. Linearly dispersive delay lines have applications in group delay equalizers, pulse compression systems, and for generating variable time delays for target simulation in phased arrays. From Table 14.1 it can be seen that MSW delay lines have favorable features in terms of time delay and bandwidth. However, at the present state of the art, the insertion loss of those devices is quite high. With improvements in MSW technology, these delay lines should be capable of meeting the requirements for the aforementioned applications.

REFERENCES

1. L. Whicker, "Selecting Ferrite Phasers for Phased Arrays," *Microwaves,* Vol. 11, August 1972, pp. 44–48.
2. W.E. Hord, "Microwave and Millimeter Wave Ferrite Phase Shifters," *Microwave J.,* Vol. 32, 1989, pp. 81–89.
3. G.P. Rodrigue, "A Generation of Microwave Ferrite Devices," *Proc. IEEE,* Vol. 76, February 1988, pp. 121–137.
4. *IEEE Trans. on Microwave Theory and Tech.,* Special Issue on Microwave Control Devices for Array Antenna Systems, Vol. MTT-22, June 1974.
5. Y. Ayasli *et al.,* "A Monolithic Single Chip X-Band Four Bit Phase Shifter," *IEEE Trans. on Microwave Theory and Tech.,* Vol. MTT-30, December 1982, pp. 2201–2206.
6. J.L. Vorhaus, R.A. Pucel, and Y. Tajima, "Monolithic Dual-Gate GaAs FET Digital Phase Shifter," *IEEE Trans. on Microwave Theory and Tech.,* Vol. MTT-30, July 1982, pp. 982–992.
7. R.S. Pengelly, "GaAs Monolithic Microwave Circuit for Phased-Array Applications," *IEE Proc.,* Vol. 127, Pt. F, 1980, pp. 301–311.
8. R.A. Pucel *et al.,* "A Multi-Chip GaAs Monolithic Transmit/Receive Module at X-Band," *IEEE Int. Microwave Symp. Digest,* 1982, pp. 489–492.
9. E. Brookner, *Radar Technology,* Artech House, 1977.
10. E. Brookner (ed.), *Aspects of Modern Radar,* Artech House, 1988.
11. D.A. Ethington, "The AN/TPQ-36 and AN/TPQ-37 Fire Finder Radar Systems," EASCON-77, Conf. Rec., 1977, pp. 4.3A–4.3F.
12. W.T. Patton, "Microwave Design for Reliability/Availability, The AN/SPY-1 Radar System," Chapter 3 in E. Brookner (ed.) *Aspects of Modern Radar,* Artech House, 1988.
13. J.S. Ajioka, "TRMS Antenna," *IEEE Int. Radar Conf.,* 1975, pp. 382–384.
14. G.R.G. Thompson, "A Modular Approach to Multi-Function Radar Design for Naval Application," *IEE Int. Conf. Proc. Radar—87,* October 1987, pp. 32–36.
15. D.G. Laighton, "Hybrid and Monolithic Microwave Integrated Circuitry (MMIC)," Chapter 4 in E. Brookner (ed.), *Aspects of Modern Radar,* Artech House, 1988.
16. C.R. Boyd, "Analog Rotary Field Ferrite Phase Shifters," *Microwave J.,* Vol. 20, December 1977, pp. 41–43.
17. W.E. Hord, "Design Considerations for Rotary Field Ferrite Phase Shifters," *Microwave J.,* Vol. 31, November 1988, pp. 105–115.
18. H.E. Schrank, "Low Sidelobe Phased Array and Reflector Antennas," Chapter 6 in E. Brookner (ed.), *Aspects of Modern Radar,* Artech House, 1988.
19. C.L. Chen *et al.* "A Low-Loss Ku-Band Monolithic Analog Phase Shifter," *IEEE Trans. on Microwave Theory and Tech.,* Vol. MTT-35, March 1987, pp. 315–320.
20. J.C. Sethares, J.M. Owens, and C.V. Smith, Jr., "MSW Non-Dispersive Electronically Tunable Time Delay Elements," *Electronics Letters,* Vol. 16, No. 22, October 1980, pp. 825–826.
21. L.R. Adkins *et al.,*" Electronically Variable Time Delays Using Cascaded Magnetostatic Delay Lines," *J. Applied Physics,* Vol. 55, No. 6, Pt. 2B, March 1984, pp. 2518–2520.

INDEX

The Artech House Microwave Library

Stripline Circuit Design, Harlan Howe, Jr.

Terrestrial Digital Microwave Communications, Ferdo Ivanek, et al.

Time-Domain Response of Multiconductor Transmission Lines: Software and User's Manual, A.R. Djordjevic, et al.

Transmission Line Design Handbook, Brian C. Waddell

www.ingramcontent.com/pod-product-compliance
Lightning Source LLC
Chambersburg PA
CBHW021429180326
41458CB00001B/191